T0205814

Introduction to
Matrix Theory
with Applications in
Economics and Engineering

Second Edition

SERIES ON CONCRETE AND APPLICABLE MATHEMATICS

ISSN: 1793-1142

Series Editor: Professor George A. Anastassiou
Department of Mathematical Sciences
University of Memphis
Memphis, TN 38152, USA

Published*

*To view the complete list of the published volumes in the series, please visit:
http://www.worldscientific/series/scam

Series on Concrete and Applicable Mathematics – Vol. 23

Introduction to
Matrix Theory
with Applications in
Economics and Engineering

Second Edition

Ferenc Szidarovszky
Corvinus University, Hungary

Sandor Molnar
Szent Istvan University, Hungary

Mark Molnar
Eötvös Lóránd University of Sciences, Hungary

World Scientific

NEW JERSEY · LONDON · SINGAPORE · BEIJING · SHANGHAI · HONG KONG · TAIPEI · CHENNAI · TOKYO

Published by

World Scientific Publishing Co. Pte. Ltd.

5 Toh Tuck Link, Singapore 596224

USA office: 27 Warren Street, Suite 401-402, Hackensack, NJ 07601

UK office: 57 Shelton Street, Covent Garden, London WC2H 9HE

Library of Congress Cataloging-in-Publication Data
Names: Szidarovszky, Ferenc, author. | Molnár, Sándor, Dr, author. |
  Molnar, Mark (Illustrator), author.
Title: Introduction to matrix theory : with applications in economics and engineering /
  Ferenc Szidarovszky, Corvinus University, Hungary Sandor Molnar, Szent Istvan University,
  Hungary, Mark Molnar, Eötvös Lóránd University of Sciences, Hungary.
Description: 2nd edition. | Singapore ; Hackensack, NJ : World Scientific Publishing Co. Pte. Ltd.,
  [2022] | Series: Series on concrete and applicable mathematics, 1793-1142 ; vol. 23 |
  Includes bibliographical references and index.
Identifiers: LCCN 2022013272 | ISBN 9789811256646 (hardcover) |
  ISBN 9789811257933 (paperback) | ISBN 9789811256653 ((ebook for institutions) |
  ISBN 9789811256660 (ebook for individuals)
Subjects: LCSH: Matrices. | AMS: Linear and multilinear algebra; matrix theory --
  Instructional exposition (textbooks, tutorial papers, etc.).
Classification: LCC QA188 .S95 2022 | DDC 512.9/434--dc23/eng20220617
LC record available at https://lccn.loc.gov/2022013272

British Library Cataloguing-in-Publication Data
A catalogue record for this book is available from the British Library.

For any available supplementary material, please visit
https://www.worldscientific.com/worldscibooks/10.1142/12849#t=suppl

Typeset by Stallion Press
Email: enquiries@stallionpress.com

Printed in Singapore

Preface

Linear algebra and matrix theory are among the most important and most frequently applied branches of mathematics. They are especially important in solving engineering and economic models, where either the model is assumed linear, or the nonlinear model is approximated by a linear model, and the resulting linear model is examined.

This book is mainly a textbook, that covers a one semester upper division course or a two semester lower division course on the subject. The book is written for students studying engineering, economics and business, however it can also be used in courses offered by the mathematics department, or by any kind of engineering and social sciences.

Each chapter consists of three major parts. The first part introduces the new concepts, discusses and proves the main theorems. The new concepts and theoretical results are always illustrated by easy-to-follow numerical examples. The second part of each chapter presents some applications of the material of the chapter. We have selected these applications from special methodology of linear algebra and matrix theory (such as block matrices, matrix exponential, singular value decomposition, pseudoinverses, etc.), linear systems theory (such as discrete and continuous systems), statistics (for example, the least squares method), numerical analysis (such as interpolation polynomials, integral equations), engineering (such as solving differential equations), as well as economic modelling (for example, oligopoly, and producer-consumer models). The last section of each chapter offers exercises to improve understanding of the material and to help students to gain experience in problem solving. In each set of exercises we have presented some simple examples which can be solved easily by using the methodology of that chapter, however in each set we also offer some more difficult problems which require deeper understanding and skill in mathematical developments.

The book is organized as follows. Chapter 1 introduces the concept of vectors and matrices, and discusses the elements of matrix algebra. Vector spaces, subspaces, linear independence, basis, and inner-product spaces are examined in Chapter 2. The most important application of linear algebra is the solution of systems of linear algebraic equations. In Chapter 3 the elimination method is introduced and we demonstrate how to use this method to determine inverses of square matrices. Determinants and their main properties are discussed in Chapter 4. Linear mappings, linear transformations, the vector space of linear mappings, and matrix representations are investigated in Chapter 5. The discussion on diagonal, triangular, and Jordan canonical forms of matrices is based on the theory of eigenvalues and eigenvectors as well as on the main properties of invariant subspaces. The fundamentals of these topics are given in Chapter 6, and special matrices are introduced and examined in Chapter 7. Here we discuss the special properties of diagonal, tridiagonal, triangular, selfadjoint, unitary, and normal matrices, and introduce different kinds of definite matrices and related special matrix classes (such as quasidefinite, quasisemidefinit, N-, P-, N-P, and P-N matrices). The last chapter of the book introduces and discusses the elements of matrix analysis including vector and matrix norms, and related topics. In a one-semester introductory course we suggest to cover Chapters 1, 2, 3 and large part of Chapters 6. In a one-semester higher level (or in a second) course on the subject we suggest to cover Chapters 4, 5, part of 6, 7, and 8.

It is our pleasure to acknowledge encouragement we have received from our colleagues and friends in both Hungary and the U.S.A. In preparing the manuscript we obtained significant help from our students, and colleagues. Our special thanks should be addressed to Dr. Zoltan Varga for providing us with helpful critique. We mention that it was through various grants from the National Science Foundation and the Joint U.S.-Hungarian Research and Technology Fund that the authors were able to enjoy the continuation of their productive and happy collaboration.

And finally special thanks are addressed to the editorial staff of World Scientific Publising Co. for the continuous support, encouragement and the careful and timely editorial work.

Ferenc Szidarovszky

Tucson, Arizona, U.S.A.

Sandor Molnar, Mark Molnar

Budapest, Hungary

Contents

List of Tables

List of Figures

Chapter 1

Vectors and Matrices

1.1 Introduction

In modeling and solving problems in engineering, economics, and in any other field of applied sciences there is often a need to present data in an organized way of a rectangular array. For example, if the prices of five items are listed as $(p_1, p_2, p_3, p_4, p_5)$ then such an array is constructed. This array has the specialty that it consists of only one row, therefore it is often called a *row vector*. Assume next, that a small firm produces three kinds of products. If the production levels are denoted by x_1, x_2 and x_3 then another type of special array

$$\begin{pmatrix} x_1 \\ x_2 \\ x_3 \end{pmatrix}$$

can be constructed, which consists of only one column. Therefore, it is sometimes called a *column vector*. Consider again the same small firm and assume that for the next week the management considers two alternative production plans. Let x_1, x_2, x_3 and y_1, y_2, y_3 denote the alternative production volumes. The data can be conveniently summarized in a rectangular array form:

$$\begin{pmatrix} x_1 & y_1 \\ x_2 & y_2 \\ x_3 & y_3 \end{pmatrix}, \qquad (1.1)$$

where the rows correspond to the different products, and the columns correspond to the alternative plans. This array consists of 3 rows and 2 columns therefore it is usually called a 3×2 *matrix* (pronounced "three by two" matrix). By constructing such arrays, a new mathematical structure is developed.

Definition 1.1. For a given (m, n) pair of positive integers, an $m \times n$ *matrix* is a rectangular array of real (or complex) numbers given as

$$\mathbf{A} = \begin{pmatrix} a_{11} & a_{12} & \cdots & a_{1n} \\ a_{21} & a_{22} & \cdots & a_{2n} \\ \cdots & \cdots & \cdots & \cdots \\ a_{m1} & a_{m2} & \cdots & a_{mn} \end{pmatrix}.$$

This matrix has m rows and n columns. Instead of saying that this matrix is $m \times n$ we may say that its *type* is $m \times n$. The numbers a_{11}, a_{12}, are called the *elements* or *entries* of the matrix. Notice, that each matrix element has two subscripts. The first subscript indicates the row in which the element is located, and the second subscript shows the column in which the element is placed. The set of all real (or complex) $m \times n$ matrices is denoted by $R^{m \times n}$ (or $C^{m \times n}$), which is the obvious generalization of the usual notation R (or C) for the set of all real (or complex) numbers.

Example 1.1. The type of matrix

$$\mathbf{A} = \begin{pmatrix} 1 & 2 & 3 \\ 4 & 5 & 6 \end{pmatrix}$$

is 2×3, since it consists of two rows and three columns, furthermore

$$a_{11} = 1, \quad a_{12} = 2, \quad a_{13} = 3,$$
$$a_{21} = 4, \quad a_{22} = 5, \quad a_{23} = 6.$$

◆

Matrices are usually denoted by bold faced capital letters[a] such as **A**, **B**, **C** and so on. Sometimes we refer to the matrix element a_{ij} as the (i, j) element or (i, j) entry of the matrix. In some applications it is convenient to use the notation $\mathbf{A} = (a_{ij})$, when a special emphasis is placed on the matrix elements. If one needs to indicate the type of matrix **A**, then the simple notation $\mathbf{A}_{m \times n}$ or the slightly more complicated $\mathbf{A} = (a_{ij})_{m \times n}$ or

$$\mathbf{A} = \left(a_{ij} \right)_{i,j=1}^{m,n}$$

can be used, which shows that the value of i (the row-index) is between 1 and m, and the value of j (the column-index) is between 1 and n.

In most applications the rows and columns of matrices refer to certain quantities, parameters, or alternatives. If the prices of different products are summarized in a row vector (as it was done previously), then the columns refer to the different products, and if the production volumes are summarized in a column vector, then the rows refer to the products. Similarly, in the case of matrix (1.1), the rows correspond to the three products and the columns refer to the two production plans. In many cases it is useful to interchange the meanings of the rows and columns. Then a new matrix is constructed in which each column is formed from the elements of the corresponding row placed in the same order. The same result is obtained, when the elements of each column are placed in the corresponding row of the new matrix.
This matrix operation can be formally defined as follows.

---

[a] Another notation for matrices used sometimes are capital underlined letters.

Definition 1.2. Let $\mathbf{A}$ be an $m \times n$ matrix, then the *transpose* of $\mathbf{A}$ is the $n \times m$ matrix, the (i, j) element of which is a_{ji}. The transpose of $\mathbf{A}$ is denoted by $\mathbf{A}^T$ and this matrix operation is called *transposition.*

Example 1.2. The transpose of a row vector is a column vector:

$$(1,2,3)^T = \begin{pmatrix} 1 \\ 2 \\ 3 \end{pmatrix},$$

the transpose of a column vector is a row vector:

$$\begin{pmatrix} 4 \\ 5 \\ 6 \end{pmatrix}^T = (4,5,6),$$

and the transpose of an $m \times n$ matrix is an $n \times m$ matrix:

$$\begin{pmatrix} 1 & 2 & 3 \\ 4 & 5 & 6 \end{pmatrix}^T = \begin{pmatrix} 1 & 4 \\ 2 & 5 \\ 3 & 6 \end{pmatrix}.$$

♦

Notice that $(\mathbf{A}^T)^T = \mathbf{A}$, since the (i, j) element of $\mathbf{A}^T$ is a_{ji}, and therefore the (i, j) element of $(\mathbf{A}^T)^T$ is a_{ij}, which is the (i, j) element of the original matrix $\mathbf{A}$.

In several cases it is convenient to emphasize in the notation if a matrix is a column vector or a row vector. Column vectors are denoted by boldface lower case letters such as $\mathbf{a}$, $\mathbf{b}$, $\mathbf{c}$ and so on. Since row vectors are the transposes of column vectors, they can be denoted as $\mathbf{a}^T$, $\mathbf{b}^T$, $\mathbf{c}^T$, and so on.

For an arbitrary matrix of the type $m \times n$, m of course, need not be equal to n. In the important special case of $m = n$, the matrix is called a *square matrix*. The common value of m and n is called the *order* of the matrix. The entries a_{11}, a_{22}, ..., a_{mm} of a square matrix of order m are called the *diagonal elements,* and they form the main *diagonal* or simply the diagonal of the matrix.

Example 1.3. Matrix

$$\begin{pmatrix} 1 & 5 & 6 \\ 7 & 2 & 8 \\ 9 & 10 & 3 \end{pmatrix}$$

is a square matrix of order 3, and the elements 1, 2 and 3 form the diagonal of the matrix.

◆

A matrix composed entirely of zeros is called the *zero* (or *null*) *matrix*, and a vector of zeros is called a *zero* (or *null*) *vector*. A zero matrix is denoted by **O**, and a zero column (or row) vector is denoted by **0** (or $\mathbf{0}^T$).

A square matrix with all off-diagonal elements equal to zero is a *diagonal* matrix. A special diagonal matrix, where all diagonal elements are equal to one, is called the *identity* matrix. The $n \times n$ identity matrix is usually denoted by $\mathbf{I}_n$. A square matrix in which all elements below the diagonal are equal to zero is called *upper triangular*, and similarly, a square matrix with all zero elements above the diagonal is called *lower triangular*.

Example 1.4. Consider matrices

$$\mathbf{A} = \begin{pmatrix} 1 & 0 & 0 \\ 0 & 2 & 0 \\ 0 & 0 & 3 \end{pmatrix}, \quad \mathbf{B} = \begin{pmatrix} 0 & 1 & 1 \\ 0 & 1 & 1 \\ 0 & 0 & 2 \end{pmatrix}, \quad \mathbf{C} = \begin{pmatrix} 0 & 0 & 0 \\ 1 & 1 & 0 \\ 1 & 1 & 2 \end{pmatrix},$$

then **A** is diagonal, **B** is upper triangular, and **C** is lower triangular. Notice, that **A** (like any other diagonal matrix) satisfies the definitions of both upper and lower triangular matrices, therefore it is also upper and lower triangular.

◆

Notice that the transpose of an upper triangular matrix is lower triangular, and the transpose of a lower triangular matrix is upper triangular. It is easy to see that the transpose of a diagonal matrix is itself.

In economic theory, triangular matrices have a special meaning. Consider an $n \times n$ real square matrix $\mathbf{A}$, and assume that element a_{ij} represents the effect of unit i towards unit j, where the rows and columns of the matrix correspond to some economic units (for example, sectors in input-output models). In the case of an upper triangular matrix, $a_{ij} = 0$ for $i + j < n + 1$; and for lower triangular matrices $a_{ij} = 0$ for $i + j > n + 1$. That is, the zero matrix elements indicate that there is no effect from unit i to units $j < n + 1 - i$ (or $j > n + 1 - i$).

Definition 1.3. An $n \times n$ matrix $\mathbf{A}$ is called *decomposable* if there is a nonempty proper subset J of $\{1, 2, ..., n\}$ such that

$$a_{ij} = 0 \quad \text{for } i \notin J \text{ and } j \in J.$$

An $n \times n$ real matrix is called *indecomposable*, if it is not decomposable and is not the 1×1 zero matrix.

It is easy to see that a matrix is decomposable if and only if its transpose is decomposable. Any decomposable matrix can be transformed into the special form

$$\begin{pmatrix} \mathbf{A}_{11} & \mathbf{A}_{12} \\ \mathbf{O} & \mathbf{A}_{22} \end{pmatrix}$$

by interchanging its rows and columns, where $\mathbf{A}_{11}$ is $k \times k$, $\mathbf{A}_{12}$ is $k \times (n - k)$, $\mathbf{A}_{22}$ is an $(n - k) \times (n - k)$ matrix, $\mathbf{O}$ is the $(n - k) \times k$ zero matrix, and set J becomes $\{1, 2, ..., k\}$. In terms of the above economic interpretation the zero block indicates that there is no effect from the units not belonging to J towards the units belonging to J.

Definition 1.4. A square matrix $\mathbf{A}$ is called *symmetric* if $\mathbf{A}^T = \mathbf{A}$, and it is called *skew-symmetric* if $\mathbf{A}^T = -\mathbf{A}$.

Notice that an $n \times n$ matrix is symmetric if and only if for $k = 1, 2, ..., n$, its kth column has the same elements as its kth row in the same order.

As a special case, all diagonal matrices are symmetric. The diagonal of a skew symmetric matrix consists of zeros.

Example 1.5. Matrix

$$\begin{pmatrix} 1 & 2 & 3 \\ 2 & 4 & 5 \\ 3 & 5 & 6 \end{pmatrix}$$

is a symmetric 3×3 matrix, and matrix

$$\begin{pmatrix} 0 & 1 & 2 \\ -1 & 0 & -3 \\ -2 & 3 & 0 \end{pmatrix}$$

is a skew symmetric 3×3 matrix.

◆

1.2 Comparison of Matrices

Matrices **A** and **B** are equal if they have the same type and the corresponding elements in the two matrices are equal. If $\mathbf{A} = (a_{ij})$ and $\mathbf{B} = (b_{ij})$, then $\mathbf{A} = \mathbf{B}$ if and only if $a_{ij} = b_{ij}$ for all i and j.

Similarly we say that for real matrices **A** and **B**, $\mathbf{A} \leq \mathbf{B}$, if they have the same type and for all i and j, $a_{ij} \leq b_{ij}$. Analogously, $\mathbf{A} < \mathbf{B}$ if they have the same type and for all i and j, $a_{ij} < b_{ij}$,

Notice that matrices can be compared in the above sense only if they have the same type.

Example 1.6. Let

$$A = \begin{pmatrix} 1 & 2 \\ 3 & 4 \end{pmatrix}, \quad B = \begin{pmatrix} 1 & 2 \\ 3 & 4 \end{pmatrix}, \quad C = \begin{pmatrix} 1 & 3 \\ 3 & 5 \end{pmatrix}, \quad D = \begin{pmatrix} 2 & 3 \\ 4 & 5 \end{pmatrix},$$

then for example, $A = B$, $A \leq C$, $A < D$. If one defines

$$E = \begin{pmatrix} 1 & 2 & 3 \\ 4 & 5 & 6 \end{pmatrix},$$

then it cannot be compared to any of matrices A, B, C or D.

♦

In comparing matrices an important comment has to be made. If a and b are two real numbers, then exactly one of the relations $a = b$, $a < b$ or $a > b$ holds. That is, any two real numbers can be compared in this way. However real matrices may not be compared even if they have the same type. For example, row vectors $a^T = (1, 2)$ and $b^T = (2, 1)$ cannot be compared, since they are different, in the first element a^T is the smaller but in the second element b^T is smaller. This phenomenon plays an important role in many fields of applied sciences. For example, in single objective optimization we are looking for a best solution, since any two values of the objective functions can be compared; however, in the case of optimizing with multiple objectives we are looking for so called efficient solutions, when none of the objective function values can be improved without worsening another one.

1.3   Elementary Matrix Algebra

In this section matrix operations will be introduced, and their main properties will be discussed.

Definition 1.5. Let A be an $m \times n$ real (or complex) matrix and let a be a real (or complex) number. The product aA is defined as the $m \times n$ matrix with (i, j) element $a \cdot a_{ij}$. That is, each element of the matrix is multiplied by a.

This definition can be briefly written as $a\mathbf{A} = \left(a \cdot a_{ij} \right)_{m \times n}$.

As a simple example assume that the elements of a matrix represent cost data, and each element is given in dollars. If someone wants to change the dimension of the data to $1000, then each matrix element has to be multiplied by the same constant 0.001.

Example 1.7. For a numerical example assume that $a = 3$ and

$$\mathbf{A} = \begin{pmatrix} 1 & 2 \\ 3 & 4 \end{pmatrix}, \quad \text{then} \quad a\mathbf{A} = \begin{pmatrix} 3 & 6 \\ 9 & 12 \end{pmatrix}.$$

◆

Notice, if $a = 0$, then $a\mathbf{A}$ is the zero matrix for all $\mathbf{A}$, since each matrix element is multiplied by zero. The real (or complex) number a is sometimes called a *scalar,* and this matrix operation is called *multiplication by scalars.*

Definition 1.6. The *sum* of matrices $\mathbf{A}$ and $\mathbf{B}$ is defined whenever $\mathbf{A}$ and $\mathbf{B}$ have the same type. Each element of $\mathbf{A} + \mathbf{B}$ equals the sum of the two corresponding elements of $\mathbf{A}$ and $\mathbf{B}$. In other words, $\mathbf{A} + \mathbf{B}$ is the matrix the (i, j) element of which is $a_{ij} + b_{ij}$ for all i, j, where a_{ij} and b_{ij} are the (i, j) elements of $\mathbf{A}$ and $\mathbf{B}$, respectively. The *difference* matrix $\mathbf{A} - \mathbf{B}$ is analogously defined to be the matrix with (i, j) elements $a_{ij} - b_{ij}$.
We can summarize this definition as

$$\mathbf{A} + \mathbf{B} = \left(a_{ij} + b_{ij} \right)_{m \times n} \quad \text{and} \quad \mathbf{A} - \mathbf{B} = \left(a_{ij} - b_{ij} \right)_{m \times n}.$$

Example 1.8. Matrices

$$\mathbf{A} = \begin{pmatrix} 1 & 2 \\ 3 & 4 \end{pmatrix} \quad \text{and} \quad \mathbf{B} = \begin{pmatrix} 1 & 2 & 3 \\ 4 & 5 & 6 \end{pmatrix}$$

cannot be added or subtracted, since they have different types. However

$$\begin{pmatrix} 1 & 2 \\ 3 & 4 \end{pmatrix} + \begin{pmatrix} 0 & 2 \\ 1 & 3 \end{pmatrix} = \begin{pmatrix} 1 & 4 \\ 4 & 7 \end{pmatrix},$$

and

$$\begin{pmatrix} 1 & 2 & 3 \\ 4 & 5 & 6 \end{pmatrix} - \begin{pmatrix} 1 & 3 & 4 \\ 2 & 5 & 5 \end{pmatrix} = \begin{pmatrix} 0 & -1 & -1 \\ 2 & 0 & 1 \end{pmatrix}.$$

◆

The above matrix operations satisfy the following properties:
(a) If $\mathbf{A}$ and $\mathbf{B}$ have the same type, then

$$\mathbf{A} + \mathbf{B} = \mathbf{B} + \mathbf{A}. \tag{1.2}$$

That is, matrix addition is *commutative*, which is a simple consequence of the fact that in adding matrices we add the corresponding matrix elements, and the addition of real (or complex) numbers is commutative.
(b) If $\mathbf{A}$, $\mathbf{B}$ and $\mathbf{C}$ have the same type, then

$$(\mathbf{A} + \mathbf{B}) + \mathbf{C} = \mathbf{A} + (\mathbf{B} + \mathbf{C}) \tag{1.3}$$

That is, matrix addition is *associative*. This property is also the simple consequence of the associativity of the addition of real (or complex) numbers. If $\mathbf{A}_1$, $\mathbf{A}_2$, ..., $\mathbf{A}_K$ are real (or complex) matrices of the same type, then their sum is defined by the recursion $\mathbf{S}_1 = \mathbf{A}_1$, and $\mathbf{S}_i = \mathbf{A}_{i-1} + \mathbf{A}_i$ for $i = 2, 3, ..., K$.
Then

$$\mathbf{A}_1 + \mathbf{A}_2 + \cdots + \mathbf{A}_K = \mathbf{S}_K.$$

(c) Let $\mathbf{A}$ be any matrix, and $\mathbf{O}$ be the zero matrix of the same order. Then

$$\mathbf{A} + \mathbf{O} = \mathbf{A}. \tag{1.4}$$

(d) If $\mathbf{A}$ and $\mathbf{B}$ have the same type, then with any scalar a,

$$a(\mathbf{A} + \mathbf{B}) = a\mathbf{A} + a\mathbf{B}, \tag{1.5}$$

and if a and b are two scalars, and $\mathbf{A}$ is any matrix, then

$$(a+b)\mathbf{A} = a\mathbf{A} + b\mathbf{A}. \tag{1.6}$$

These two equations are called the *distributivity* properties. Equation (1.5) shows distributivity with respect to the addition of matrices, and (1.6) is the distributive property with respect to the addition of scalars.

(e) If $\mathbf{A}$ and $\mathbf{B}$ have the same type, then

$$(\mathbf{A} + \mathbf{B})^T = \mathbf{A}^T + \mathbf{B}^T. \tag{1.7}$$

Assume that both $\mathbf{A}$ and $\mathbf{B}$ are $m \times n$, then both sides of this equation are $n \times m$ matrices, and the (i, j) elements are the same: $a_{ji} + b_{ji}$ for all i and j.

(f) For positive integers k,

$$\mathbf{A} + \mathbf{A} + \ldots + \mathbf{A} = k\mathbf{A}, \tag{1.8}$$

where on the left-hand side we have k terms, each of them equals the same matrix $\mathbf{A}$.

(g) For all matrices $\mathbf{A}$ and $\mathbf{B}$ of the same type,

$$\mathbf{A} - \mathbf{B} = \mathbf{A} + (-1) \cdot \mathbf{B}, \tag{1.9}$$

since for all i and j,

$$a_{ij} - b_{ij} = a_{ij} + (-1) \cdot b_{ij}.$$

Multiplication of matrices will be defined next. For simplifying the discussion, particular cases will be introduced before presenting the general definition.

As the first special case we define the product of row vectors by column vectors. Let $\mathbf{a}^T = (a_i)$ and $\mathbf{b} = (b_j)$ be a row and a column vector, respectively. The product $\mathbf{a}^T \mathbf{b}$ is defined only when the two vectors have the same number of elements, and in this case

$$\mathbf{a}^T \mathbf{b} = a_1 b_1 + a_2 b_2 + \ldots + a_n b_n = \sum_{i=1}^{n} a_i b_i,$$

where n is the common "length" of the vectors.

Example 1.9. The product

$$(1,2,3)\begin{pmatrix} 4 \\ 5 \end{pmatrix}$$

cannot be defined, but

$$(1,2,3)\begin{pmatrix} 2 \\ 3 \\ 4 \end{pmatrix} = 1 \times 2 + 2 \times 3 + 3 \times 4 = 20.$$

♦

Notice that $\mathbf{a}^T \mathbf{b}$ is always a scalar, which can also be considered as a 1×1 matrix. This multiplication can be illustrated and explained by the simple economic example, when a firm produces 3 items, the sale prices of which form the row vector $\mathbf{p}^T = (p_1, p_2, p_3)$ and the produced quantities are given in a column vector

$$\mathbf{x} = \begin{pmatrix} x_1 \\ x_2 \\ x_3 \end{pmatrix}.$$

Then the revenue (or total sale value) by selling the products is given by the product

$$\mathbf{p}^T \mathbf{x} = p_1 x_1 + p_2 x_2 + p_3 x_3.$$

Assume next that **A** is an $m \times n$ matrix, and **x** is a column vector. The product **Ax** is defined only if the length of **x** equals the length of the rows of **A** (that is, when **x** has n elements), the product is an m-element column vector the ith entry of which is obtained as the product of the ith row of **A** by the column vector **x**.

That is, the ith element of **Ax** equals $\sum\limits_{j=1}^{n} a_{ij} x_j$.

Example 1.10. The product

$$\begin{pmatrix} 1 & 2 \\ 3 & 4 \end{pmatrix} \begin{pmatrix} 1 \\ 2 \\ 3 \end{pmatrix}$$

is not defined, but

$$\begin{pmatrix} 1 & 2 \\ 3 & 4 \end{pmatrix} \begin{pmatrix} 1 \\ 2 \end{pmatrix} = \begin{pmatrix} 1 \times 1 + 2 \times 2 \\ 3 \times 1 + 4 \times 2 \end{pmatrix} = \begin{pmatrix} 5 \\ 11 \end{pmatrix}.$$

♦

Assume next that $\mathbf{x}^T$ is a row vector and **A** is an $m \times n$ matrix. The product $\mathbf{x}^T \mathbf{A}$ is defined only if the length of $\mathbf{x}^T$ equals the length of the columns of **A** (that is, when **x** has m-elements), the product is an n-element row vector, the ith entry of which is obtained as the product of $\mathbf{x}^T$ by the ith column of **A**. That is, the ith element of $\mathbf{x}^T \mathbf{A}$ equals

$$\sum\limits_{j=1}^{m} x_j a_{ji}.$$

Example 1.11. The product

$$(1,2,3) \begin{pmatrix} 1 & 2 \\ 3 & 4 \end{pmatrix}$$

cannot be defined, but

$$(1, 2)\begin{pmatrix} 1 & 2 \\ 3 & 4 \end{pmatrix} = (1 \times 1 + 2 \times 3, \ 1 \times 2 + 2 \times 4) = (7, 10)$$

♦

We are ready now to consider the general case of matrix multiplication.

Definition 1.7. Let **A** and **B** be two matrices. The product **AB** can be defined only if the rows of **A** have the same length as the columns of **B**, and then the (i, j) element of **AB** equals the product of the ith row of **A** by the jth column of **B**. If **A** is $m \times n$ and **B** is $p \times q$, then **AB** is defined only if $n = p$, its type is $m \times q$, and its (i, j) element is obtained as

$$\sum_{k=1}^{n} a_{ik} b_{kj}.$$

Example 1.12. The product

$$\begin{pmatrix} 1 & 2 \\ 3 & 4 \end{pmatrix} \begin{pmatrix} 1 & 4 \\ 2 & 5 \\ 3 & 6 \end{pmatrix}$$

is not defined, but

$$\begin{pmatrix} 1 & 2 \\ 3 & 4 \end{pmatrix} \begin{pmatrix} 1 & -1 \\ 2 & -2 \end{pmatrix} = \begin{pmatrix} 5 & -5 \\ 11 & -11 \end{pmatrix},$$

since

$$(1,2)\begin{pmatrix} 1 \\ 2 \end{pmatrix} = 1 \times 1 + 2 \times 2 = 5,$$

$$(1,2)\begin{pmatrix} -1 \\ -2 \end{pmatrix} = 1 \times (-1) + 2 \times (-2) = -5,$$

$$(3,4)\begin{pmatrix} 1 \\ 2 \end{pmatrix} = 3 \times 1 + 4 \times 2 = 11,$$

and

$$(3,4)\begin{pmatrix} -1 \\ -2 \end{pmatrix} = 3 \times (-1) + 4 \times (-2) = -11.$$

♦

As we have seen before, a row vector can be multiplied by a column vector only if they have the same length, and the product is always a scalar. However, a column vector can always be multiplied by a row vector even if they have different lengths, and the product is always a matrix, which is called a *dyad*. If $\mathbf{x}$ is an m-element column vector and $\mathbf{y}^T$ is an n element row vector, then

$$\mathbf{x}\mathbf{y}^T = \begin{pmatrix} x_1 \\ x_2 \\ \vdots \\ x_m \end{pmatrix}(y_1, y_2, ..., y_n) = \begin{pmatrix} x_1 y_1 & x_1 y_2 & \cdots & x_1 y_n \\ x_2 y_1 & x_2 y_2 & \cdots & x_2 y_n \\ \cdots & \cdots & \cdots & \cdots \\ x_m y_1 & x_m y_2 & \cdots & x_m y_n \end{pmatrix},$$

since the ith row of $\mathbf{x}$ is the scalar x_i and the jth column of $\mathbf{y}^T$ is the number y_j, and their product is $x_i y_j$.

If one selects two arbitrary matrices $\mathbf{A}$ and $\mathbf{B}$, then in their multiplication he/she should face one of the following possibilities:

(i) Neither $\mathbf{AB}$, nor $\mathbf{BA}$ exists. Such an example is provided by matrices

$$A = \begin{pmatrix} 1 & 2 \\ 3 & 4 \end{pmatrix} \quad \text{and} \quad B = \begin{pmatrix} 1 & 2 & 3 \\ 1 & 1 & 1 \\ 2 & 1 & 0 \end{pmatrix}.$$

(ii) Exactly one of the products **AB** and **BA** exists. For example, select

$$A = \begin{pmatrix} 1 & 2 \\ 1 & 1 \end{pmatrix} \quad \text{and} \quad B = \begin{pmatrix} 1 & 0 & 1 \\ 1 & 1 & 1 \end{pmatrix},$$

then **AB** does exist and

$$AB = \begin{pmatrix} 1 & 2 \\ 1 & 1 \end{pmatrix}\begin{pmatrix} 1 & 0 & 1 \\ 1 & 1 & 1 \end{pmatrix} = \begin{pmatrix} 3 & 2 & 3 \\ 2 & 1 & 2 \end{pmatrix},$$

but

$$BA = \begin{pmatrix} 1 & 0 & 1 \\ 1 & 1 & 1 \end{pmatrix}\begin{pmatrix} 1 & 2 \\ 1 & 1 \end{pmatrix}$$

cannot be defined, since the rows of **B** have 3 elements and the columns of **A** have only 2.

(iii) Both **AB** and **BA** exist, but they have different types. As an example, select

$$A = (1,1) \quad \text{and} \quad B = \begin{pmatrix} 2 \\ 2 \end{pmatrix},$$

then

$$AB = (1,1)\begin{pmatrix} 2 \\ 2 \end{pmatrix} = 1 \times 2 + 1 \times 2 = 4$$

is a scalar and

$$\mathbf{BA} = \begin{pmatrix} 2 \\ 2 \end{pmatrix}(1,1) = \begin{pmatrix} 2 & 2 \\ 2 & 2 \end{pmatrix}$$

is a dyad.

(iv) Both **AB** and **BA** exist, they have the same type but the products are not equal. Select

$$\mathbf{A} = \begin{pmatrix} 1 & 1 \\ -1 & -1 \end{pmatrix} \quad \text{and} \quad \mathbf{B} = \begin{pmatrix} 1 & 1 \\ 1 & 1 \end{pmatrix},$$

then

$$\mathbf{AB} = \begin{pmatrix} 1 & 1 \\ -1 & -1 \end{pmatrix}\begin{pmatrix} 1 & 1 \\ 1 & 1 \end{pmatrix} = \begin{pmatrix} 2 & 2 \\ -2 & -2 \end{pmatrix}$$

and

$$\mathbf{BA} = \begin{pmatrix} 1 & 1 \\ 1 & 1 \end{pmatrix}\begin{pmatrix} 1 & 1 \\ -1 & -1 \end{pmatrix} = \begin{pmatrix} 0 & 0 \\ 0 & 0 \end{pmatrix}.$$

This example also shows a different problem in matrix multiplication. In the second case **BA** is the zero matrix, however neither **A** nor **B** is zero, and furthermore there is no zero element in **A** or **B**. This shows a different phenomenon than the one we used to have and apply in the case of real numbers. The product of real (or complex) numbers is zero only if at least one of the factors equals zero. This idea is used when one solves real equations by factorization. Unfortunately, this method cannot be used in solving matrix equations.

(v) Both **AB** and **BA** exist, they have the same type and are equal. Such a special case can be illustrated by matrices

$$\mathbf{A} = \begin{pmatrix} 1 & 1 \\ 1 & 1 \end{pmatrix} \quad \text{and} \quad \mathbf{B} = \begin{pmatrix} 2 & 2 \\ 2 & 2 \end{pmatrix},$$

when

$$\mathbf{AB} = \begin{pmatrix} 1 & 1 \\ 1 & 1 \end{pmatrix} \begin{pmatrix} 2 & 2 \\ 2 & 2 \end{pmatrix} = \begin{pmatrix} 4 & 4 \\ 4 & 4 \end{pmatrix}$$

and

$$\mathbf{BA} = \begin{pmatrix} 2 & 2 \\ 2 & 2 \end{pmatrix} \begin{pmatrix} 1 & 1 \\ 1 & 1 \end{pmatrix} = \begin{pmatrix} 4 & 4 \\ 4 & 4 \end{pmatrix}.$$

Hence, the multiplication of matrices is not a commutative operation in general. However, it satisfies the following properties:

(a) If the product $(\mathbf{AB}) \cdot \mathbf{C}$ exists, then $\mathbf{A} \cdot (\mathbf{BC})$ also exists and

$$(\mathbf{AB}) \cdot \mathbf{C} = \mathbf{A} \cdot (\mathbf{BC}). \tag{1.10}$$

That is, matrix multiplication is an *associative* operation. This property can be proved as follows. Introduce the notation $\mathbf{D} = \mathbf{AB}$ and $\mathbf{E} = \mathbf{BC}$, then the (i, j) element of $\mathbf{D}$ is given as

$$d_{ij} = \sum_l a_{il} b_{lj},$$

and therefore the (i, j) element of the left-hand side of (1.10) is as follows:

$$\sum_k d_{ik} c_{kj} = \sum_k \left(\sum_l a_{il} b_{lk} \right) c_{kj}.$$

On the other hand, the (i, j) element of $\mathbf{E}$ is

$$e_{ij} = \sum_k b_{ik} c_{kj},$$

and therefore the (i, j) element of the right-hand side of (1.10) equals

$$\sum_l a_{il} e_{lj} = \sum_l a_{il} \left(\sum_k b_{lk} c_{kj} \right).$$

Since a_{il} does not depend on the summation variable k, this expression gives the same value as the (i, j) element of the left-hand side of (1.10).

(b) If one of the matrices $(\mathbf{A} + \mathbf{B}) \cdot \mathbf{C}$ and $\mathbf{AC} + \mathbf{BC}$ exists, then the other matrix is also defined and they are equal:

$$(\mathbf{A} + \mathbf{B}) \cdot \mathbf{C} = \mathbf{AC} + \mathbf{BC}. \tag{1.11}$$

Similarly, if one of the matrices $\mathbf{A} \cdot (\mathbf{B} + \mathbf{C})$ and $\mathbf{AB} + \mathbf{AC}$ exists, then the other matrix is also defined and they are equal:

$$\mathbf{A} \cdot (\mathbf{B} + \mathbf{C}) = \mathbf{AB} + \mathbf{AC}. \tag{1.12}$$

These properties show that matrix multiplication is *distributive*, and their proof is similar to the one presented above for associativity.

(c) If $\mathbf{A}$ is $m \times n$, and $\mathbf{I}_m$ and $\mathbf{I}_n$ denote the $m \times m$ and $n \times n$ identity matrices, respectively, then

$$\mathbf{I}_m \cdot \mathbf{A} = \mathbf{A} \qquad \text{and} \qquad \mathbf{A} \cdot \mathbf{I}_n = \mathbf{A}. \tag{1.13}$$

These properties can also be proved in an easy way, the proofs are left as an exercise. Relations (1.13) show that multiplying by identity matrices leaves matrices unchanged. The same holds for real (or complex) numbers, when we multiply them by 1. Therefore, identity matrices can be considered as the matrix-versions of the real number 1.

(d) If $\mathbf{AB}$ exists, then $\mathbf{B}^T\mathbf{A}^T$ also exists, furthermore

$$(\mathbf{AB})^T = \mathbf{B}^T\mathbf{A}^T\dots. \tag{1.14}$$

A simple proof of this equation can be given by comparing the (i, j) elements of the two sides of the equality. The (i, j) element of $\mathbf{AB}$ equals $\sum_k a_{ik} b_{kj}$, therefore the (i, j) element of $(\mathbf{AB})^T$ is obtained by interchanging i and j: $\sum_k a_{jk} b_{ki} = \sum_k b_{ki} a_{jk}$.

This is the (i, j) element of the right-hand side, since b_{ki} is the (i, k) element of $\mathbf{B}^T$, and a_{jk} is the (k, j) element of $\mathbf{A}^T$.

(e) For any matrix $\mathbf{A}$ and zero matrix $\mathbf{O}$,

$$\mathbf{AO} = \mathbf{O} \quad \text{and} \quad \mathbf{OA} = \mathbf{O} \tag{1.15}$$

assuming that the left-hand sides are defined. Notice that if $\mathbf{A}$ is $m \times n$ and $\mathbf{O}$ is $n \times p$, then $\mathbf{AO}$ is the $m \times p$ zero matrix, and if $\mathbf{A}$ is $m \times n$ and $\mathbf{O}$ is $p \times m$, then $\mathbf{OA}$ is the $p \times n$ zero matrix.

Let $\mathbf{A}_1$, $\mathbf{A}_2$, ..., $\mathbf{A}_k$ be real (or complex) matrices of the types $m_1 \times n_1$, $m_2 \times n_2$, ..., $m_k \times n_k$, respectively, and assume that $n_1 = m_2$, $n_2 = m_3$, ..., $n_{k-1} = m_k$. The product of these matrices is defined by the recursion $\mathbf{P}_1 = \mathbf{A}_1$ and $\mathbf{P}_i = \mathbf{P}_{i-1}\mathbf{A}_i$ for $i = 2, 3, ..., k$ and letting $\mathbf{A}_1 \cdot \mathbf{A}_2 ... \mathbf{A}_k = \mathbf{P}_k$. In the special case, when $\mathbf{A}$ is an $n \times n$ square matrix we may select $\mathbf{A}_1 = \mathbf{A}_2 = ... = \mathbf{A}_k = \mathbf{A}$, and the product $\mathbf{A} \cdot \mathbf{A} ... \mathbf{A}$ can be simply denoted by $\mathbf{A}^k$. For convenience, we define $\mathbf{A}^0 = \mathbf{I}_n$ for all $n \times n$ square matrices $\mathbf{A}$.

Example 1.13. Select

$$\mathbf{A} = \begin{pmatrix} 1 & 2 \\ 1 & 2 \end{pmatrix},$$

then

$$\mathbf{A}^2 = \mathbf{AA} = \begin{pmatrix} 1 & 2 \\ 1 & 2 \end{pmatrix}\begin{pmatrix} 1 & 2 \\ 1 & 2 \end{pmatrix} = \begin{pmatrix} 3 & 6 \\ 3 & 6 \end{pmatrix},$$

$$\mathbf{A}^3 = \mathbf{A}^2\mathbf{A} = \begin{pmatrix} 3 & 6 \\ 3 & 6 \end{pmatrix}\begin{pmatrix} 1 & 2 \\ 1 & 2 \end{pmatrix} = \begin{pmatrix} 9 & 18 \\ 9 & 18 \end{pmatrix},$$

and so on.

◆

Assume next that

$$p(x) = a_0 + a_1 x + a_2 x^2 + ... + a_m x^m$$

is a single-variable polynomial with real (or complex) coefficients, and $\mathbf{A}$ is an $n \times n$ square matrix. The *matrix-polynomial* $p(\mathbf{A})$ is defined as

$$p(\mathbf{A}) = a_0 \mathbf{I}_n + a_1 \mathbf{A} + a_2 \mathbf{A}^2 + ... + a_m \mathbf{A}^m, \tag{1.16}$$

where $\mathbf{I}_n$ is the $n \times n$ identity matrix.

Example 1.14. Let $p(x) = 2 + 2x + x^2$ and **A** as in the previous example, then

$$p(\mathbf{A}) = 2\begin{pmatrix} 1 & 0 \\ 0 & 1 \end{pmatrix} + 2\begin{pmatrix} 1 & 2 \\ 1 & 2 \end{pmatrix} + \begin{pmatrix} 3 & 6 \\ 3 & 6 \end{pmatrix}$$

$$= \begin{pmatrix} 2 & 0 \\ 0 & 2 \end{pmatrix} + \begin{pmatrix} 2 & 4 \\ 2 & 4 \end{pmatrix} + \begin{pmatrix} 3 & 6 \\ 3 & 6 \end{pmatrix} = \begin{pmatrix} 7 & 10 \\ 5 & 12 \end{pmatrix}.$$

♦

1.4 Inverse of a Matrix

We start this section with the definition of the inverse of a square matrix.

Definition 1.8. Let **A** be an $n \times n$ square matrix. The *inverse* of **A** is the $n \times n$ matrix **X** which satisfies the relation

$$\mathbf{AX} = \mathbf{XA} = \mathbf{I}_n, \tag{1.17}$$

where $\mathbf{I}_n$ is the $n \times n$ identity matrix. If such an **X** exists, then **A** is called an *invertible* matrix, and the inverse of **A** is denoted by $\mathbf{A}^{-1}$.

These relations show that inverse matrices generalize the concept of the reciprocal of a real (or complex) number, since for all real (or complex) $a \neq 0$,

$$aa^{-1} = a\frac{1}{a} = 1 \quad \text{and} \quad a^{-1}a = \frac{1}{a}a = 1.$$

If $a \neq 0$, then a^{-1} exists. Unfortunately, in the case of matrices the situation is more complicated, as it is illustrated in the following example.

Example 1.15. Consider matrix

$$\mathbf{A} = \begin{pmatrix} 1 & 1 \\ 1 & 1 \end{pmatrix}.$$

We will now prove that this matrix has no inverse. Assume that it has, and let

$$\mathbf{A}^{-1} = \begin{pmatrix} a_{11} & a_{12} \\ a_{21} & a_{22} \end{pmatrix}.$$

Then relation

$$\mathbf{A}\mathbf{A}^{-1} = \begin{pmatrix} 1 & 1 \\ 1 & 1 \end{pmatrix} \begin{pmatrix} a_{11} & a_{12} \\ a_{21} & a_{22} \end{pmatrix} = \begin{pmatrix} 1 & 0 \\ 0 & 1 \end{pmatrix} = \mathbf{I}_2$$

implies that

$$a_{11} + a_{21} = 1$$
$$a_{12} + a_{22} = 0$$
$$a_{11} + a_{21} = 0$$
$$a_{12} + a_{22} = 1,$$

where we equated the (1,1), (1,2), (2,1), and (2,2) elements of the left hand and right-hand sides, respectively. Notice that the first and third equations contradict to each other, since $a_{11} + a_{21}$ must not have two different values at the same time. (Similar contradiction is obtained from the second and fourth equations.) In this case $\mathbf{A} \neq \mathbf{O}$, and the matrix has even no zero element. If we change the sign of only one element of $\mathbf{A}$ to get matrix

$$\mathbf{B} = \begin{pmatrix} 1 & 1 \\ -1 & 1 \end{pmatrix},$$

then

$$\mathbf{B}^{-1} = \begin{pmatrix} \dfrac{1}{2} & -\dfrac{1}{2} \\ \dfrac{1}{2} & \dfrac{1}{2} \end{pmatrix},$$

which can be verified by simple calculation:

$$\mathbf{BB}^{-1} = \begin{pmatrix} 1 & 1 \\ -1 & 1 \end{pmatrix} \begin{pmatrix} \dfrac{1}{2} & -\dfrac{1}{2} \\ \dfrac{1}{2} & \dfrac{1}{2} \end{pmatrix} = \begin{pmatrix} 1 & 0 \\ 0 & 1 \end{pmatrix}$$

and

$$\mathbf{B}^{-1}\mathbf{B} = \begin{pmatrix} \dfrac{1}{2} & -\dfrac{1}{2} \\ \dfrac{1}{2} & \dfrac{1}{2} \end{pmatrix} \begin{pmatrix} 1 & 1 \\ -1 & 1 \end{pmatrix} = \begin{pmatrix} 1 & 0 \\ 0 & 1 \end{pmatrix}.$$

$\blacklozenge$

If $\mathbf{A}$ is the $n \times n$ zero matrix, then $\mathbf{AX} = \mathbf{O}$ for all $n \times n$ matrices $\mathbf{X}$, therefore $\mathbf{A}$ has no inverse. If $\mathbf{A} = \mathbf{I}_n$ then $\mathbf{A}^{-1} = \mathbf{I}_n$, since $\mathbf{I}_n \mathbf{I}_n = \mathbf{I}_n$.

In this moment we do not see an easy way to check if a given matrix has an inverse or not. In later chapters of this book we will introduce simple, practical conditions to check the existence of the inverse of a matrix. However, we can easily show that in the case of the existence of an inverse of a given square matrix the inverse must be unique. Assume in contrary that both $\mathbf{X}$ and $\mathbf{Y}$ are inverses of a matrix $\mathbf{A}$. Then

$$\mathbf{X} = \mathbf{XI}_n = \mathbf{X}(\mathbf{AY}) = (\mathbf{XA})\mathbf{Y} = \mathbf{I}_n\mathbf{Y} = \mathbf{Y};$$

hence $\mathbf{X}$ and $\mathbf{Y}$ are necessarily equal to each other. It is also easy to see that if both matrices $\mathbf{A}$ and $\mathbf{B}$ are invertible $n \times n$ matrices, then the inverse of their product also exists and

$$(\mathbf{AB})^{-1} = \mathbf{B}^{-1}\mathbf{A}^{-1}. \tag{1.18}$$

This relation can be verified by simple calculation:

$$(\mathbf{AB})(\mathbf{B}^{-1}\mathbf{A}^{-1}) = \mathbf{A}(\mathbf{BB}^{-1})\mathbf{A}^{-1} = \left(\mathbf{AI}\right)\mathbf{A}^{-1} = \mathbf{AA}^{-1} = \mathbf{I},$$

and

$$(\mathbf{B}^{-1}\mathbf{A}^{-1})(\mathbf{AB}) = \mathbf{B}^{-1}(\mathbf{A}^{-1}\mathbf{A})\mathbf{B} = \left(\mathbf{B}^{-1}\mathbf{I}\right)\mathbf{B} = \mathbf{B}^{-1}\mathbf{B} = \mathbf{I}.$$

1.5 Further Examples and Applications

In this section some additional examples and applications of matrix algebra will be outlined.

1. Our first example is the *algebra of block matrices*. Assume that the $m \times n$ real (or complex) matrix $\mathbf{A}$ is divided into blocks as

$$\mathbf{A} = \begin{pmatrix} \mathbf{A}_{11} & \mathbf{A}_{12} & \cdots & \mathbf{A}_{1s} \\ \mathbf{A}_{21} & \mathbf{A}_{22} & \cdots & \mathbf{A}_{2s} \\ \cdots & \cdots & \cdots & \cdots \\ \mathbf{A}_{r1} & \mathbf{A}_{r2} & \cdots & \mathbf{A}_{rs} \end{pmatrix},$$

where $\mathbf{A}_{ij}$ is an $m_i \times n_i$ matrix. It is assumed that

$$m = \sum_{i=1}^{r} m_i \quad \text{and} \quad n = \sum_{j=1}^{s} n_j.$$

Suppose that matrix $\mathbf{B}$ has the same size as $\mathbf{A}$, and it is also divided into blocks as

$$\mathbf{B} = \begin{pmatrix} \mathbf{B}_{11} & \mathbf{B}_{12} & \cdots & \mathbf{B}_{1s} \\ \mathbf{B}_{21} & \mathbf{B}_{22} & \cdots & \mathbf{B}_{2s} \\ \cdots & \cdots & \cdots & \cdots \\ \mathbf{B}_{r1} & \mathbf{B}_{r2} & \cdots & \mathbf{B}_{rs} \end{pmatrix},$$

where for all i and j, blocks $\mathbf{A}_{ij}$ and $\mathbf{B}_{ij}$ have the same size. Since $\mathbf{A}$ and $\mathbf{B}$ are added and subtracted element-wise,

$$\mathbf{A} + \mathbf{B} = \begin{pmatrix} \mathbf{A}_{11} + \mathbf{B}_{11} & \mathbf{A}_{12} + \mathbf{B}_{12} & \cdots & \mathbf{A}_{1s} + \mathbf{B}_{1s} \\ \mathbf{A}_{21} + \mathbf{B}_{21} & \mathbf{A}_{22} + \mathbf{B}_{22} & \cdots & \mathbf{A}_{2s} + \mathbf{B}_{2s} \\ \cdots & \cdots & \cdots & \cdots \\ \mathbf{A}_{r1} + \mathbf{B}_{r1} & \mathbf{A}_{r2} + \mathbf{B}_{r2} & \cdots & \mathbf{A}_{rs} + \mathbf{B}_{rs} \end{pmatrix}, \qquad (1.19)$$

and

$$\mathbf{A} - \mathbf{B} = \begin{pmatrix} \mathbf{A}_{11} - \mathbf{B}_{11} & \mathbf{A}_{12} - \mathbf{B}_{12} & \cdots & \mathbf{A}_{1s} - \mathbf{B}_{1s} \\ \mathbf{A}_{21} - \mathbf{B}_{21} & \mathbf{A}_{22} - \mathbf{B}_{22} & \cdots & \mathbf{A}_{2s} - \mathbf{B}_{2s} \\ \cdots & \cdots & \cdots & \cdots \\ \mathbf{A}_{r1} - \mathbf{B}_{r1} & \mathbf{A}_{r2} - \mathbf{B}_{r2} & \cdots & \mathbf{A}_{rs} - \mathbf{B}_{rs} \end{pmatrix}. \qquad (1.20)$$

If the sizes of the corresponding blocks of $\mathbf{A}$ and $\mathbf{B}$ are different, we cannot add or subtract the corresponding blocks, since their sum and difference are defined only if they have the same size. However, if matrices $\mathbf{A}$ and $\mathbf{B}$ have the same size and in their block divisions the sizes of the corresponding blocks are different, then $\mathbf{A} + \mathbf{B}$ can be obtained only by adding the respective elements of the matrices. In this case we can assume that each block of $\mathbf{A}$ and $\mathbf{B}$ is 1×1.

Example 1.16. Let

$$\mathbf{A} = \left(\begin{array}{cc|cc} 1 & 1 & 2 & 2 \\ 1 & 2 & 3 & 1 \\ \hline 0 & 1 & 1 & 1 \\ 1 & -1 & 1 & 2 \end{array} \right)$$

be divided into four 2×2 blocks, and assume that

$$\mathbf{B} = \begin{pmatrix} 2 & 1 & -1 & 3 \\ \hline 1 & 0 & 0 & 0 \\ 2 & 2 & 2 & 1 \\ 1 & 1 & 1 & 1 \end{pmatrix}$$

is divided also into four blocks, but their sizes are 1×2, 1×2, 3×2, 3×2, respectively. Notice that both $\mathbf{A} + \mathbf{B}$ and $\mathbf{A} - \mathbf{B}$ exist, however neither of them can be obtained by adding or subtracting the corresponding blocks.

♦

Matrices can also be multiplied block-wise if the division of both matrices into blocks satisfy certain compatibility conditions. Assume that $\mathbf{A}$ is $m \times n$ and it is divided into blocks as before. Assume that $\mathbf{B}$ is an $n \times p$ matrix with block-form

$$\mathbf{B} = \begin{pmatrix} \mathbf{B}_{11} & \mathbf{B}_{12} & \cdots & \mathbf{B}_{1t} \\ \mathbf{B}_{21} & \mathbf{B}_{22} & \cdots & \mathbf{B}_{2t} \\ \cdots & \cdots & \cdots & \cdots \\ \mathbf{B}_{s1} & \mathbf{B}_{s2} & \cdots & \mathbf{B}_{st} \end{pmatrix}$$

where the size of block $\mathbf{B}_{ij}$ is $n_i \times p_j$ with $n = \sum_{i=1}^{s} n_i$ and $p = \sum_{j=1}^{t} p_j$. For $i = 1, 2, \ldots, r$ and $j = 1, 2, \ldots, t$ define

$$\mathbf{C}_{ij} = \sum_{l=1}^{s} \mathbf{A}_{il} \mathbf{B}_{lj} \tag{1.21}$$

which is the "formal product" of the ith block-row

$$\left(\mathbf{A}_{i1}, \mathbf{A}_{i2}, \ldots, \mathbf{A}_{is} \right)$$

of matrix $\mathbf{A}$ by the jth block-column

$$\begin{pmatrix} \mathbf{B}_{1j} \\ \mathbf{B}_{2j} \\ \dots \\ \mathbf{B}_{sj} \end{pmatrix}$$

of matrix $\mathbf{B}$. Then it is easy to see that the product $\mathbf{C} = \mathbf{A} \cdot \mathbf{B}$ can be divided into blocks as follows:

$$\mathbf{C} = \begin{pmatrix} \mathbf{C}_{11} & \mathbf{C}_{12} & \dots & \mathbf{C}_{1t} \\ \mathbf{C}_{21} & \mathbf{C}_{22} & \dots & \mathbf{C}_{2t} \\ \dots & \dots & \dots & \dots \\ \mathbf{C}_{r1} & \mathbf{C}_{r2} & \dots & \mathbf{C}_{rt} \end{pmatrix}.$$

assuming that all products in (1.21) are defined.

Example 1.17. Select

$$\mathbf{A} = \left(\begin{array}{cc|cc} 1 & 1 & 1 & 1 \\ 1 & 2 & 0 & 0 \\ \hline 1 & 1 & 1 & 1 \\ 0 & 0 & 1 & 2 \end{array} \right) \quad \text{and} \quad \mathbf{B} = \left(\begin{array}{cc|cc} 1 & 0 & 1 & 1 \\ 0 & 1 & 1 & 1 \\ \hline 1 & 1 & 1 & 0 \\ 1 & 1 & 0 & 0 \end{array} \right).$$

If both matrices are divided into 2×2 blocks as shown above, we have the blocks

$$\mathbf{A}_{11} = \begin{pmatrix} 1 & 1 \\ 1 & 2 \end{pmatrix}, \quad \mathbf{A}_{12} = \begin{pmatrix} 1 & 1 \\ 0 & 0 \end{pmatrix}, \quad \mathbf{A}_{21} = \begin{pmatrix} 1 & 1 \\ 0 & 0 \end{pmatrix}, \quad \mathbf{A}_{22} = \begin{pmatrix} 1 & 1 \\ 1 & 2 \end{pmatrix},$$

and

$$\mathbf{B}_{11} = \begin{pmatrix} 1 & 0 \\ 0 & 1 \end{pmatrix}, \quad \mathbf{B}_{12} = \begin{pmatrix} 1 & 1 \\ 1 & 1 \end{pmatrix}, \quad \mathbf{B}_{21} = \begin{pmatrix} 1 & 1 \\ 1 & 1 \end{pmatrix}, \quad \mathbf{B}_{22} = \begin{pmatrix} 1 & 0 \\ 0 & 0 \end{pmatrix}.$$

Then the 2×2 blocks of the product $\mathbf{C} = \mathbf{A} \cdot \mathbf{B}$ can be obtained as follows:

$$\mathbf{C}_{11} = \mathbf{A}_{11}\mathbf{B}_{11} + \mathbf{A}_{12}\mathbf{B}_{21} = \begin{pmatrix} 1 & 1 \\ 1 & 2 \end{pmatrix}\begin{pmatrix} 1 & 0 \\ 0 & 1 \end{pmatrix} + \begin{pmatrix} 1 & 1 \\ 0 & 0 \end{pmatrix}\begin{pmatrix} 1 & 1 \\ 1 & 1 \end{pmatrix}$$

$$= \begin{pmatrix} 1 & 1 \\ 1 & 2 \end{pmatrix} + \begin{pmatrix} 2 & 2 \\ 0 & 0 \end{pmatrix} = \begin{pmatrix} 3 & 3 \\ 1 & 2 \end{pmatrix},$$

$$\mathbf{C}_{12} = \mathbf{A}_{11}\mathbf{B}_{12} + \mathbf{A}_{12}\mathbf{B}_{22} = \begin{pmatrix} 1 & 1 \\ 1 & 2 \end{pmatrix}\begin{pmatrix} 1 & 1 \\ 1 & 1 \end{pmatrix} + \begin{pmatrix} 1 & 1 \\ 0 & 0 \end{pmatrix}\begin{pmatrix} 1 & 0 \\ 0 & 0 \end{pmatrix}$$

$$= \begin{pmatrix} 2 & 2 \\ 3 & 3 \end{pmatrix} + \begin{pmatrix} 1 & 0 \\ 0 & 0 \end{pmatrix} = \begin{pmatrix} 3 & 2 \\ 3 & 3 \end{pmatrix},$$

$$\mathbf{C}_{21} = \mathbf{A}_{21}\mathbf{B}_{11} + \mathbf{A}_{22}\mathbf{B}_{21} = \begin{pmatrix} 1 & 1 \\ 0 & 0 \end{pmatrix}\begin{pmatrix} 1 & 0 \\ 0 & 1 \end{pmatrix} + \begin{pmatrix} 1 & 1 \\ 1 & 2 \end{pmatrix}\begin{pmatrix} 1 & 1 \\ 1 & 1 \end{pmatrix}$$

$$= \begin{pmatrix} 1 & 1 \\ 0 & 0 \end{pmatrix} + \begin{pmatrix} 2 & 2 \\ 3 & 3 \end{pmatrix} = \begin{pmatrix} 3 & 3 \\ 3 & 3 \end{pmatrix},$$

and

$$\mathbf{C}_{22} = \mathbf{A}_{21}\mathbf{B}_{12} + \mathbf{A}_{22}\mathbf{B}_{22} = \begin{pmatrix} 1 & 1 \\ 0 & 0 \end{pmatrix}\begin{pmatrix} 1 & 1 \\ 1 & 1 \end{pmatrix} + \begin{pmatrix} 1 & 1 \\ 1 & 2 \end{pmatrix}\begin{pmatrix} 1 & 0 \\ 0 & 0 \end{pmatrix}$$

$$= \begin{pmatrix} 2 & 2 \\ 0 & 0 \end{pmatrix} + \begin{pmatrix} 1 & 0 \\ 1 & 0 \end{pmatrix} = \begin{pmatrix} 3 & 2 \\ 1 & 0 \end{pmatrix}.$$

Therefore,

$$\mathbf{A} \cdot \mathbf{B} = \mathbf{C} = \left(\begin{array}{cc|cc} 3 & 3 & 3 & 2 \\ 1 & 2 & 3 & 3 \\ \hline 3 & 3 & 3 & 2 \\ 3 & 3 & 1 & 0 \end{array} \right).$$

The direct multiplication of matrices $\mathbf{A}$ and $\mathbf{B}$ has to give the same answer, that can be easily checked.

♦

If the division of **A** and **B** into blocks does not satisfy the above conditions then **A** and **B** cannot be multiplied by using blocks even in cases when **A** · **B** exists. In such cases the original definition of matrix multiplication by using the matrix elements can be used.

As the conclusion of this example we will examine inverses of block matrices. Assume that the $n \times n$ matrix **A** is divided into blocks as

$$\mathbf{A} = \begin{pmatrix} \mathbf{P} & \mathbf{Q} \\ \mathbf{R} & \mathbf{S} \end{pmatrix},$$

where **P** is $m \times m$, **Q** is $m \times (n-m)$, **R** is $(n-m) \times m$, and **S** is an $(n-m) \times (n-m)$ matrix. We will determine the inverse of **A** in a similar block-form

$$\mathbf{A}^{-1} = \begin{pmatrix} \mathbf{X} & \mathbf{Y} \\ \mathbf{U} & \mathbf{V} \end{pmatrix},$$

where **X** is $m \times m$, **Y** is is $m \times (n-m)$ **U** is $(n-m) \times m$, and **V** is $(n-m) \times (n-m)$. The definition of inverse matrices implies that

$$\begin{pmatrix} \mathbf{P} & \mathbf{Q} \\ \mathbf{R} & \mathbf{S} \end{pmatrix} \begin{pmatrix} \mathbf{X} & \mathbf{Y} \\ \mathbf{U} & \mathbf{V} \end{pmatrix} = \begin{pmatrix} \mathbf{I}_m & \mathbf{O} \\ \mathbf{O} & \mathbf{I}_{n-m} \end{pmatrix}.$$

Comparing the corresponding blocks of the two sides of this equation gives the relations

$$\begin{aligned} \mathbf{PX} + \mathbf{QU} &= \mathbf{I}_m \\ \mathbf{PY} + \mathbf{QV} &= \mathbf{O} \\ \mathbf{RX} + \mathbf{SU} &= \mathbf{O} \\ \mathbf{RY} + \mathbf{SV} &= \mathbf{I}_{n-m}. \end{aligned} \qquad (1.22)$$

Assuming that **P** is invertible, the second equation implies that

$$\mathbf{Y} = -\mathbf{P}^{-1}\mathbf{QV} \qquad (1.23)$$

and substituting this relation into the fourth equation gives an equation for block **V**:

$$\left(-\mathbf{R}\mathbf{P}^{-1}\mathbf{Q}+\mathbf{S}\right)\mathbf{V} = \mathbf{I}_{n-m},$$

that is, $\mathbf{S} - \mathbf{R}\mathbf{P}^{-1}\mathbf{Q}$ must be invertible, and

$$\mathbf{V} = \left(\mathbf{S} - \mathbf{R}\mathbf{P}^{-1}\mathbf{Q}\right)^{-1}. \tag{1.24}$$

From the third equation of (1.22) we have

$$\mathbf{U} = -\mathbf{S}^{-1}\mathbf{R}\mathbf{X} \tag{1.25}$$

assuming that **S** is invertible. Substitute this equation into the first equation of (1.22) to see that $\mathbf{P} - \mathbf{Q}\mathbf{S}^{-1}\mathbf{R}$ must be invertible, and

$$\mathbf{X} = \left(\mathbf{P} - \mathbf{Q}\mathbf{S}^{-1}\mathbf{R}\right)^{-1}. \tag{1.26}$$

Notice that equations (1.26), (1.25), (1.24), (1.23) can be used to recover the unknown blocks **X**, **U**, **V** and **Y** of the inverse matrix $\mathbf{A}^{-1}$.

Example 1.18. We will now invert matrix

$$\mathbf{A} = \begin{pmatrix} 1 & 0 & 1 \\ 0 & 1 & 1 \\ 1 & 1 & 1 \end{pmatrix}.$$

In this case we may select

$$\mathbf{P} = \begin{pmatrix} 1 & 0 \\ 0 & 1 \end{pmatrix}, \quad \mathbf{Q} = \begin{pmatrix} 1 \\ 1 \end{pmatrix}, \quad \mathbf{R} = (1,1), \quad \text{and} \quad \mathbf{S} = (1).$$

Equation (1.26) implies that

$$\mathbf{X} = \left(\begin{pmatrix} 1 & 0 \\ 0 & 1 \end{pmatrix} - \begin{pmatrix} 1 \\ 1 \end{pmatrix} \cdot 1 \cdot (1,1) \right)^{-1} = \begin{pmatrix} 0 & -1 \\ -1 & 0 \end{pmatrix}^{-1} = \begin{pmatrix} 0 & -1 \\ -1 & 0 \end{pmatrix}$$

since $\mathbf{S}^{-1} = (1)$. From equation (1.25) we have

$$\mathbf{U} = -1 \cdot (1,1) \begin{pmatrix} 0 & -1 \\ -1 & 0 \end{pmatrix} = (1,1).$$

Equation (1.24) is then applied to find $\mathbf{V}$:

$$\mathbf{V} = \left(1 - (1,1) \begin{pmatrix} 1 & 0 \\ 0 & 1 \end{pmatrix} \begin{pmatrix} 1 \\ 1 \end{pmatrix} \right)^{-1} = (1-2)^{-1} = \frac{1}{-1} = -1,$$

where we used the fact that the inverse of the identity matrix is itself. And finally, from equation (1.23) we get

$$\mathbf{Y} = -\begin{pmatrix} 1 & 0 \\ 0 & 1 \end{pmatrix} \cdot \begin{pmatrix} 1 \\ 1 \end{pmatrix} \cdot (-1) = \begin{pmatrix} 1 \\ 1 \end{pmatrix}.$$

Hence,

$$\mathbf{A}^{-1} = \begin{pmatrix} 0 & -1 & 1 \\ -1 & 0 & 1 \\ 1 & 1 & -1 \end{pmatrix}.$$

♦

2. In many applications a *certain part of a matrix* (a row, column, element, or even a block of the matrix) is needed for further computation. In this example of matrix algebra we will show how to obtain such matrix parts by using only matrix operations. Let $\mathbf{A}$ be a given $m \times n$ real (or complex) matrix with (i, j) element a_{ij}.

Let $\mathbf{e}_j^{(n)}$ denote the n-element column vector the jth element of which is one, and all other elements are equal to zero. Then

$$\mathbf{A} \cdot \mathbf{e}_j^{(n)} = \begin{pmatrix} a_{11} & a_{12} & \cdots & a_{1j} & \cdots & a_{1n} \\ a_{21} & a_{22} & \cdots & a_{2j} & \cdots & a_{2n} \\ \cdots & \cdots & \cdots & \cdots & \cdots & \cdots \\ a_{m1} & a_{m2} & \cdots & a_{mj} & \cdots & a_{mn} \end{pmatrix} \begin{pmatrix} 0 \\ 0 \\ \vdots \\ 1 \\ \vdots \\ 0 \end{pmatrix} \leftarrow j^{\text{th}} \text{ element.}$$

$$\uparrow$$
$$j^{\text{th}} \text{ column}$$

For $i = 1, 2, \ldots, m$, the ith element of the product is obtained by multiplying the ith row of $\mathbf{A}$ by the column vector $\mathbf{e}_j^{(n)}$:

$$\left(a_{i1}, a_{i2}, \ldots, a_{ij}, \ldots, a_{in} \right) \begin{pmatrix} 0 \\ 0 \\ \vdots \\ 1 \\ \vdots \\ 0 \end{pmatrix} \leftarrow j^{\text{th}} \text{ element} = a_{ij}$$

$$\uparrow$$
$$j^{\text{th}} \text{ element}$$

since all other terms equal zero. Hence

$$\mathbf{A} \cdot \mathbf{e}_j^{(n)} = \begin{pmatrix} a_{1j} \\ a_{2j} \\ \vdots \\ a_{ij} \\ \vdots \\ a_{mj} \end{pmatrix},$$

which is the jth column of $\mathbf{A}$.

Let now $\mathbf{e}_i^{(m)T}$ denote the m-element row vector the ith element of which is equal to one and all other elements are zeros. Then

$$\mathbf{e}_i^{(m)T} \mathbf{A} = \left(0,0,...,1,...0\right) \cdot \begin{pmatrix} a_{11} & a_{12} & \cdots & a_{1n} \\ a_{21} & a_{22} & \cdots & a_{2n} \\ \cdots & \cdots & \cdots & \cdots \\ a_{i1} & a_{i2} & \cdots & a_{in} \\ \cdots & \cdots & \cdots & \cdots \\ a_{m1} & a_{m2} & \cdots & a_{mn} \end{pmatrix} \leftarrow i^{\text{th}} \text{ row}$$

i^{th} element

For $j = 1, 2, \ldots, n$, the jth element of the product is obtained by multiplying $\mathbf{e}_i^{(m)T}$ by the jth column of $\mathbf{A}$:

$$\left(0,0,...,1,...,0\right) \begin{pmatrix} a_{1j} \\ a_{2j} \\ \vdots \\ a_{ij} \\ \vdots \\ a_{mj} \end{pmatrix} \leftarrow i^{\text{th}} \text{ element} = a_{ij}$$

i^{th} element

since all other terms are equal to zero. Therefore

$$\mathbf{e}_i^{(m)T} \mathbf{A} = \left(a_{i1}, a_{i2},..., a_{ij},..., a_{in}\right),$$

which is the ith row of $\mathbf{A}$.

It is also easy to see that for all i and j,

$$a_{ij} = \mathbf{e}_i^{(m)T} \mathbf{A} \mathbf{e}_j^{(n)},$$

since

$$\mathbf{e}_i^{(m)T}\left(\mathbf{Ae}_j^{(n)}\right) = \left(0, 0, ..., 1, ..., 0\right)\begin{pmatrix} a_{1j} \\ a_{2j} \\ \vdots \\ a_{ij} \\ \vdots \\ a_{mj} \end{pmatrix} \leftarrow i^{\text{th}}\text{ element} = a_{ij}$$

$$\uparrow$$
$$i^{\text{th}}\text{ element}$$

since all other terms equal zero.

The above relations can be presented in a much more general framework. Let $1 \le i_1 < i_2 < ... < i_r \le m$ and $1 \le j_1 < j_2 < ... < j_s \le n$ be arbitrary integers. Consider matrix

$$\mathbf{A}_1 = \begin{pmatrix} a_{i_1 j_1} & a_{i_1 j_2} & \cdots & a_{i_1 j_s} \\ a_{i_2 j_1} & a_{i_2 j_2} & \cdots & a_{i_2 j_s} \\ \cdots & \cdots & \cdots & \cdots \\ a_{i_r j_1} & a_{i_r j_2} & \cdots & a_{i_r j_s} \end{pmatrix}$$

which can be obtained from $\mathbf{A}$ by deleting all rows except rows i_1, i_2, ..., i_r and all columns except columns $j_1, j_2, ..., j_s$. Define the $r \times m$ matrix $\mathbf{U}$ and the $n \times s$ matrix $\mathbf{V}$ such that

$$u_{1i_1} = u_{2i_2} = ... = u_{ri_r} = 1, \quad \text{all other} \quad u_{ij} = 0;$$

and

$$v_{j_1 1} = v_{j_2 2} = ... = v_{j_s s} = 1, \quad \text{all other} \quad v_{ij} = 0.$$

Then

$$\mathbf{A}_1 = \mathbf{UAV}. \tag{1.27}$$

In the particular case, when $\mathbf{A}$ is a square matrix, $r = s$, and $i_1 = j_1$, $i_2 = j_2$, ..., $i_r = j_r$, then matrix $\mathbf{A}_1$ called a *principal submatrix* of $\mathbf{A}$. Notice that all principal submatrices are square matrices, and if $\mathbf{A}$ is

$n \times n$, then there are $\binom{n}{r}$ $r \times r$ principal submatrices of $\mathbf{A}$.

Using a similar idea as before, some further vector characteristics can be derived, that have significant applications in statistics.

Let x_1, ..., x_n be sample elements. They can be summarized as a column vector

$$\mathbf{x} = \begin{pmatrix} x_1 \\ x_2 \\ \vdots \\ x_n \end{pmatrix}.$$

By introducing the n-element row vector

$$\mathbf{1}^T = \left(1, 1, ..., 1\right)$$

it is easy to see that

$$\mathbf{1}^T \mathbf{x} = x_1 + x_2 + ... + x_n,$$

therefore, the sample mean $(\overline{x})$ can be obtained as

$$\overline{x} = \frac{1}{n}\mathbf{1}^T \mathbf{x}.$$

If we notice that $\mathbf{1}^T \mathbf{1} = n$, then we also have

$$\overline{x} = \frac{\mathbf{1}^T \mathbf{x}}{\mathbf{1}^T \mathbf{1}}.$$

The sample variance can be expressed as

$$S_x^2 = \frac{1}{n-1} \sum_{k=1}^{n} \left(x_k - \overline{x} \right)^2.$$

Notice now that

$$\mathbf{x} - \overline{x} \cdot \mathbf{1} = \begin{pmatrix} x_1 - \overline{x} \\ x_2 - \overline{x} \\ \cdots \\ x_n - \overline{x} \end{pmatrix},$$

therefore

$$S_x^2 = \frac{1}{n-1} \left(\mathbf{x} - \overline{x} \cdot \mathbf{1} \right)^T \left(\mathbf{x} - \overline{x} \cdot \mathbf{1} \right) = \frac{\left(\mathbf{x} - \overline{x} \cdot \mathbf{1} \right)^T \left(\mathbf{x} - \overline{x} \cdot \mathbf{1} \right)}{n-1}$$

$$= \frac{\left(\mathbf{x} - \dfrac{\mathbf{1}^T \mathbf{x}}{\mathbf{1}^T \mathbf{1}} \cdot \mathbf{1} \right)^T \left(\mathbf{x} - \dfrac{\mathbf{1}^T \mathbf{x}}{\mathbf{1}^T \mathbf{1}} \cdot \mathbf{1} \right)}{\mathbf{1}^T \mathbf{1} - 1}.$$

Consider next two *n*-element samples, $x_1, x_2, \ldots, x_n$ and $y_1, y_2, \ldots, y_n$. The covariance between these samples can be written as follows:

$$Cov\left(\mathbf{x}, \mathbf{y}\right) = \frac{1}{n} \sum_{k=1}^{n} \left(x_k - \overline{x} \right)\left(y_k - \overline{y} \right) = \frac{1}{n}\left(\mathbf{x} - \overline{x} \cdot \mathbf{1} \right)^T \left(\mathbf{y} - \overline{y} \cdot \mathbf{1} \right)$$

$$= \frac{\left(\mathbf{x} - \dfrac{\mathbf{1}^T \mathbf{x}}{\mathbf{1}^T \mathbf{1}} \mathbf{1} \right)^T \left(\mathbf{y} - \dfrac{\mathbf{1}^T \mathbf{y}}{\mathbf{1}^T \mathbf{1}} \mathbf{1} \right)}{\mathbf{1}^T \mathbf{1}}.$$

The correlation between the two samples has the general form

$$r = \frac{Cov\left(\mathbf{x}, \mathbf{y}\right)}{S_x \cdot S_y},$$

it can also be expressed by using vector operations if we substitute the above expressions for the covariance and the two variances.

3. Time invariant linear *dynamic systems with discrete* time scale can be generally formulated as the difference equation

$$\mathbf{x}(t+1) = \mathbf{A}\mathbf{x}(t) + \mathbf{b}, \tag{1.28}$$

where the n-element vector $\mathbf{x}$ is the state variable, $\mathbf{A}$ is a given $n \times n$ real matrix, and $\mathbf{b}$ is a given n-element real vector. The initial state $\mathbf{x}_0 = \mathbf{x}(0)$ is also assumed to be known. An elementary problem of systems theory is to find $\mathbf{x}(t)$ for all future times as easily as possible. In this application we will suggest a solution for this problem. Substitute $t = 0$, 1 and 2 into equation (1.28) to see that

$$\mathbf{x}(1) = \mathbf{A}\mathbf{x}(0) + \mathbf{b} = \mathbf{A}\mathbf{x}_0 + \mathbf{b},$$
$$\mathbf{x}(2) = \mathbf{A}\mathbf{x}(1) + \mathbf{b} = \mathbf{A}(\mathbf{A}\mathbf{x}_0 + \mathbf{b}) + \mathbf{b} = \mathbf{A}^2\mathbf{x}_0 + (\mathbf{A} + \mathbf{I})\mathbf{b},$$
$$\mathbf{x}(3) = \mathbf{A}\mathbf{x}(2) + \mathbf{b} = \mathbf{A}(\mathbf{A}^2\mathbf{x}_0 + (\mathbf{A} + \mathbf{I})\mathbf{b}) + \mathbf{b} = \mathbf{A}^3\mathbf{x}_0 + (\mathbf{A}^2 + \mathbf{A} + \mathbf{I})\mathbf{b}.$$

These initial solution vectors suggest that in general,

$$\mathbf{x}(t) = \mathbf{A}^t\mathbf{x}_0 + (\mathbf{A}^{t-1} + \mathbf{A}^{t-2} + \dots + \mathbf{A} + \mathbf{I})\mathbf{b} = \mathbf{A}^t\mathbf{x}_0 + \left(\sum_{l=0}^{t-1}\mathbf{A}^l\right)\mathbf{b} \tag{1.29}$$

where we use that fact that $\mathbf{I} = \mathbf{A}^0$. This solution formula can be proved by finite induction. For $t = 1$, 2 and 3 the formula is valid as it is shown from the above initial values of the state variable. Assume that the formula is valid for an integer $t > 0$. Then from equation (1.28) we conclude that

$$\mathbf{x}(t+1) = \mathbf{A}\mathbf{x}(t) + \mathbf{b} = \mathbf{A}\left(\mathbf{A}^t\mathbf{x}_0 + \left(\sum_{l=0}^{t-1}\mathbf{A}^l\right)\mathbf{b}\right) + \mathbf{b} = \mathbf{A}^{t+1}\mathbf{x}_0 + \left(\sum_{l=0}^{t}\mathbf{A}^l\right)\mathbf{b},$$

that is, the formula remains valid for $t + 1$. Hence it holds for all $t \geq 1$.

In applying the solution formula (1.29), we need a fast method to find powers of $\mathbf{A}$. Later, in Chapter 6 we will show a general method for the efficient computation of $\mathbf{A}^t$ for all $t \geq 1$. In many special cases the power matrix $\mathbf{A}^t$ can be determined by calculating some initial powers $\mathbf{A}^2$, $\mathbf{A}^3$, $\mathbf{A}^4$ and observing the general formula, which has to be than proved by finite induction. Matrix $\mathbf{A}^t$ is called the state transition matrix of system (1.28).

Example 1.19. Consider matrix

$$\mathbf{A} = \begin{pmatrix} 1 & 1 \\ 2 & 2 \end{pmatrix}.$$

Then

$$\mathbf{A}^2 = \begin{pmatrix} 1 & 1 \\ 2 & 2 \end{pmatrix}\begin{pmatrix} 1 & 1 \\ 2 & 2 \end{pmatrix} = \begin{pmatrix} 3 & 3 \\ 6 & 6 \end{pmatrix} = 3 \cdot \mathbf{A},$$
$$\mathbf{A}^3 = \mathbf{A}^2 \cdot \mathbf{A} = (3 \cdot \mathbf{A})\mathbf{A} = 3 \cdot \mathbf{A}^2 = 3(3 \cdot \mathbf{A}) = 3^2 \cdot \mathbf{A},$$
$$\mathbf{A}^4 = \mathbf{A}^3 \cdot \mathbf{A} = (3^2 \cdot \mathbf{A})\mathbf{A} = 3^2 \cdot \mathbf{A}^2 = 3^2(3 \cdot \mathbf{A}) = 3^3 \cdot \mathbf{A}.$$

By using finite induction, it is easy to show that in general,

$$\mathbf{A}^t = 3^{t-1}\mathbf{A}.$$

We will next solve the difference equation

$$\mathbf{x}(t+1) = \begin{pmatrix} 1 & 1 \\ 2 & 2 \end{pmatrix}\mathbf{x}(t) + \begin{pmatrix} 1 \\ 0 \end{pmatrix}, \quad \mathbf{x}(0) = \begin{pmatrix} 1 \\ 1 \end{pmatrix}.$$

From Equation (1.29) we have

$$\mathbf{x}(t) = 3^{t-1}\mathbf{A}\mathbf{x}_0 + \left(\mathbf{I} + \left(\sum_{l=1}^{t-1} 3^{l-1}\right)\mathbf{A}\right)\mathbf{b} = 3^{t-1}\mathbf{A}\mathbf{x}_0 + \left(\mathbf{I} + \frac{3^{t-1}-1}{3-1}\mathbf{A}\right)\mathbf{b}.$$

Notice that for $\mathbf{A}^0$ we must not use the general formula of $\mathbf{A}^t$ since it holds usually only for $t \geq 1$. We have to use the fact that $\mathbf{A}^0 = \mathbf{I}$.

Substituting the actual form of $\mathbf{A}, \mathbf{x}_0$ and $\mathbf{b}$ into the above equation gives the solution:

$$\mathbf{x}(t) = 3^{t-1} \begin{pmatrix} 1 & 1 \\ 2 & 2 \end{pmatrix} \begin{pmatrix} 1 \\ 1 \end{pmatrix} + \left(\begin{pmatrix} 1 & 0 \\ 0 & 1 \end{pmatrix} + \frac{1}{2}(3^{t-1}-1) \begin{pmatrix} 1 & 1 \\ 2 & 2 \end{pmatrix} \right) \begin{pmatrix} 1 \\ 0 \end{pmatrix}$$

$$= 3^{t-1} \begin{pmatrix} 2 \\ 4 \end{pmatrix} + \begin{pmatrix} 1 \\ 0 \end{pmatrix} + \frac{3^{t-1}-1}{2} \begin{pmatrix} 1 \\ 2 \end{pmatrix} = \begin{pmatrix} \dfrac{5}{2} \cdot 3^{t-1} + \dfrac{1}{2} \\ 5 \cdot 3^{t-1} - 1 \end{pmatrix}.$$

◆

Time invariant linear *dynamic systems with continuous* time scale can be generally formulated as the differential equation

$$\dot{\mathbf{x}}(t) = \mathbf{A}\mathbf{x}(t) + \mathbf{b} \tag{1.30}$$

where $\mathbf{x}$ is the state variable, matrix $\mathbf{A}$ is $n \times n$, and vector $\mathbf{b}$ has n elements. It is well known from linear systems theory (see, for example, Szidarovszky and Bahill, 1992) that the solution is given as

$$\mathbf{x}(t) = e^{\mathbf{A} \cdot t} \mathbf{x}_0 + \int_0^t e^{\mathbf{A} \cdot (t-\tau)} \mathbf{b} \, d\tau. \tag{1.31}$$

In Chapter 6 we will show a general method to compute matrix $e^{\mathbf{A}t}$, which is defined as the sum of the infinite series

$$e^{\mathbf{A}t} = \mathbf{I} + \frac{\mathbf{A} \cdot t}{1!} + \frac{\mathbf{A}^2 t^2}{2!} + \frac{\mathbf{A}^3 t^3}{3!} + \dots. \tag{1.32}$$

It is also well known that this series is convergent for all real t and arbitrary $n \times n$ real matrices. In special cases, a general formula may be derived for $\mathbf{A}^k$, and this general formula can be substituted into equality (1.32) to get a closed form for $e^{\mathbf{A}t}$. This matrix is called matrix exponential, or the state transition matrix of system (1.30).

Example 1.20. Consider again the 2 × 2 matrix

$$\mathbf{A} = \begin{pmatrix} 1 & 1 \\ 2 & 2 \end{pmatrix}.$$

In Example 1.19 we have seen that for all $k \geq 1$,

$$\mathbf{A}^k = 3^{k-1} \cdot \mathbf{A} = 3^{k-1} \begin{pmatrix} 1 & 1 \\ 2 & 2 \end{pmatrix}.$$

Therefore

$$e^{\mathbf{A}t} = \mathbf{I} + \sum_{k=1}^{\infty} \frac{3^{k-1} \mathbf{A} t^k}{k!} = \mathbf{I} + \mathbf{A} \cdot \sum_{k=1}^{\infty} \frac{3^{k-1} t^k}{k!}$$

$$= \mathbf{I} + \mathbf{A} \cdot \frac{1}{3} \sum_{k=1}^{\infty} \frac{(3t)^k}{k!} = \mathbf{I} + \mathbf{A} \cdot \frac{1}{3}\left(e^{3t} - 1\right)$$

$$= \begin{pmatrix} 1 & 0 \\ 0 & 1 \end{pmatrix} + \frac{e^{3t}-1}{3}\begin{pmatrix} 1 & 1 \\ 2 & 2 \end{pmatrix} = \begin{pmatrix} \frac{1}{3}\left(e^{3t}+2\right) & \frac{1}{3}\left(e^{3t}-1\right) \\ \frac{2}{3}\left(e^{3t}-1\right) & \frac{1}{3}\left(2e^{3t}+1\right) \end{pmatrix}.$$

We will next solve the initial value-problem

$$\dot{\mathbf{x}}(t) = \begin{pmatrix} 1 & 1 \\ 2 & 2 \end{pmatrix} \mathbf{x}(t) + \begin{pmatrix} 1 \\ 0 \end{pmatrix}, \quad \mathbf{x}(0) = \begin{pmatrix} 1 \\ 1 \end{pmatrix},$$

that is, we will derive a formula for $\mathbf{x}(t)$ at all future times $t > 0$ by assuming that the initial state at $t = 0$ is given as before. From equation (1.31) we have

$$\mathbf{x}(t) = \begin{pmatrix} \frac{1}{3}\left(e^{3t}+2\right) & \frac{1}{3}\left(e^{3t}-1\right) \\ \frac{2}{3}\left(e^{3t}-1\right) & \frac{1}{3}\left(2e^{3t}+1\right) \end{pmatrix} \begin{pmatrix} 1 \\ 1 \end{pmatrix}$$

$$+ \int_0^t \begin{pmatrix} \frac{1}{3}\left(e^{3(t-\tau)}+2\right) & \frac{1}{3}\left(e^{3(t-\tau)}-1\right) \\ \frac{2}{3}\left(e^{3(t-\tau)}-1\right) & \frac{1}{3}\left(2e^{3(t-\tau)}+1\right) \end{pmatrix} \begin{pmatrix} 1 \\ 0 \end{pmatrix} d\tau$$

$$= \begin{pmatrix} \frac{1}{3}\left(2e^{3t}+1\right) \\ \frac{1}{3}\left(4e^{3t}-1\right) \end{pmatrix} + \int_0^t \begin{pmatrix} \frac{1}{3}\left(e^{3(t-\tau)}+2\right) \\ \frac{2}{3}\left(e^{3(t-\tau)}-1\right) \end{pmatrix} d\tau.$$

Notice that

$$\int_0^t \frac{1}{3}\left(e^{3(t-\tau)}+2\right)d\tau = \frac{1}{3}\left[\frac{e^{3(t-\tau)}}{-3}+2\tau\right]_{\tau=0}^t$$

$$= \frac{1}{3}\left[\left(-\frac{1}{3}+2t\right)-\left(-\frac{1}{3}e^{3t}\right)\right] = \frac{1}{9}\left(e^{3t}+6t-1\right)$$

and

$$\int_0^t \frac{2}{3}\left(e^{3(t-\tau)}-1\right)d\tau = \frac{2}{3}\left[\frac{e^{3(t-\tau)}}{-3}-\tau\right]_{\tau=0}^t$$

$$= \frac{2}{3}\left[\left(-\frac{1}{3}-t\right)-\left(-\frac{1}{3}e^{3t}\right)\right] = \frac{2}{9}\left(e^{3t}-3t-1\right),$$

therefore

$$\mathbf{x}(t) = \begin{pmatrix} \dfrac{1}{9}\left(7e^{3t} + 6t + 2\right) \\ \dfrac{1}{9}\left(14e^{3t} - 6t - 5\right) \end{pmatrix}.$$

◆

4. A special linear dynamic economic system will be introduced next, which is called the *dynamic Cournot oligopoly model*. Assume that n firms produce a homogeneous good and sell it on the same market. Assume that the market demand function $d(p)$ is decreasing and linear, where p is the sale price. Then its inverse is also decreasing and linear assuming that demand equals total supply:

$$p(d) = a \cdot d + b \quad (a < 0, \quad b > 0).$$

The constant b shows the highest price that the consumers would be willing to pay if the product is not available on the market directly. The coefficient a shows the decrease in the price if the quantity of available products increases by unity. Assume that the production cost of firm $k(1 \leq k \leq n)$ is also linear:

$$C_k(x_k) = b_k x_k + c_k \quad (b_k > 0, \quad c_k \geq 0),$$

where x_k denotes the production level of firm k. Here c_k shows the fixed cost, and b_k is the marginal cost (that is, the cost increase by the increase of the production level by unity). The profit of firm k is therefore given as

$$\phi_k(x_1, \ldots, x_n) = x_k\left(a\left(\sum_{l=1}^{n} x_l\right) + b\right) - \left(b_k x_k + c_k\right).$$

Consider next the following *discrete dynamic extension* of this model. Let $x_1(0), \ldots, x_n(0)$ denote the production levels of the firms at the initial time period $t = 0$. Assume that at each further time period $t + 1$ $(t \geq 0)$, each firm maximizes its profit under the assumption that the competitors do not change their production levels from the previous time period. That is, firm k $(k = 1, 2, ..., n)$ maximizes its profit

$$x_k \left(a \left(x_k + \sum_{l \neq k} x_l(t) \right) + b \right) - \left(b_k x_k + c_k \right).$$

Assuming interior optimum, simple differentiation shows that the optimal solution for x_k is given as

$$x_k(t+1) = -\frac{1}{2} \sum_{l \neq k} x_l(t) + \frac{b_k - b}{2a}. \tag{1.33}$$

This equation is the special case of the difference equation (1.28) with

$$\mathbf{A} = \begin{pmatrix} 0 & -\frac{1}{2} & -\frac{1}{2} & \cdots & -\frac{1}{2} & -\frac{1}{2} \\ -\frac{1}{2} & 0 & -\frac{1}{2} & \cdots & -\frac{1}{2} & -\frac{1}{2} \\ \cdots & \cdots & \cdots & \cdots & \cdots & \cdots \\ -\frac{1}{2} & -\frac{1}{2} & -\frac{1}{2} & \cdots & -\frac{1}{2} & 0 \end{pmatrix} \quad \text{and} \quad \mathbf{b} = \begin{pmatrix} \dfrac{b_1 - b}{2a} \\ \dfrac{b_2 - b}{2a} \\ \cdots \\ \dfrac{b_n - b}{2a} \end{pmatrix}.$$

The *continuous dynamic extension* of the model can be formulated in the following way. Assume again that $x_1(0), ..., x_n(0)$ are the initial production levels. At each time period $t \geq 0$, each firm adjusts its production level proportionally to its marginal profit, that is, for all k,

$$\dot{x}_k(t) = m_k \left(2a x_k(t) + a \sum_{l \neq k} x_l(t) + b - b_k \right) \tag{1.34}$$

where $m_k > 0$ is a given constant. Notice that the marginal profit of firm k is the derivative of its profit with respect to x_k, and equation (1.34) requires that x_k increases if the profit increases in x_k, and x_k decreases if the profit decreases in x_k. These equations can be summarized as the differential equation (1.30) with

$$
\mathbf{A} = \begin{pmatrix} m_1 & & & \\ & m_2 & & \\ & & \ddots & \\ & & & m_n \end{pmatrix} \begin{pmatrix} 2a & a & \dots & a & a \\ a & 2a & \dots & a & a \\ \dots & \dots & \dots & \dots & \dots \\ a & a & \dots & a & 2a \end{pmatrix}
$$

and

$$
\mathbf{b} = \begin{pmatrix} m_1 (b - b_1) \\ m_2 (b - b_2) \\ \dots \\ m_n (b - b_n) \end{pmatrix}.
$$

5. Our next application deals with *dynamic producer-consumer models*. Consider a market where a commodity or a service is supplied by n competing firms. Let x_k, denote the output of firm k ($k = 1, 2, \dots, n$), and assume that $C_k(x_k) = B_k x_k^2 + b_k x_k + c_k$ is its cost function ($B_k, b_k, c_k > 0$). At each time period, firm k maximizes its expected profit given its price prediction $p_k^E(t+1)$. The expected profit of this firm is the following:

$$
x_k p_k^E (t+1) - \left(B_k x_k^2 + b_k x_k + c_k \right),
$$

and assuming interior optimum, simple differentiation shows that the profit maximizing output is given as

$$
x_k (t+1) = -\frac{1}{2B_k} \left(b_k - p_k^E (t+1) \right).
$$

If firm k believes that the price does not change from time period t to $t + 1$, then this firm selects $p_k^E (t+1) = p(t)$ and the profit maximizing output is given as

$$x_k(t+1) = -\frac{1}{2B_k}(b_k - p(t)). \tag{1.35}$$

Let $d(t) = Dp(t) + d$ denote the market demand function, from which the price $p(t)$ can be obtained as

$$p(t) = \frac{1}{D}(d(t) - d). \tag{1.36}$$

If we assume that at each time period, the supply equals the demand, then $d(t) = \sum_{l=1}^{n} x_l(t)$, and substituting this equation and (1.36) into (1.35) the following difference equation is obtained:

$$
\begin{aligned}
x_k(t+1) &= -\frac{1}{2B_k}\left(b_k - \frac{1}{D}\left(\sum_{l=1}^{n} x_l(t) - d\right)\right) \\
&= \sum_{l=1}^{n} \frac{1}{2B_k D} x_l(t) - \frac{1}{2B_k}\left(b_k + \frac{d}{D}\right),
\end{aligned}
\tag{1.37}
$$

which is the special case of the difference equations (1.28) with

$$
\mathbf{A} = \begin{pmatrix}
\dfrac{1}{2B_1 D} & \dfrac{1}{2B_1 D} & \cdots & \dfrac{1}{2B_1 D} \\[2mm]
\dfrac{1}{2B_2 D} & \dfrac{1}{2B_2 D} & \cdots & \dfrac{1}{2B_2 D} \\[2mm]
\cdots & \cdots & \cdots & \cdots \\[2mm]
\dfrac{1}{2B_n D} & \dfrac{1}{2B_n D} & \cdots & \dfrac{1}{2B_n D}
\end{pmatrix},
\quad \text{and} \quad
\mathbf{b} = -\begin{pmatrix}
\dfrac{1}{2B_1}\left(b_1 + \dfrac{d}{D}\right) \\[2mm]
\dfrac{1}{2B_2}\left(b_2 + \dfrac{d}{D}\right) \\[2mm]
\cdots \\[2mm]
\dfrac{1}{2B_n}\left(b_n + \dfrac{d}{D}\right)
\end{pmatrix}.
$$

The continuous counterpart of this model can be introduced as follows. Assume that at each time period $t \geq 0$, each firm adjusts its production output proportionally to its expected marginal profit:

$$\dot{x}_k(t) = m_k\left(p_k^E(t) - 2B_k x_k(t) - b_k\right),$$

where $p_k^E(t)$ is its price prediction for time period t. Assuming again that $p_k^E(t) = p(t)$, equation (1.36) holds, $d(t) = \sum_{l=1}^{n} r_l(t)$, then an ordinary differential equation is obtained for x_k:

$$\dot{x}_k(t) = m_k\left(\frac{1}{D}\left(\sum_{l=1}^{n} x_l(t) - d\right) - 2B_k x_k(t) - b_k\right)$$

$$= \left(\frac{m_k}{D} - 2B_k m_k\right) x_k(t) + \frac{m_k}{D}\sum_{l \neq k} x_l(t) + \left(-\frac{dm_k}{D} - m_k b_k\right)$$

which is a special case of the differential equation (1.30) with

$$\mathbf{A} = \begin{pmatrix} m_1\left(\dfrac{1}{D} - 2B_1\right) & \dfrac{m_1}{D} & \dfrac{m_1}{D} & \cdots & \dfrac{m_1}{D} \\[2ex] \dfrac{m_2}{D} & m_2\left(\dfrac{1}{D} - 2B_2\right) & \dfrac{m_2}{D} & \cdots & \dfrac{m_2}{D} \\[2ex] \cdots & \cdots & \cdots & \cdots & \cdots \\[2ex] \dfrac{m_n}{D} & \dfrac{m_n}{D} & \dfrac{m_n}{D} & \cdots & m_n\left(\dfrac{1}{D} - 2B_n\right) \end{pmatrix}$$

and

$$\mathbf{b} = \begin{pmatrix} -m_1\left(\dfrac{d}{D} + b_1\right) \\ -m_2\left(\dfrac{d}{D} + b_2\right) \\ \cdots \\ -m_n\left(\dfrac{d}{D} + b_n\right) \end{pmatrix}. \tag{1.38}$$

6. Matrix multiplications can be illustrated by the simple example of a bakery. Assume that the bakery makes three types of biscuits. The recipes are summarized in the following table:

	Flour	Sugar	Margarine
Type 1	0.5 (kg)	0.2 (kg)	0.3 (kg)
Type 2	0.55 (kg)	0.25 (kg)	0.2 (kg)
Type 3	0.6 (kg)	0.15 (kg)	0.25 (kg)

The quantities of the ingredients are given for one kilogram of each type. Let p_1, p_2, p_3 denote the prices per kg of the row materials, then the material cost per kg of each biscuit type can be calculated as

$$c_1 = 0.5p_1 + 0.2p_2 + 0.3p_3$$
$$c_2 = 0.55p_1 + 0.25p_2 + 0.2p_3$$
$$c_3 = 0.6p_1 + 0.15p_2 + 0.25p_3.$$

By introducing the notation

$$\mathbf{A} = \begin{pmatrix} 0.5 & 0.2 & 0.3 \\ 0.55 & 0.25 & 0.2 \\ 0.6 & 0.15 & 0.25 \end{pmatrix}, \quad \mathbf{p} = \begin{pmatrix} p_1 \\ p_2 \\ p_3 \end{pmatrix}, \quad \mathbf{c} = \begin{pmatrix} c_1 \\ c_2 \\ c_3 \end{pmatrix}$$

these relations can be rewritten as

$$\mathbf{c} = \mathbf{Ap}.$$

Assume next that the company sells three kinds of boxes of assorted biscuits as given in Table 1.1.

Table 1.1. Data for Application 1.6

	Type 1	Type 2	Type 3
Assortment 1	0.5 (kg)	0.7 (kg)	0.8 (kg)
Assortment 2	1.2 (kg)	0.3 (kg)	0.5 (kg)
Assortment 3	1 (kg)	0.5 (kg)	0.5 (kg)

The numbers show the amount of each type of biscuits in each type of assortment. Notice that each assortment contains 2 kg, and the costs of the boxes are

$$k_1 = 0.5c_1 + 0.7c_2 + 0.8c_3$$
$$k_2 = 1.2c_1 + 0.3c_2 + 0.5c_3$$
$$k_3 = c_1 + 0.5c_2 + 0.5c_3.$$

Let

$$\mathbf{B} = \begin{pmatrix} 0.5 & 0.7 & 0.8 \\ 1.2 & 0.3 & 0.5 \\ 1 & 0.5 & 0.5 \end{pmatrix}, \text{ and } \mathbf{k} = \begin{pmatrix} k_1 \\ k_2 \\ k_3 \end{pmatrix},$$

then we have

$$\mathbf{k} = \mathbf{Bc}.$$

Since $\mathbf{c} = \mathbf{Ap}$, this relation can be rewritten as

$$\mathbf{k} = \mathbf{BAp} = (\mathbf{BA})\mathbf{p} \tag{1.39}$$

giving a direct relation between the prices of the ingredients and the costs of the assortments.

7. Matrices are used in describing directed graphs, which model, for example, material flows or many network problems. A directed graph is defined by a finite set of elements, $P_1, P_2, \ldots, P_n$, together with a finite collection of ordered pairs, (P_i, P_j), of distinct elements where no ordered pair is repeated. The elements $P_1, P_2, \ldots, P_n$ are called the *vertices*, and the ordered pairs are called the *directed arcs*.

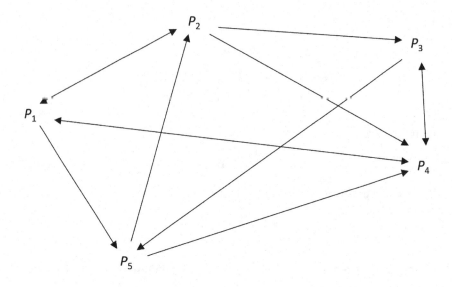

Figure 1.1. A directed graph

Figure 1.1 shows a directed graph with vertices P_1, ..., P_5 and arcs (P_1, P_2), (P_1, P_4), (P_1, P_5), (P_2, P_1), (P_2, P_3), (P_2, P_4), (P_3, P_4), (P_3, P_5), (P_4, P_1), (P_4, P_3), (P_5, P_2), and (P_5, P_4). With a directed graph with n vertices, we may associate an $n \times n$ real matrix $\mathbf{A}$ with elements

$$a_{ij} = \begin{cases} 1, & \text{if } \left(P_i, P_j\right) \text{ is a directed arc} \\ 0 & \text{otherwise}. \end{cases}$$

This matrix is called the vertex matrix of the directed graph.

For example, in the previous example we select $n = 5$, and

$$\mathbf{A} = \begin{pmatrix} 0 & 1 & 0 & 1 & 1 \\ 1 & 0 & 1 & 1 & 0 \\ 0 & 0 & 0 & 1 & 1 \\ 1 & 0 & 1 & 0 & 0 \\ 0 & 1 & 0 & 1 & 0 \end{pmatrix}.$$

For any square matrix having 0 and 1 elements with all diagonal elements are zero, there exists a directed graph such that the given matrix is its vertex matrix. The elements of the vertex matrix show the direct connections from each vertex to the other vertices.

Let now $r \geq 1$ be a positive integer, and let $a_{ij}^{(r)}$ denote the (i, j)-element of the matrix $\mathbf{A}^r$. Then it is easy to show that $a_{ij}^{(r)}$ equals the number of r-step connections from P_i to P_j. In the case of the above matrix,

$$\mathbf{A}^2 = \begin{pmatrix} 0 & 1 & 0 & 1 & 1 \\ 1 & 0 & 1 & 1 & 0 \\ 0 & 0 & 0 & 1 & 1 \\ 1 & 0 & 1 & 0 & 0 \\ 0 & 1 & 0 & 1 & 0 \end{pmatrix}\begin{pmatrix} 0 & 1 & 0 & 1 & 1 \\ 1 & 0 & 1 & 1 & 0 \\ 0 & 0 & 0 & 1 & 1 \\ 1 & 0 & 1 & 0 & 0 \\ 0 & 1 & 0 & 1 & 0 \end{pmatrix} = \begin{pmatrix} 2 & 1 & 2 & 2 & 0 \\ 1 & 1 & 1 & 2 & 2 \\ 1 & 1 & 1 & 1 & 0 \\ 0 & 1 & 0 & 2 & 2 \\ 2 & 0 & 2 & 1 & 0 \end{pmatrix}.$$

The $(1,1)$-element with value 2 shows that there are two 2-step connections from vertex 1 to itself. From the graph we see that $P_1 \rightarrow P_2 \rightarrow P_1$ and $P_1 \rightarrow P_4 \rightarrow P_1$ are the two such connections.

The number of at most r-step connections from P_i to P_j is given as $a_{ij} + a_{ij}^{(2)} + \ldots + a_{ij}^{(r)}$. For example, the number of at most two-step connections are summarized in matrix

$$\mathbf{A} + \mathbf{A}^2 = \begin{pmatrix} 2 & 2 & 2 & 3 & 1 \\ 2 & 1 & 2 & 3 & 2 \\ 1 & 1 & 1 & 2 & 1 \\ 1 & 1 & 1 & 2 & 2 \\ 2 & 1 & 2 & 2 & 0 \end{pmatrix}.$$

Consider as an example, the (1,4) element of this matrix. The direct connection from P_1 to P_4 gives the only 1-step connection, and the two 2-step connections are $P_1 \to P_2 \to P_4$ and $P_1 \to P_5 \to P_4$.

1.6 Exercises

1. Specify a 3×4 matrix which has only positive entries.

2. Specify 5 different matrices which have entries equal to 1.

3. Compute

$$2 \cdot \begin{pmatrix} 1 & 2 & 3 \\ 1 & 1 & 1 \\ -1 & -1 & 0 \end{pmatrix} - 3 \begin{pmatrix} 1 & 2 & 2 \\ 1 & -1 & -1 \\ 0 & 0 & 1 \end{pmatrix}.$$

4. Prove that, if **A**, **B**, **C** are matrices of the same size, then

$$(\mathbf{A} + \mathbf{B}) - (\mathbf{A} + \mathbf{C}) = \mathbf{B} - \mathbf{C}.$$

5. Solve equation

$$\mathbf{X} + \begin{pmatrix} 1 & 1 & 1 \\ 2 & 2 & 2 \end{pmatrix} = \begin{pmatrix} 3 & 3 & 3 \\ 4 & 4 & 4 \end{pmatrix}.$$

    What is the size of **X**?

6. Solve equation

$$\begin{pmatrix} 1 & 1 & 1 \\ 2 & 2 & 2 \end{pmatrix} + \mathbf{X} = \begin{pmatrix} 3 & 3 & 3 \\ 4 & 4 & 4 \end{pmatrix}.$$

    What is the size of **X**? Explain why the results of this and the previous Exercises coincide.

7. Verify relations (1.11), (1.12), and (1.13).
8. Find values of a, b and c such that

$$\begin{pmatrix} 1+a & b \\ b & c \end{pmatrix} + \begin{pmatrix} b & a \\ c & a \end{pmatrix} = \begin{pmatrix} 1 & 0 \\ 1 & 1 \end{pmatrix}.$$

9. Let

$$\mathbf{A} = \begin{pmatrix} 1 & 1 \\ 1 & 1 \end{pmatrix}, \quad \mathbf{B} = \begin{pmatrix} 2 & 2 \\ 2 & 2 \end{pmatrix}, \quad \mathbf{C} = \begin{pmatrix} 3 & 3 \\ 3 & 3 \end{pmatrix}.$$

Determine the results of the following operations:

a) $3\mathbf{A} + \mathbf{B} - 2\mathbf{C}$,

b) $4\mathbf{A} - \mathbf{B} + \mathbf{C}$,

c) $-\mathbf{A} - \mathbf{B} + 6\mathbf{C}$.

10. Solve equation

$$2(\mathbf{X} + \mathbf{A}) + \frac{1}{2}(\mathbf{X} + \mathbf{B} - \mathbf{C}) = \mathbf{X}$$

for $\mathbf{X}$, where $\mathbf{A}$, $\mathbf{B}$ and $\mathbf{C}$ are the same as given in the previous Exercise.

11. Let

$$\mathbf{A} = \begin{pmatrix} 1 & 1 \\ 1 & 1 \end{pmatrix} \quad \text{and} \quad \mathbf{B} = \begin{pmatrix} 1 & 2 \\ -1 & -1 \end{pmatrix}.$$

Compute the results of the following operations

a) $\mathbf{AB}$

b) **BA**

c) **(AB)A**

d) **((AB)A)B**

e) **((AB)B)A**

f) **B(AB)A**.

12. Show that $(A + B)^2 = A^2 + 2AB + B^2$ if and only if $AB = BA$.

13. Show that $(A + B)(A - B) = A^2 - B^2$ if and only if $AB = BA$.

14. Show that if $AB = BA$ then $(A + B)^n = \sum_{k=0}^{n} \binom{n}{k} A^k B^{n-k}$.

15. Verify that matrix $A = \begin{pmatrix} 0 & 1 \\ 0 & 0 \end{pmatrix}$ satisfies equation $A^2 = O$.

16. Characterize all 2×2 real matrices such that $A^2 = \begin{pmatrix} 1 & 1 \\ 1 & 1 \end{pmatrix}$.

17. Show that for all $k \geq 1$ and $n \times n$ matrices A,

$$\left(I_n - A\right)\left(I_n + A + A^2 + \ldots + A^k\right) = I_n - A^{k+1}.$$

18. Find a matrix satisfying equations

a) $X^2 - 3X + 2I = O$

b) $X^3 - I = O$

c) $X^3 - X^2 - X = O$.

19. Assume that $ad - bc \neq 0$. Prove that

$$\begin{pmatrix} a & b \\ c & d \end{pmatrix}^{-1} = \frac{1}{ad - bc} \begin{pmatrix} d & -b \\ -c & a \end{pmatrix}.$$

20. Find $\begin{pmatrix} 1 & 2 \\ 2 & 2 \end{pmatrix}^{-1}$ by using the result of the previous Exercise.

21. Let $\mathbf{A} = \begin{pmatrix} 1 & 2 \\ 2 & 2 \end{pmatrix}$.

    Find matrices $\mathbf{B}, \mathbf{C}$ such that $\mathbf{B} > \mathbf{A}$ and $\mathbf{C} > \mathbf{B}$.

22. Let $\mathbf{A} = \begin{pmatrix} 1 & 2 \\ 2 & 2 \end{pmatrix}$ and $\mathbf{B} = \begin{pmatrix} 2 & 3 \\ 1 & 1 \end{pmatrix}$.

    Find a matrix $\mathbf{C}$ such that $\mathbf{C} > \mathbf{A} - \mathbf{B}$.

23. Show that every $n \times n$ real matrix can be written as the sum of a symmetric and a skew symmetric matrix.

24. Show that for all $n \times n$ real matrices $\mathbf{A}$ and any n-element column vector $\mathbf{x}$ the quadratic form $\mathbf{x}^T\mathbf{Ax}$ can be written as $\mathbf{x}^T\mathbf{Sx}$, where $\mathbf{S}$ is a symmetric matrix.

25. Let $\mathbf{A}$ be an $m \times n$ real matrix such that $\mathbf{A}^T\mathbf{A} = \mathbf{O}$. Show that all elements of $\mathbf{A}$ equal zero.

26. Is $(\mathbf{A} + \mathbf{B})^{-1} = \mathbf{A}^{-1} + \mathbf{B}^{-1}$?

27. Let $\mathbf{p}^T = (p_1, p_2, ..., p_n)$ be a price vector, and let **1** be the *n*-element vector all elements of which are equal to 1. Show that the average price can be expressed as $\dfrac{1}{n}\mathbf{p}^T\mathbf{1}$.

28. Find $\mathbf{A}^t$ for matrix $\mathbf{A} = \begin{pmatrix} 1 & 1 \\ 0 & 1 \end{pmatrix}$.

29. Find $e^{\mathbf{A}t}$ for matrix $\mathbf{A} = \begin{pmatrix} 1 & 1 \\ 0 & 1 \end{pmatrix}$.

30. Assume that all elements of the $n \times n$ real matrix $\mathbf{A}(t)$ are continously differentiable and $\mathbf{A}(t)$ is invertible for all t. Find the derivative matrix of $\mathbf{A}^{-1}(t)$.

31. In the theory of dynamic systems matrices of the form

$$\mathbf{H}(s) = \mathbf{C}(\mathbf{A}-s\mathbf{I})^{-1}\mathbf{B}$$

have special importance, where $\mathbf{A}$ is $n \times n$, $\mathbf{B}$ is $n \times m$ and $\mathbf{C}$ is $p \times n$ constant matrix. $\mathbf{H}(s)$ is known as the transfer function of the system. Based on the result of the previous Exercise find $\dfrac{d}{ds}\mathbf{H}(s)$.

Chapter 2

Vector Spaces and Inner-Product Spaces

2.1 Introduction

In this chapter some structural properties of the set of real (or complex) vectors will be introduced and analysed. Our analysis will be based on a new concept known as vector spaces which can be defined as follows:

Definition 2.1. Let V be a set of arbitrary elements and assume that the operation of addition is defined for all pairs $\mathbf{x}, \mathbf{y} \in V$, and multiplication by real (or complex) scalars is defined for all scalars a and $\mathbf{x} \in V$ and the following properties are satisfied:

(i) $\mathbf{x} + \mathbf{y} = \mathbf{y} + \mathbf{x}$ for all $\mathbf{x}, \mathbf{y} \in V$ (commutativity);

(ii) $(\mathbf{x} + \mathbf{y}) + \mathbf{z} = \mathbf{x} + (\mathbf{y} + \mathbf{z})$ for all $\mathbf{x}, \mathbf{y}, \mathbf{z} \in V$ (associativity);

(iii) there exists a zero element $\mathbf{0} \in V$ such that for all $\mathbf{x} \in V$, $\mathbf{x} + \mathbf{0} = \mathbf{x}$;

(iv) for all $\mathbf{x} \in V$ there is an element denoted by $(-\mathbf{x})$ such that
$$\mathbf{x} + (-\mathbf{x}) = \mathbf{0}.$$

(v) $a(\mathbf{x} + \mathbf{y}) = a\mathbf{x} + a\mathbf{y}$ for all scalars a, and $\mathbf{x}, \mathbf{y} \in V$ (distributivity);

(vi) $(a + b)\mathbf{x} = a\mathbf{x} + b\mathbf{x}$ for all scalars a and b, and $\mathbf{x} \in V$ (distributivity);

(vii) $(ab)\mathbf{x} = a(b\mathbf{x})$ for all scalars a and b, and $\mathbf{x} \in V$ (associativity);

(viii) $1 \cdot \mathbf{x} = \mathbf{x}$ for all $\mathbf{x} \in V$.

Then V is called a *vector space,* and its elements are usually called *vectors.* If scalars are defined as real numbers, then the vector space is called *real,* and if scalars are complex numbers then the vector space is called *complex.*

Notice that in properties (iii) and (iv) we did not assume the uniqueness of the zero element $\mathbf{0}$ and the negative $(-\mathbf{x})$ of any $\mathbf{x}$. Their uniqueness follows from the other properties, which can be demonstrated in the following way. Assume that **0** and **0′** are both zero elements in a vector space. Then properties (iii) and (i) imply that

$$\mathbf{0} + \mathbf{0}' = \mathbf{0}$$

(when **x** is selected as **0**), and

$$\mathbf{0} + \mathbf{0}' = \mathbf{0}' + \mathbf{0} = \mathbf{0}'$$

(when **x** is selected as **0′**). Therefore **0** = **0′**.

Assume next that both (−**x**) and (−**x**)′ are negatives of an **x**. Consider the element (−**x**) + **x** + (−**x**)′. Since

$$(-\mathbf{x}) + \mathbf{x} + (-\mathbf{x})' = \big((-\mathbf{x}) + \mathbf{x}\big) + (-\mathbf{x})' = \big(\mathbf{x} + (-\mathbf{x})\big) + (-\mathbf{x})'$$
$$= \mathbf{0} + (-\mathbf{x})' = (-\mathbf{x})' + \mathbf{0} = (-\mathbf{x})',$$

and on the other hand

$$(-\mathbf{x}) + \mathbf{x} + (-\mathbf{x})' = (-\mathbf{x}) + \big(\mathbf{x} + (-\mathbf{x})'\big) = (-\mathbf{x}) + \mathbf{0} = (-\mathbf{x}),$$

the negatives are necessarily equal to each other.

We will next show that the zero element of V can be obtained by multiplying any **x** ∈ V by zero. Introduce the notation $0 \cdot \mathbf{x} = \mathbf{z}$ Then **x** = **x** + **z**, since

$$\mathbf{x} = 1 \cdot \mathbf{x} = (1 + 0) \cdot \mathbf{x} = 1 \cdot \mathbf{x} + 0 \cdot \mathbf{x} = \mathbf{x} + \mathbf{z}.$$

Add the element (−**x**) to both sides, to see that

$$\mathbf{0} = \mathbf{x} + (-\mathbf{x}) = \mathbf{x} + \mathbf{z} + (-\mathbf{x}) = \mathbf{x} + (-\mathbf{x}) + \mathbf{z}$$
$$= \mathbf{0} + \mathbf{z} = \mathbf{z} + \mathbf{0} = \mathbf{z}.$$

This property implies that for any scalar a, $a \cdot \mathbf{0} = \mathbf{0}$, since with any **x** ∈ V,

$$a \cdot \mathbf{0} = a \cdot (0 \cdot \mathbf{x}) = (a \cdot 0)\mathbf{x} = 0\mathbf{x} = \mathbf{0}.$$

Notice that the selection (−**x**) = (−1) · **x** satisfies property (iv), since

$$\mathbf{x} + (-\mathbf{x}) = 1 \cdot \mathbf{x} + (-1) \cdot \mathbf{x} = \big(1 + (-1)\big) \cdot \mathbf{x} = 0 \cdot \mathbf{x} = \mathbf{0}.$$

The uniqueness of $(-\mathbf{x})$ implies that necessarily $(-\mathbf{x}) = (-1) \cdot \mathbf{x}$.

Before analysing the further properties of vector spaces, some important examples are presented.

Example 2.1. Let V be the set of all n-element real column vectors, and define vector addition and multiplication by real scalars in the usual way (as given in Chapter 1). Then properties (i), (ii), (v), (vi), (vii), and (viii) are obviously satisfied. If $\mathbf{0}$ denotes the n-element zero column vector, and $(-\mathbf{x})$ is defined as $(-1) \cdot \mathbf{x}$, then (iii) and (iv) are also satisfied. Hence, the set of all n-element real vectors is a vector space, which is usually denoted by R^n. One may show in a similar way that the set of all complex n-element column vectors also form a vector space with the usual addition and multiplication by real or complex scalars. The set of all n-element complex column vectors is usually denoted by C^n.

◆

Example 2.2. Let now V denote the set of all real (or complex) $m \times n$ matrices, where m and n are fixed. Define the addition and multiplication by real (or complex) scalars in the way as it was shown in Chapter 1. If $\mathbf{O}$ is the $m \times n$ zero matrix, the element $(-\mathbf{A})$ is defined by multiplying $\mathbf{A}$ by (-1), then V is a vector space. The set of all real (or complex) $m \times n$ matrices is usually denoted by $R^{m \times n}$ (or $C^{m \times n}$), as it was introduced in Chapter 1.

◆

Example 2.3. Define next V as the set of all single variable real polynomials of degree at most n, where n is fixed. Let p and q be two polynomials:

$$p(t) = a_0 + a_1 t + \ldots + a_n t^n$$

and

$$q(t) = b_0 + b_1 t + \ldots + b_n t^n,$$

then the sum of p and q is defined in the following way:

$$p(t) + q(t) = (a_0 + b_0) + (a_1 + b_1)t + \ldots + (a_n + b_n)t^n.$$

If the degree of either p or q is less then n, then the "missing" coefficients up to a_n or b_n are assumed to be zeros. If a is a scalar, then multiplication by a is defined as

$$a \cdot p(t) = (a \cdot a_0) + (a \cdot a_1)t + \ldots + (a \cdot a_n)t^n.$$

The zero polynomial has all zero coefficients, and $(-p)$ is given by changing the sings all of coefficients of p:

$$(-p)(t) = (-a_0) + (-a_1)t + \ldots + (-a_n)t^n.$$

Then V is a vector space.

$\blacklozenge$

Example 2.4. Let V be the set of all continous real functions defined on a closed interval $[\alpha, \beta]$. The sum $f + g$ and scalar multiple $a \cdot f$ of real functions are defined in the usual way:

$$(f + g)(x) = f(x) + g(x)$$

and

$$(a \cdot f)(x) = a \cdot f(x)$$

for all $x \in [\alpha, \beta]$. It is easy to see that V is a vector space, when $(-f)(x) = -f(x)$, and the zero function has zero values for all $x \in [\alpha, \beta]$. This vector space is usually denoted by $C[\alpha, \beta]$.

$\blacklozenge$

Example 2.5. Let next V be the set of all infinite real sequences (a_n). The sum of sequences is defined as

$$(a_n) + (b_n) = (a_n + b_n),$$

that is, all corresponding terms are added. Scalar multiples of sequences are defined by multiplying each term by the scalar:

$$a \cdot (a_n) = (a \cdot a_n).$$

If one defines the sequence $-(a_n)$ as $(-a_n)$ and the zero sequence as $a_n = 0$ for all $n \geq 0$, then V is a vector space. For example,

$$(1, 1, 1, ...) + (1, 2, 3, ...) = (2, 3, 4, ...),$$

$$2 \cdot (1, 1, 1, ...) = (2, 2, 2, ...), \quad \text{and} \quad -(1, 1, 1, ...) = (-1, -1, -1, ...)$$

♦

2.2 Subspaces

Before introducing the concept of subspaces in general, a simple example is presented. Consider the vector space of all n dimensional real column vectors with the same operations as in Example 2.1. Let $\mathbf{z}$ be a nonzero vector in $V = R^n$, and define V_1 as the set of all real-multiples of $\mathbf{z}$:

$$V_1 = \{c \cdot \mathbf{z} \mid c \in R\}.$$

Here R denotes the set of real numbers. We can easily show, that V_1 satisfies all properties of a vector space. Assume that $\mathbf{x}$ and $\mathbf{y} \in V_1$, then $\mathbf{x} = a_1 \cdot \mathbf{z}$ and $\mathbf{y} = a_2 \cdot \mathbf{z}$ with some scalars a_1 and a_2. Then

$$\mathbf{x} + \mathbf{y} = a_1\mathbf{z} + a_2\mathbf{z} = (a_1 + a_2)\mathbf{z}$$

and for all scalars a,

$$a \cdot \mathbf{x} = a \cdot (a_1 \cdot \mathbf{z}) = (a \cdot a_1)\mathbf{z}$$

that is, both $\mathbf{x} + \mathbf{y}$ and $a \cdot \mathbf{x}$ are in V_1. Similarly, $\mathbf{0} = 0 \cdot \mathbf{z} \in V_1$ and for all $\mathbf{x} \in V_1$, $\mathbf{x} = a \cdot \mathbf{z}$ with some scalar a, therefore

$$(-\mathbf{x}) = (-1) \cdot \mathbf{x} = (-1) \cdot (a \cdot \mathbf{z}) = (-a) \cdot \mathbf{z} \in V_1.$$

Hence the zero element and $(-\mathbf{x})$ for all $\mathbf{x} \in V_1$ are in V_1. All properties (commutativity, associativity, and distributivity) of the operations also hold in V_1, since they are true in the larger set V.

Definition 2.2. A subset V_1 of a vector space V is called a *subspace*, if V_1 itself is a vector space.

Notice that the entire vector space V also can be considered as a subspace of itself. Similarly, the set containing only the zero element of V also satisfies the defining properties of a vector space, therefore it is also a subspace in V. These two special subspaces are called the *trivial subspaces* of V.

In verifying the conditions of vector spaces for any subset V_1 we do not need to check the operation properties (i), (ii), (v), (vi), (vii), and (viii), since they are true for all elements of V, and since $V_1 \subseteq V$, they remain true for all elements of V_1 as well.

Therefore, we only have to check that for all $\mathbf{x}$ and $\mathbf{y} \in V_1$, the sums $\mathbf{x} + \mathbf{y}$ are in V_1, and for all scalars a and $\mathbf{x} \in V_1$, the products $a \cdot \mathbf{x}$ are in V_1. Notice that for all $\mathbf{x}$, $0 \cdot \mathbf{x} = \mathbf{0}$, and $(-1) \cdot \mathbf{x} = (-\mathbf{x})$ therefore the zero element and the negative of all $\mathbf{x} \in V_1$ are necessarily in V_1. Since properties (iii) and (iv) hold in the entire V, they are valid in the subset V_1. This observation can be summarized as follows.

Theorem 2.1. If V is a vector space and $V_1 \subseteq V$, then V_1 is a subspace of V if and only if
(a) $\mathbf{x} + \mathbf{y} \in V_1$ for all $\mathbf{x}, \mathbf{y} \in V_1$;
(b) $a \cdot \mathbf{x} \in V_1$ for all scalars a and $\mathbf{x} \in V_1$.
The application of the theorem can be illustrated by the following important example.

Example 2.6. Let V be a vector space, and x_1, x_2, ..., x_k be given elements from V. Define

$$V_1 = \left\{ a_1 \cdot \mathbf{x}_1 + a_2 \cdot \mathbf{x}_2 + \ldots + a_k \cdot \mathbf{x}_k \,\middle|\, a_1, \, a_2, \, \ldots, \, a_k \in R \right\}.$$

Notice that the introductory example of this section is the special case of this example, when one selects $k = 1$. We will now verify that V_1 is a subspace. If $\mathbf{x}, \mathbf{y} \in V_1$, then with some scalars $a_1, \ldots, a_k$ and $b_1, \ldots, b_k$

$$\mathbf{x} = a_1 \cdot \mathbf{x}_1 + a_2 \cdot \mathbf{x}_2 + \ldots + a_k \cdot \mathbf{x}_k$$

and

$$\mathbf{y} = b_1 \cdot \mathbf{x}_1 + b_2 \cdot \mathbf{x}_2 + \ldots + b_k \cdot \mathbf{x}_k,$$

and therefore

$$\mathbf{x} + \mathbf{y} = (a_1 + b_1) \cdot \mathbf{x}_1 + (a_2 + b_2) \cdot \mathbf{x}_2 + \ldots + (a_k + b_k) \cdot \mathbf{x}_k.$$

Hence, $\mathbf{x} + \mathbf{y} \in V_1$. Similarly, for all scalars a,

$$a \cdot \mathbf{x} = a \cdot \left(a_1 \cdot \mathbf{x}_1 + a_2 \cdot \mathbf{x}_2 + \ldots + a_k \cdot \mathbf{x}_k \right)$$
$$= (a \cdot a_1) \cdot \mathbf{x}_1 + (a \cdot a_2) \cdot \mathbf{x}_2 + \ldots + (a \cdot a_k) \cdot \mathbf{x}_k \in V_1.$$

Therefore, both conditions of the theorem are satisfied showing that V_1 is a vector space.

♦

For any $\mathbf{x}_1, \ldots, \mathbf{x}_k \in V$ and scalars $a_1, \ldots, a_k$ the element $a_1 \cdot \mathbf{x}_1 + \ldots + a_k \cdot \mathbf{x}_k \in V$ is called a *linear combination* of the given vectors. Example 2.6 can be reformulated by saying that for any given finite set of the elements of a vector space, the set of all linear combinations of the given elements always forms a subspace.

In verifying that a subset V_1 of a given vector space is a subspace, the following theorem is applied most frequently.

Theorem 2.2. If V is a vector space and $V_1 \subseteq V$, then V_1 is a subspace if and only if for all $\mathbf{x}, \mathbf{y} \in V_1$ and scalars a and b,

$$a \cdot \mathbf{x} + b \cdot \mathbf{y} \in V_1.$$

Proof. Assume first, that V_1 is a subspace. Then for all a and $\mathbf{x}$, $a \cdot \mathbf{x} \in V_1$ and for all b and $\mathbf{y}$ and $b \cdot \mathbf{y} \in V_1$. Since the sum of the elements of V_1 must be in V_1,

$$a \cdot \mathbf{x} + b \cdot \mathbf{y} \in V_1.$$

Assume next that V_1 satisfies the condition of the theorem. Select an arbitrary a and $\mathbf{x}$ and define $b = 0$. Then

$$a \cdot \mathbf{x} + b \cdot \mathbf{y} = a \cdot \mathbf{x} \in V_1.$$

Choose next $a = b = 1$, then

$$a \cdot \mathbf{x} + b \cdot \mathbf{y} = \mathbf{x} + \mathbf{y} \in V_1,$$

therefore, both conditions of Theorem 2.1 are satisfied, and hence V_1 is a subspace.

♣

The advantage of this result compared to Theorem 2.1. is the fact that only one condition has to be verified to show that a subset of a vector space is a subspace.

Some important properties of the subspaces of a given vector space are presented next.

Theorem 2.3. Let V be a vector space, and $\{V_\alpha\}$ ($\alpha \in A$ with some index set A) be a finite or infinite set of (not necessarily all) subspaces of V. Then the intersection of the given subspaces is also a subspace.

Proof. Let V_1 denote the intersection of the subspaces V_α. We will show that V_1 satisfies the condition of Theorem 2.2. Let $\mathbf{x}$ and $\mathbf{y} \in V_1$ and let a and b be two scalars. The definition of V_1 implies that for all α, $\mathbf{x}$ and $\mathbf{y}$ are in V_α, and since V_α is a subspace, $a \cdot \mathbf{x} + b \cdot \mathbf{y} \in V_\alpha$ for all α. Therefore $a \cdot \mathbf{x} + b \cdot \mathbf{y}$ has to belong to the intersection of sets V_α. That is,

$$a \cdot \mathbf{x} + b \cdot \mathbf{y} \in V_1,$$

which completes the proof.

♣

Definition 2.3. Let S be an arbitrary (finite or infinite) subset of a vectorspace V. The subspace *generated* by set S is the intersection of all subspaces of V containing S.

Since V itself contains S and it is a trivial subspace in itself, there is at least one subspace containing S. From Theorem 2.3 we know that the intersection of all such subspaces is also a subspace. We will use the notation $V(S)$ for the subspace generated by S. From the definition we may conclude that if V_1 is any subspace of V containing S, then $V(S)$ is also contained in V_1, therefore $V(S)$ is the *smallest* subspace containing S. The construction of $V(S)$ is given in the following result.

Theorem 2.4. Let V be a vector space and let S be an arbitrary non-empty subset of V. Then $V(S)$ is the set of all linear combinations of finitely many elements of S:

$$V(S) = \left\{ \sum_{i=1}^{N} a_i \mathbf{u}_i \; \middle| \; \begin{array}{l} N \text{ is a positive integer, } \mathbf{u}_i \in S, \; i = 1, 2, ..., N \\ \text{and } a_i \text{ is a scalar for all } i \end{array} \right\} (2.1)$$

Proof. First we will show that the set of all linear combinations of finitely many elements of S is a subspace in V. For simple reference, let $L(S)$ denote this set. It is easy to verify that $L(S)$ satisfies the condition of Theorem 2.2. Let $\mathbf{x}$ and $\mathbf{y}$ be two elements of $L(S)$, and let a and b be two scalars. Then we may assume that

$$\mathbf{x} = a_1 \mathbf{x}_1 + ... + a_k \mathbf{x}_k + b_1 \mathbf{y}_1 + ... + b_l \mathbf{y}_l$$

and

$$\mathbf{y} = c_1 \mathbf{x}_1 + ... c_k \mathbf{x}_k + d_1 \mathbf{z}_1 + ... + d_m \mathbf{z}_m$$

where $\mathbf{x}_1, ..., \mathbf{x}_k, \mathbf{y}_1, ..., \mathbf{y}_l, \mathbf{z}_1, ..., \mathbf{z}_m \in S$ and the common elements in the linear combinations are denoted by $\mathbf{x}_1, ..., \mathbf{x}_k$. Simple calculation shows that

$$a \cdot \mathbf{x} + b \cdot \mathbf{y} = (a \cdot a_1) \cdot \mathbf{x}_1 + \ldots + (a \cdot a_k) \cdot \mathbf{x}_k + (a \cdot b_1) \cdot \mathbf{y}_1 + \ldots + (a \cdot b_l) \cdot \mathbf{y}_l$$
$$+ (b \cdot c_1) \cdot \mathbf{x}_1 + \ldots + (b \cdot c_k) \cdot \mathbf{x}_k + (b \cdot d_1) \cdot \mathbf{z}_1 + \ldots + (b \cdot d_m) \cdot \mathbf{z}_m$$
$$= (a \cdot a_1 + b \cdot c_1) \cdot \mathbf{x}_1 + \ldots + (a \cdot a_k + b \cdot c_k) \mathbf{x}_k$$
$$+ (a \cdot b_1) \cdot \mathbf{y}_1 + \ldots + (a \cdot b_l) \cdot \mathbf{y}_l + (b \cdot d_1) \cdot \mathbf{z}_1 + \ldots + (b \cdot d_m) \cdot \mathbf{z}_m,$$

which is the linear combination of the elements $\mathbf{x}_1, \ldots, \mathbf{x}_k, \mathbf{y}_1, \ldots, \mathbf{y}_l, \mathbf{z}_1, \ldots, \mathbf{z}_m$, therefore it is in $L(S)$. Hence $L(S)$ is a vector space. Notice that for all $\mathbf{x} \in S$, $\mathbf{x} = 1 \cdot \mathbf{x} \in L(S)$ consequently $S \subseteq L(S)$.

Assume next that V_1 is a subspace of V such that $S \subseteq V_1$. Let $\mathbf{x}_1, \ldots, \mathbf{x}_k \in S$ be arbitrary elements and let $a_1, \ldots, a_k$ be arbitrary scalars. Since V_1 is a subspace containing $\mathbf{x}_1, \ldots, \mathbf{x}_k$, the element $\mathbf{y}_1 = a_1\mathbf{x}_1 + a_2\mathbf{x}_2$ must belong to V_1. Then $\mathbf{y}_2 = 1\mathbf{y}_1 + a_3\mathbf{x}_3 = a_1\mathbf{x}_1 + a_2\mathbf{x}_2 + a_3\mathbf{x}_3$ also belongs to V_1. Similarly, $\mathbf{y}_3 = 1\mathbf{y}_2 + a_4\mathbf{x}_4 = a_1\mathbf{x}_1 + a_2\mathbf{x}_2 + a_3\mathbf{x}_3 + a_4\mathbf{x}_4 \in V_1$.

Repeating the same reasoning $k-1$ times we conclude that $\mathbf{y}_{k-1} = 1\mathbf{y}_{k-2} + a_k\mathbf{x}_k = a_1\mathbf{x}_1 + a_2\mathbf{x}_2 + a_3\mathbf{x}_3 + \ldots + a_{k-1}\mathbf{x}_{k-1} + a_k\mathbf{x}_k \in V_1$. In other words, all linear combinations of finitely many elements of S necessarily belong to V_1.

That is, $L(S) \subseteq V_1$. Hence $L(S)$ is the smallest subspace containing S, therefore it is the subspace generated by S.

♣

The assertion of the theorem is illustrated by the following examples.

Example 2.7. Consider the set R^3 of real 3-element column vectors, and let $\mathbf{x} \neq \mathbf{0}$ be an element of R^3. Then the subspace generated by vector $\mathbf{x}$ is the set of all real multiples of $\mathbf{x}$:

$$V(\{\mathbf{x}\}) = \{a \cdot \mathbf{x} \mid a \text{ is a real scalar}\}.$$

forming a straight line which passes through the origin, since $0\mathbf{x} = \mathbf{0}$.

♦

Example 2.8. Consider again the vector space R^3, and let $\mathbf{x}$ and $\mathbf{y}$ be two nonzero vectors such that $\mathbf{y}$ is not a constant multiple of $\mathbf{x}$. Then the set

$$V\left(\{\mathbf{x},\mathbf{y}\}\right) = \left\{a\mathbf{x} + b\mathbf{y} \mid a,b \text{ are real scalars}\right\}$$

is a plane containing the origin. Figure 2.1 illustrates this subspace, which is the plane containing the origin and the endpoints of vectors $\mathbf{x}$ and $\mathbf{y}$.

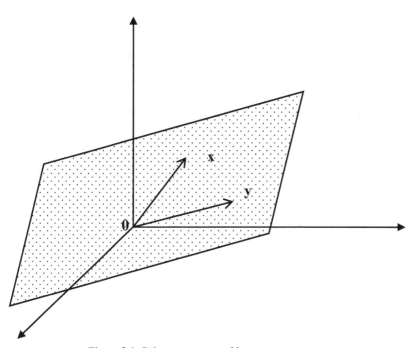

Figure 2.1. Subspace generated by two vectors

♦

Example 2.9. Let V be the set of all real variable, real valued functions defined on the interval $(-\infty, \infty)$, and let real numbers be considered as scalars. For a given integer $n \geq 1$, select the functions

$$p_0(x) = 1, \ p_1(x) = x, \ p_2(x) = x^2, \ \ldots, \ p_n(x) = x^n.$$

Then the subspace generated by these functions is the set of all linear combinations of these functions, which is the set of all real polynomials of degree of maximum n:

$$V\left(\{p_0, p_1, \ldots, p_n\}\right) = \left\{a_0 + a_1 r + \ldots + a_n r^n \mid a_0, a_1, \ldots, a_n \text{ are real scalars}\right\}.$$

                                                              ◆

2.3   Linear Independence, Basis

In the previous section we have seen that any element of a subspace generated by a subset of a vector space can be obtained as a linear combination of finitely many elements from that subset. The description of the generated subspace becomes simpler, if the necessary number of the elements in the linear combinations is smaller. In this section we will find the minimum number of such elements and the corresponding minimal realization of subspaces. Our analysis will be based on the concept of linear independence, which will be defined next.

Definition 2.4. Let V be a vector space. The elements $\mathbf{x}_1, \ldots, \mathbf{x}_n \in V$ are called *linearly independent*, if

$$a_1\mathbf{x}_1 + a_2\mathbf{x}_2 + \ldots + a_n\mathbf{x}_n = \mathbf{0}$$

implies that $a_1 = a_2 = \ldots = a_n = 0$. If the elements are not linearly independent, then we say that they are *linearly dependent*.

That is, a linear combination of linearly independent vectors is zero only if all coefficients are equal to zero.

Example 2.10. Consider the vector space R^n, and select the vectors:

$$\mathbf{e}_1 = \begin{pmatrix} 1 \\ 0 \\ 0 \\ \vdots \\ 0 \\ 0 \end{pmatrix}, \; \mathbf{e}_2 = \begin{pmatrix} 0 \\ 1 \\ 0 \\ \vdots \\ 0 \\ 0 \end{pmatrix}, \; \ldots, \; \mathbf{e}_n = \begin{pmatrix} 0 \\ 0 \\ 0 \\ \vdots \\ 0 \\ 1 \end{pmatrix}.$$

Notice that for $k = 1, 2, \ldots, n$, the kth entry of $\mathbf{e}_k$ equals 1, and all other elements are equal to zero. We will next prove that these vectors are linearly independent. A linear combination of these vectors is zero, when with some constants $a_1, \ldots, a_n$,

$$a_1 \begin{pmatrix} 1 \\ 0 \\ 0 \\ \vdots \\ 0 \\ 0 \end{pmatrix} + a_2 \begin{pmatrix} 0 \\ 1 \\ 0 \\ \vdots \\ 0 \\ 0 \end{pmatrix} + \ldots + a_n \begin{pmatrix} 0 \\ 0 \\ 0 \\ \vdots \\ 0 \\ 1 \end{pmatrix} = \begin{pmatrix} 0 \\ 0 \\ 0 \\ \vdots \\ 0 \\ 0 \end{pmatrix}.$$

The first element of the left-hand side vector is

$$a_1 \cdot 1 + a_2 \cdot 0 + \ldots + a_n \cdot 0,$$

which is zero only if $a_1 = 0$. The second element of the left-hand side vector equals

$$a_1 \cdot 0 + a_2 \cdot 1 + \ldots + a_n \cdot 0,$$

which is zero only if $a_2 = 0$. In general, the kth ($k = 1, 2, \ldots, n$) element of the left-hand side vector is

$$a_1 \cdot 0 + \ldots + a_{k-1} \cdot 0 + a_k \cdot 1 + a_{k+1} \cdot 0 + \ldots + a_n \cdot 0,$$

which is zero only if $a_k = 0$. Thus, all coefficients are necessarily zero.

♦

Example 2.11. Let next V be the set of all continous real functions defined for all $x \in (-\infty, \infty)$. Select the functions

$$p_0(x) = 1, \ p_1(x) = x, \ p_2(x) = x^2, \ \dots, \ p_n(x) = x^n$$

from V. We will now prove that these functions are linearly independent. If their linear combination is the zero element in V, then for all real x,

$$a_0 + a_1 x + a_2 x^2 + \dots + a_n x^n = 0,$$

which is possible only if $a_0 = a_1 = \dots = a_n = 0$.

Contrary to this, assume that at least one coefficient is nonzero. Assume that $a_k \neq 0$ and $a_{k+1} = a_{k+2} = \dots = a_n = 0$. If $k = 0$ then the polynomial is a nonzero constant. Otherwise, for all real $x \neq 0$,

$$a_0 + a_1 x + a_2 x^2 + \dots + a_k x^k = x^k \left(a_k + \frac{a_{k-1}}{x} + \dots + \frac{a_1}{x^{k-1}} + \frac{a_0}{x^k} \right).$$

If $x \to \infty$, then $x^k \to \infty$, and all terms $\dfrac{a_{k-1}}{x}, \dots, \dfrac{a_1}{x^{k-1}}, \dfrac{a_0}{x^k}$ converge to zero. Therefore, this polynomial converges to ∞ if $a_k > 0$ and it converges to $-\infty$ if $a_k < 0$. Hence the polynomial cannot have zero value for all real x.

$$\blacklozenge$$

Some elementary properties of linearly independent elements are listed below:

1. If $\mathbf{0}$ is among the selected elements, then they are linearly dependent. Assume that $\mathbf{0}, \mathbf{x}_2, \dots, \mathbf{x}_k$ are selected. Then

$$1 \cdot \mathbf{0} + 0 \cdot \mathbf{x}_2 + \dots + 0 \cdot \mathbf{x}_k = \mathbf{0}$$

showing that a linear combination equals zero with a nonzero coefficient. Hence these elements are linearly dependent.

2. If among $\mathbf{x}_1, \mathbf{x}_2, \dots, \mathbf{x}_k$ at least one element is repeated, then they are linearly dependent.

Assume now that $\mathbf{x}_1 = \mathbf{x}_2$ otherwise we have to renumber the elements. Then

$$1 \cdot \mathbf{x}_1 + (-1) \cdot \mathbf{x}_2 + 0 \cdot \mathbf{x}_3 + \ldots + 0 \cdot \mathbf{x}_k = \mathbf{0},$$

and there are two nonzero coefficients.

3. Any subset of the set of linearly independent elements contains only linearly independent vectors.
Assume that $\mathbf{x}_1, \mathbf{x}_2, \ldots, \mathbf{x}_k$ are linearly independent, and let $\mathbf{x}_1, \mathbf{x}_2, \ldots, \mathbf{x}_l$ $(l < k)$ be a subset of the original set of elements. In contrary to the assertion assume that the subset has linearly dependent vectors. Then with some scalars $a_1, a_2, \ldots, a_l$,

$$a_1 \cdot \mathbf{x}_1 + \ldots + a_l \cdot \mathbf{x}_l = \mathbf{0}$$

and there is at least one nonzero coefficient a_i $(1 \le i \le l)$.
Consider now the linear combination

$$a_1 \cdot \mathbf{x}_1 + \ldots + a_l \cdot \mathbf{x}_l + 0 \cdot \mathbf{x}_{l+1} + \ldots + 0 \cdot \mathbf{x}_k$$

of the original elements, which is the zero vector, but the coefficient a_i is nonzero. Therefore, the original elements are also linearly dependent, which contradicts the assumption. Hence, the subset contains only linearly independent elements.

4. Let $\mathbf{x} \ne \mathbf{0}$ be an element of a vector space. Then it forms a linearly independent (one element) set.

In contrary to the assertion assume that $a \cdot \mathbf{x} = \mathbf{0}$ with some $a \ne 0$. Then $\dfrac{1}{a}$ exists and

$$\mathbf{x} = 1 \cdot \mathbf{x} = \left(\frac{1}{a} \cdot a \right) \cdot \mathbf{x} = \frac{1}{a} \cdot (a \cdot \mathbf{x}) = \frac{1}{a} \cdot \mathbf{0} = \mathbf{0},$$

which is an obvious contradiction to the assumption that $\mathbf{x}$ is nonzero.

Theorem 2.5. The elements x_1, x_2, ..., x_n of a vector space V are linearly dependent if and only if at least one of them can be expressed as the linear combination of the others.

Proof. Assume first that x_1, x_2, ..., x_n are linearly dependent. Then with some scalars a_1, a_2, ..., a_n,

$$a_1 \, x_1 + a_2 \, x_2 + \ldots + a_n \, x_n = 0,$$

where at least one coefficient is nonzero, say $a_i \neq 0$.
Then

$$a_i x_i = -a_1 x_1 - \ldots - a_{i-1} x_{i-1} - a_{i+1} x_{i+1} - \ldots - a_n x_n,$$

therefore

$$x_i = \left(-\frac{a_1}{a_i}\right) x_1 + \ldots + \left(-\frac{a_{i-1}}{a_i}\right) x_{i-1} + \left(-\frac{a_{i+1}}{a_i}\right) x_{i+1} + \ldots + \left(-\frac{a_n}{a_i}\right) x_n$$

where the right hand side is a linear combination of the elements x_1, ..., x_{i-1}, x_{i+1}, ..., x_n.

Assume next that at least one element, say x_i equals the linear combination of the others. Then

$$x_i = a_1 \cdot x_1 + \ldots + a_{i-1} \cdot x_{i-1} + a_{i+1} \cdot x_{i+1} + \ldots + a_n \cdot x_n,$$

which can be rewritten as

$$a_1 \cdot x_1 + \ldots + a_{i-1} \cdot x_{i-1} + (-1) \cdot x_i + a_{i+1} \cdot x_{i+1} + \ldots + a_n \cdot x_n = 0.$$

The left-hand side is a linear combination of all of the elements, and the coefficient of x_i is nonzero, therefore the elements are linearly dependent.

♣

In the case when all selected elements are nonzero, this theorem can be reformulated as follows.

Theorem 2.6. Assume that x_1, x_2, ..., x_n are nonzero elements of a vector space V. Then they are linearly dependent if and only if there exists an i ($i = 2, 3, ..., n$) such that x_i is the linear combination of x_1, ..., x_{i-1}.

Proof. Assume first that the vectors are linearly dependent. This assumption implies that

$$a_1 \cdot x_1 + a_2 \cdot x_2 + ... + a_n \cdot x_n = 0$$

where at least one of the coefficients is nonzero. Let i be the largest integer ($1 \leq i \leq n$) such that $a_i \neq 0$. Then

$$a_1 \cdot x_1 + ... + a_{i-1}x_{i-1} + a_i \cdot x_i = 0,$$

since $a_{i+1} = a_{i+2} = ... = a_n = 0$. This equation implies that

$$a_i \cdot x_i = -a_1 x_1 - ... - a_{i-1} x_{i-1},$$

that is,

$$x_i = \left(-\frac{a_1}{a_i} \right) \cdot x_1 + ... + \left(-\frac{a_{i-1}}{a_i} \right) x_{i-1}.$$

Assume next that x_i can be obtained as a linear combination of x_1, ..., x_{i-1}. Then it is the linear combination of x_1, ..., x_{i-1}, x_{i+1}, ..., x_n, where x_{i+1}, ..., x_n are multiplied by zero. Therefore Theorem 2.5 implies that x_1, ..., x_n are linearly dependent.

♣

Example 2.12. The vectors

$$x_1 = \begin{pmatrix} 1 \\ 1 \\ 1 \end{pmatrix}, x_2 = \begin{pmatrix} 2 \\ 1 \\ 1 \end{pmatrix}, \text{ and } x_3 = \begin{pmatrix} 4 \\ 3 \\ 3 \end{pmatrix}$$

are linearly dependent, since $x_3 = 2 \cdot x_1 + 1 \cdot x_2$, which can be shown by simple calculation:

$$2 \cdot \mathbf{x}_1 + 1 \cdot \mathbf{x}_2 = 2 \cdot \begin{pmatrix} 1 \\ 1 \\ 1 \end{pmatrix} + 1 \cdot \begin{pmatrix} 2 \\ 1 \\ 1 \end{pmatrix} = \begin{pmatrix} 2 \\ 2 \\ 2 \end{pmatrix} + \begin{pmatrix} 2 \\ 1 \\ 1 \end{pmatrix} = \begin{pmatrix} 4 \\ 3 \\ 3 \end{pmatrix} = \mathbf{x}_3.$$

♠

Theorem 2.7. The elements $\mathbf{x}_1$, ..., $\mathbf{x}_n$ of a vector space V are linearly independent if and only if every element of the subspace generated by these vectors can be uniquely expressed as the linear combination of these vectors.

Proof. Assume first that an element of the generated subspace can be expressed as two different linear combinations:

$$\mathbf{x} = a_1\mathbf{x}_1 + \ldots + a_n\mathbf{x}_n = b_1\mathbf{x}_1 + \ldots + b_n\mathbf{x}_n.$$

This equation implies that

$$\left(a_1 - b_1\right) \cdot \mathbf{x}_1 + \ldots + \left(a_n - b_n\right) \cdot \mathbf{x}_n = \mathbf{0},$$

where for at least one i, $a_i - b_i \neq 0$. This implies that $\mathbf{x}_1$, ..., $\mathbf{x}_n$ are linearly dependent.

Assume next that each element of V ($\{\mathbf{x}_1, \ldots, \mathbf{x}_n\}$) can be uniquely expressed as the linear combination of $\mathbf{x}_1$, ..., $\mathbf{x}_n$. Since

$$0 \cdot \mathbf{x}_1 + 0 \cdot \mathbf{x}_2 + \ldots + 0 \cdot \mathbf{x}_n = \mathbf{0},$$

there is no other linear combination of these elements that equals 0. Therefore $\mathbf{x}_1$, ..., $\mathbf{x}_n$ are linearly independent.

♣

Before defining the basis and dimension of vector spaces and subspaces, one more concept should be introduced.

Definition 2.5. Let V be a vector space and let S be a subset of V. We say that S is a *generating system* of V, if $L(S) = V$ that is, if V coincides with the set of all linear combinations of finitely many elements of S.

Every vector space V has generating systems, since V itself is a generating system. (Notice that each element $\mathbf{x} \in V$ is a linear combination of itself: $1 \cdot \mathbf{x}$). Vector spaces (or subspaces) with finite generating systems are called *finitely generated*.

Definition 2.6. A vector space $V \neq \{\mathbf{0}\}$ is said to be *n-dimensional*, if it has at least one *n*-element generating system and any set with less than n elements is not a generating system of V. The dimension of V is denoted by $\dim(V)$. If a vector space does not have finite generating system, then it is said to be *infinite dimensional*.
This concept is illustrated next.

Example 2.13. Vector space $V = R^n$ is finite dimensional, since the vectors

$$\mathbf{e}_1 = \begin{pmatrix} 1 \\ 0 \\ 0 \\ \vdots \\ 0 \\ 0 \end{pmatrix}, \ \mathbf{e}_2 = \begin{pmatrix} 0 \\ 1 \\ 0 \\ \vdots \\ 0 \\ 0 \end{pmatrix}, \ ..., \ \mathbf{e}_n = \begin{pmatrix} 0 \\ 0 \\ 0 \\ \vdots \\ 0 \\ 1 \end{pmatrix}$$

form a generating system. Let

$$\mathbf{a} = \begin{pmatrix} a_1 \\ a_2 \\ a_3 \\ \vdots \\ a_{n-1} \\ a_n \end{pmatrix} \in V$$

be an arbitrary vector, then it is easy to see by comparing the entries of the vectors of the two sides of the equality, that

$$\mathbf{a} = a_1 \cdot \mathbf{e}_1 + a_2 \cdot \mathbf{e}_2 + ... + a_n \cdot \mathbf{e}_n.$$

Hence, arbitrary $\mathbf{a} \in V$ can be expressed as the linear combination of the vectors $\mathbf{e}_1, \ldots, \mathbf{e}_n$.

◆

Example 2.14. Let V be the set of all single-variable real polynomials. Then V is infinite-dimensional, since there is no finite generating system. Assume to the contrary that a finite set of polynomials $p_1, p_2, \ldots, p_N$ is a generating system. Let $n_1, n_2, \ldots, n_N$ denote the degrees of these polynomials, respectively, and define $n = \max\{n_1, n_2, \ldots, n_N\}$, which is the largest among the degrees. Since every linear combination of these polynomials is a polynomial of degree at most n, no polynomial of higher degree can be obtained as a linear combination. Therefore, polynomials $p_1, p_2, \ldots, p_N$ do not form a generating system.

◆

Definition 2.7. Let V be a finite dimensional vector space. A finite set of elements $\mathbf{x}_1, \ldots, \mathbf{x}_n$ is called a *basis* of V, if
    (i) they are linearly independent;
    (ii) they form a generating system of V.

Example 2.15. Combining the results of Examples 2.10 and 2.13 we conclude that vectors $\mathbf{e}_1, \mathbf{e}_2, \ldots, \mathbf{e}_n$ form a basis in R^n.

◆

Example 2.16. Let V be the set of single variable, real polynomials of degree at most n. Then

$$p_0(t) = 1,\ p_1(t) = t,\ p_2(t) = t^2, \ldots, p_n(t) = t^n$$

is a basis of V.

◆

Example 2.17. Let V be an arbitrary vector space, and let $\mathbf{0}$ denote the zero element of V. Assume that S consists of only the zero element: $S = \{\mathbf{0}\}$. As we have seen earlier, this is a trivial subspace of V. Notice that S has no basis, since equality $1 \cdot \mathbf{0} = \mathbf{0}$ shows that the zero vector itself forms a linearly dependent (one-element) set. That is, no independent generating system exists. We will say that $\dim(S) = 0$. The dimension of any other subspace of V is given in Definition 2.6.

♦

Some important properties of generating system are given in the following theorems.

Theorem 2.8. Assume that $\{x_1, x_2, \ldots, x_N\}$ is a generating system of a vector space V, and for some i, x_i, can be expressed as the linear combination of elements $x_1, \ldots, x_{i-1}, x_{i+1}, \ldots, x_N$. Then $\{x_1, \ldots, x_{i-1}, x_{i+1}, \ldots, x_N\}$ is also a generating system of V.

Proof. We will show that any arbitrary element $x \in V$ can be expressed as a linear combination of $x_1, \ldots, x_{i-1}, x_{i+1}, \ldots, x_N$, that is, x_i can be deleted from the generating system. For any x,

$$x = a_1 x_1 + a_2 x_2 + \ldots + a_{i-1} x_{i-1} + a_i x_i + a_{i+1} x_{i+1} + \ldots + a_N x_N. \quad (2.2)$$

Since x_i is a linear combination of the other elements, with some scalars $c_1, \ldots, c_{i-1}, c_{i+1}, \ldots, c_N$,

$$x_i = c_1 x_1 + \ldots + c_{i-1} x_{i-1} + c_{i+1} x_{i+1} + \ldots + c_N x_N.$$

Substituting this equality into (2.2) gives the relation

$$x = a_1 x_1 + \ldots + a_{i-1} x_{i-1} + a_i \left(c_1 x_1 + \ldots + c_{i-1} x_{i-1} + c_{i+1} x_{i+1} + \ldots + c_N x_N \right)$$
$$+ a_{i+1} x_{i+1} + \ldots + a_N x_N$$
$$= \left(a_1 + a_i c_1 \right) x_1 + \ldots + \left(a_{i-1} + a_i c_{i-1} \right) x_{i-1} + \left(a_{i+1} + a_i c_{i+1} \right) x_{i+1} + \ldots + \left(a_N + a_i c_N \right) x_N,$$

which is a linear combination of only $x_1, \ldots, x_{i-1}, x_{i+1}, \ldots, x_N$. Hence, the proof is completed.

♣

Corollary. Let S be a finite generating system of a vector space $V \neq \{0\}$. Then it contains a basis of V.

Proof. If the elements of S are linearly independent, then they form a basis. Otherwise, at least one element of S can be deleted such that the remaining set is still a generating system. If it has only linearly independent vectors, then a basis is obtained. Otherwise another element can be deleted, and so on. The process has to continue until only linearly independent vectors are obtained.

♣

Theorem 2.9. Let V be a finite dimensional vector space, and assume that $\{x_1, \ldots, x_n\}$ is a generating system of V, and let $\{y_1, \ldots, y_m\}$ be a linearly independent set in V. Then $m \le n$.

Proof. Assumption that $\{x_1, \ldots, x_n\}$ is a generating system implies that y_1 is a linear combination of these elements. Therefore $\{y_1, x_1, \ldots, x_n\}$ is a linearly dependent set.

From Theorem 2.6 we know that there exists an x_i ($1 \le i \le n$) which is a linear combination of $y_1, x_1, \ldots, x_{i-1}$, and so $\{y_1, x_1, \ldots, x_{i-1}, x_{i+1}, \ldots, x_n\}$ is also a generating system of V, and thus y_2 is a linear combination of these elements. Therefore set $\{y_1, y_2, x_1, \ldots, x_{i-1}, x_{i+1}, \ldots, x_n\}$ consists of linearly dependent elements, therefore there is an index j ($j = 1, \ldots, i-1, i + 1, \ldots, n$) such that x_j is a linear combination of the preceding elements. Notice that the linear independence of $\{y_1, \ldots, y_m\}$ implies that this element cannot be y_2. Therefore, x_j can be deleted from the generating system. Continue this process until all elements $\{y_1, \ldots, y_m\}$ are entered into the generating system by deleting one x_k at each step. Hence the resulting generating system will consist of n elements and will contain all of $y_1, \ldots, y_m$ implying that $m \le n$.

♣

Corollary 1. In a finite dimensional vector space, any basis consists of the same number of elements, which is the dimension of the vector space. This coincides with our earlier discussion when we defined the dimension of the subspace $\{0\}$ as zero, since this subspace has no basis.

Corollary 2. Let $x_1, \ldots, x_k$ be linearly independent elements in an n-dimensional vector space V, where $k < n$. Then there are certain elements $y_{k+1}, \ldots, y_n \in V$ such that $\{x_1, \ldots, x_k, y_{k+1}, \ldots, y_n\}$ is a basis of V. That is, in finite dimensional vector spaces each set of linearly independent elements can be completed to form a basis.

Proof. If $\{x_1, \ldots, x_k\}$ is a generating system, then it is a basis. Otherwise there is an element y_{k+1}, which cannot be expressed as the linear combination of $x_1, \ldots, x_k$. Therefore $x_1, \ldots, x_k, y_{k+1}$ are linearly independent. If these vectors generate the entire vector space, then they form a basis. Otherwise, there is an y_{k+2}, which cannot be expressed as a linear combination of these vectors, so $x_1, \ldots, x_k, y_{k+1}, y_{k+2}$, are linearly independent. Since the vector space is n-dimensional, after $n-k$ steps we obtain a set $\{x_1, \ldots, x_k, y_{k+1}, \ldots, y_n\}$ of linearly independent vectors that generates V, which is therefore a basis.

♣

Corollary 3. Let V be an n-dimensional vector space and assume that $\{x_1, \ldots, x_n\}$ is a linearly independent set. Then it is a basis of V.

Example 2.18. The vector space R^n is n-dimensional, since from Examples 2.10 and 2.13 we know that vectors $e_1, \ldots, e_n$ form a basis. This basis is usually called the *natural basis* of R^n.

♦

Example 2.19. For given positive integers m and n, let V be the set of all real or complex $m \times n$ matrices. For $i = 1, \ldots, m$ and $j = 1, \ldots, n$, introduce matrices E_{ij}, the (i, j)-entry of which equals 1 and all other elements equal 0. Similar to the case of R^n one may easily show that these matrices form a basis of V, therefore this vector space is $(m \cdot n)$-dimensional.

♦

Example 2.20. Assume next that V is the set of all real polynomials of degree at most n. Then this vector space is $(n+1)$-dimensional, since from Examples 2.9 and 2.11 we conclude that $p_0(x) = 1$, $p_1(x) = x, \ldots, p_n(x) = x^n$ form a basis in V.

$\blacklozenge$

2.4 Inner-Product Spaces

Let V be a real vector space.

Definition 2.8. An *inner product* on V is a bivariable function, which assigns to each ordered pair $\mathbf{x}$, $\mathbf{y}$ of elements of V a real number, denoted by $(\mathbf{x}, \mathbf{y})$, such that the following conditions are satisfied:
(i) $(\mathbf{x}, \mathbf{x}) \geq 0$ for all $\mathbf{x} \in V$ and $(\mathbf{x}, \mathbf{x}) = 0$ if and only if $\mathbf{x} = \mathbf{0}$;
(ii) $(\mathbf{x}, \mathbf{y}) = (\mathbf{y}, \mathbf{x})$ for all $\mathbf{x}, \mathbf{y} \in V$;
(iii) $(\mathbf{x}_1 + \mathbf{x}_2, \mathbf{y}) = (\mathbf{x}_1, \mathbf{y}) + (\mathbf{x}_2, \mathbf{y})$ for all $\mathbf{x}_1, \mathbf{x}_2, \mathbf{y} \in V$;
(iv) $(c \cdot \mathbf{x}, \mathbf{y}) = c \cdot (\mathbf{x}, \mathbf{y})$ for all $\mathbf{x}, \mathbf{y} \in V$ and $c \in \mathbb{R}$.

Example 2.21. Let $V = R^n$ and for all $\mathbf{x}, \mathbf{y} \in V$, define

$$(\mathbf{x}, \mathbf{y}) = \mathbf{x}^T \mathbf{y}$$

(which is the product of $\mathbf{x}$ as a row vector by the column vector $\mathbf{y}$). If x_i and y_i denote the elements of $\mathbf{x}$ and $\mathbf{y}$, respectively, then

$$(\mathbf{x}, \mathbf{y}) = x_1 y_1 + \ldots + x_n y_n.$$

We will now show that this inner product satisfies all above properties:
(i) $(\mathbf{x}, \mathbf{x}) = x_1^2 + x_2^2 + \ldots + x_n^2$ is always nonnegative, and since all terms are nonnegative, it equals zero if and only if all terms equal zero, which occurs if and only if $x_i = 0$ for all i;
(ii) $(\mathbf{x}, \mathbf{y}) = x_1 y_1 + \ldots + x_n y_n = y_1 x_1 + \ldots + y_n x_n = (\mathbf{y}, \mathbf{x})$;
(iii) $(\mathbf{x}_1 + \mathbf{x}_2, \mathbf{y}) = (x_{11} + x_{21}) y_1 + \ldots + (x_{1n} + x_{2n}) y_n$

$\quad = (x_{11} y_1 + \ldots + x_{1n} y_n) + (x_{21} y_1 + \ldots + x_{2n} y_n) = (\mathbf{x}_1, \mathbf{y}) + (\mathbf{x}_2, \mathbf{y})$,

$\quad$ where x_{1i} and x_{2i} denote the entries of $\mathbf{x}_1$ and $\mathbf{x}_2$, respectively;

(iv) $(c\mathbf{x}, \mathbf{y}) = (cx_1)y_1 + \ldots + (cx_n)y_n$
$= c(x_1 y_1) + \ldots + c(x_n y_n) = c(x_1 y_1 + \ldots + x_n y_n) = c(\mathbf{x}, \mathbf{y}).$

♦

The inner product of two vectors introduced in this example plays an important role in optimization theory. One important application is the following.

Let $\mathbf{x}$ and $\mathbf{y}$ be two vectors with nonnegative entries. Then obviously

$$(\mathbf{x}, \mathbf{y}) = x_1 y_1 + \ldots + x_n y_n \geq 0,$$

and

$$(\mathbf{x}, \mathbf{y}) = 0$$

if and only if for all $k = 1, 2, \ldots, n$, either $x_k = 0$ or $y_k = 0$. This property is known as the *complementarity* condition.

Example 2.22. Let now V be the set of all continuous real functions on an interval $[a, b]$. As in Example 2.4 this vector space is usually denoted by $C[a, b]$. Let f and g be two functions from this set. Define the inner product as

$$(f, g) = \int_a^b f(x)g(x)dx.$$

Notice, that the integral exists, since both functions are continuous. We will now prove that this inner product satisfies the conditions of Definition 2.8.

(i) $(f, f) = \int_a^b f^2(x)dx \geq 0$, since $f^2(x) \geq 0$ for all $x \in [a, b]$. Assume next, that $(f, f) = 0$, that is,

$$\int_a^b f^2(x)dx = 0.$$

Assume that for some $x_0 \in [a, b]$, $f(x_0) \neq 0$. Then $f^2(x_0) > 0$, and the continuity of function f implies that f^2 is also continuous, and therefore there is a neighborhood of x_0 such that $f^2(x) > 0$ for every x in this neighborhood. Since for all $x \in [a, b]$, $f^2(x) \geq 0$, the integral of f^2 must be positive contradicting the assumption. Therefore $f(x) - 0$ for all $x \in [a, b]$.

(ii) $(f, g) = \int_a^b f(x)g(x)dx = \int_a^b g(x)f(x)dx = (g, f);$

(iii) $(f_1 + f_2, g) = \int_a^b (f_1(x) + f_2(x))g(x)dx = \int_a^b (f_1(x)g(x) + f_2(x)g(x))dx$

$= \int_a^b f_1(x)g(x)dx + \int_a^b f_2(x)g(x)dx = (f_1, g) + (f_2, g);$

(iv) $(cf, g) = \int_a^b cf(x)g(x)dx = c \int_a^b f(x)g(x)dx = c(f, g).$

$\blacklozenge$

Definition 2.8 implies the additional properties of the inner product:

(v) $(\mathbf{x}, \mathbf{y}_1 + \mathbf{y}_2) = (\mathbf{x}, \mathbf{y}_1) + (\mathbf{x}, \mathbf{y}_2)$

since

$(\mathbf{x}, \mathbf{y}_1 + \mathbf{y}_2) = (\mathbf{y}_1 + \mathbf{y}_2, \mathbf{x}) = (\mathbf{y}_1, \mathbf{x}) + (\mathbf{y}_2, \mathbf{x}) = (\mathbf{x}, \mathbf{y}_1) + (\mathbf{x}, \mathbf{y}_2);$

(vi) $(\mathbf{x}, c\mathbf{y}) = c(\mathbf{x}, \mathbf{y})$ for all $c \in R$, which can be proven as follows:

$(\mathbf{x}, c\mathbf{y}) = (c\mathbf{y}, \mathbf{x}) = c(\mathbf{y}, \mathbf{x}) = c(\mathbf{x}, \mathbf{y}).$

(vii) $(c_1\mathbf{x}, c_2\mathbf{y}) = c_1 c_2 (\mathbf{x}, \mathbf{y})$ for all $c_1, c_2 \in R$, which is the consequence of the previous property:

$(c_1\mathbf{x}, c_2\mathbf{y}) = c_1(\mathbf{x}, c_2\mathbf{y}) = c_1 c_2 (\mathbf{x}, \mathbf{y}).$

(viii) $(\mathbf{x}, \mathbf{0}) = 0$ for all $\mathbf{x}$, since

$(\mathbf{x}, \mathbf{0}) = (\mathbf{x}, \mathbf{0} + \mathbf{0}) = (\mathbf{x}, \mathbf{0}) + (\mathbf{x}, \mathbf{0}) = 2(\mathbf{x}, \mathbf{0}),$

from which the assertion follows immediately.

Definition 2.9. A vector space V is called an *inner-product space*, if an inner-product is defined on V.

The previous examples show that both R^n and $C\,[a,\,b]$ are inner-product spaces.

Definition 2.10. Let $\mathbf{x}$, $\mathbf{y}$ be two elements of an inner-product space V. We say that $\mathbf{x}$ and $\mathbf{y}$ are orthogonal, if $(\mathbf{x},\,\mathbf{y}) = 0$.

Example 2.23. In R^n, the natural basis vectors $\mathbf{e}_1$, $\mathbf{e}_2$, ..., $\mathbf{e}_n$ are pair-wise orthogonal, since for all k and l,

$$\left(\mathbf{e}_k,\mathbf{e}_l\right) = \begin{cases} 0 & \text{if } k \neq l \\ 1 & \text{if } k = l. \end{cases}$$

♦

Example 2.24. In the interval $[0,\,2\pi]$ the functions $f_0(x) = 1$, $f_k(x) = \sin(kx)$ $(k \geq 1)$, $g_k(x) = \cos(kx)$, $(k \geq 1)$ are pair-wise orthogonal. This property has an important role in the theory of Fourier-series, and can be proved by simple integration:

$$\int_0^{2\pi} f_0(x) f_k(x)dx = \int_0^{2\pi} \sin(kx)dx = \left[\frac{-\cos(kx)}{k}\right]_0^{2\pi} = 0 \ (k \geq 1),$$

$$\int_0^{2\pi} f_0(x) g_k(x)dx = \int_0^{2\pi} \cos(kx)dx = \left[\frac{\sin(kx)}{k}\right]_0^{2\pi} = 0 \ (k \geq 1),$$

$$\int_0^{2\pi} f_k(x) f_l(x)dx = \int_0^{2\pi} \sin(kx)\sin(lx)dx = \frac{1}{2}\int_0^{2\pi}\left[\cos\big((k-l)x\big) - \cos\big((k+l)x\big)\right]dx$$

$$= \frac{1}{2}\int_0^{2\pi} \cos\big((k-l)x\big)dx - \frac{1}{2}\int_0^{2\pi} \cos\big((k+l)x\big)dx = 0 \text{ for } k \neq l,$$

since both terms equal zero;

$$\int_0^{2\pi} g_k(x)g_l(x)dx = \int_0^{2\pi} \cos(kx)\cos(lx)dx = \frac{1}{2}\int_0^{2\pi}\left[\cos\big((k-l)x\big)+\cos\big((k+l)x\big)\right]dx$$

$$= \frac{1}{2}\int_0^{2\pi}\cos\big((k-l)x\big)\,dx + \frac{1}{2}\int_0^{2\pi}\cos\big((k+l)x\big)\,dx = 0 \text{ for } k \neq l,$$

since both terms equal zero again, and finally, for all k and l,

$$\int_0^{2\pi} f_k(x)g_l(x)dx = \int_0^{2\pi} \sin(kx)\cos(lx)dx$$

$$= \frac{1}{2}\int_0^{2\pi}\left[\sin\big((k+l)x\big)+\sin\big((k-l)x\big)\right]dx$$

$$= \frac{1}{2}\int_0^{2\pi}\sin\big((k+l)x\big)\,dx + \frac{1}{2}\int_0^{2\pi}\sin\big((k-l)x\big)\,dx = 0,$$

since both terms are equal to zero even if $k = l$. This set of orthogonal functions is usually called the *trigonometric system*.

♦

Definition 2.11. A set of elements in an inner-product space is called an *orthogonal system,* if it consists of pair-wise orthogonal elements.

For example, the natural basis in R^n, and system $\{1; \sin kx \cos kx, \ k \geq 1\}$ in $C[0, 2\pi]$ are orthogonal systems.

Definition 2.12. Let $\mathbf{x}$ be an element of an inner-product space. Then the *length* of $\mathbf{x}$ is defined as $\sqrt{(\mathbf{x}, \mathbf{x})}$, and is denoted by $|\mathbf{x}|$.

For example, if $\mathbf{x} \in R^n$ then

$$|\mathbf{x}| = \sqrt{x_1^2 + \ldots + x_n^2},$$

assuming that the inner product is defined the same way as in Example 2.21.

If $f \in C[a, b]$ and the inner product is defined as in Example 2.22, then

$$|f| = \sqrt{\int_a^b f^2(x)dx}.$$

Definition 2.13. A set of elements in an inner-product space is called an *orthonormal system,* if it consists of pair-wise orthogonal elements and each element from the set has unit length.

For example, the natural basis in R^n with the inner product given above is an orthonormal system, since it is an orthogonal system and for all k, $|\mathbf{e}_k| = 1$.

The next theorem is known as the *Cauchy-Schwarz inequality*, and it is considered as one of the most fundamental properties of inner-product spaces.

Theorem 2.10. Let $\mathbf{x}$, $\mathbf{y}$ be arbitrary elements of an inner-product space V. Then

$$|(\mathbf{x}, \mathbf{y})| \le |\mathbf{x}| \cdot |\mathbf{y}|. \tag{2.3}$$

Proof. Notice first that if $\mathbf{x} = 0$ or/and $\mathbf{y} = 0$, then both sides of the inequality are equal to zero, consequently the assertion holds. Assume next that $\mathbf{x} \ne 0$ and $\mathbf{y} \ne 0$. Let λ be a real number, and consider the square of the length of the element $\mathbf{x} + \lambda\mathbf{y}$:

$$\begin{aligned}
|\mathbf{x} + \lambda\mathbf{y}|^2 &= (\mathbf{x} + \lambda\mathbf{y}, \mathbf{x} + \lambda\mathbf{y}) = (\mathbf{x}, \mathbf{x} + \lambda\mathbf{y}) + (\lambda\mathbf{y}, \mathbf{x} + \lambda\mathbf{y}) \\
&= (\mathbf{x}, \mathbf{x}) + (\mathbf{x}, \lambda\mathbf{y}) + (\lambda\mathbf{y}, \mathbf{x}) + (\lambda\mathbf{y}, \lambda\mathbf{y}) \\
&= (\mathbf{x}, \mathbf{x}) + \lambda(\mathbf{x}, \mathbf{y}) + \lambda(\mathbf{y}, \mathbf{x}) + \lambda^2(\mathbf{y}, \mathbf{y}) = \lambda^2(\mathbf{y}, \mathbf{y}) + 2\lambda(\mathbf{x}, \mathbf{y}) + (\mathbf{x}, \mathbf{x}).
\end{aligned}$$

Observe that this is a quadratic function of λ, the graph of which is a parabola that opens up and its value is always nonnegative (being the square of the length of an element). Therefore, this quadratic function has at most one real root, otherwise between the roots it would have negative values. Therefore, the discriminant is zero or negative:

$$4(\mathbf{x},\mathbf{y})^2 - 4 \cdot (\mathbf{x},\mathbf{x}) \cdot (\mathbf{y},\mathbf{y}) \le 0,$$

that is

$$(\mathbf{x},\mathbf{y})^2 \le (\mathbf{x},\mathbf{x}) \cdot (\mathbf{y},\mathbf{y}).$$

Recall that $(\mathbf{x},\mathbf{x}) = |\mathbf{x}|^2$ and $(\mathbf{y},\mathbf{y}) = |\mathbf{y}|^2$ to have the assertion.

♣

Corollary. For all $\mathbf{x}$, $\mathbf{y}$ from an inner-product space,

$$|\mathbf{x} + \mathbf{y}| \le |\mathbf{x}| + |\mathbf{y}|,$$

which relation is called the *triangle inequality*. This name originates from the geometric representation of the sum of two-dimensional real vectors as shown in Figure 2.2, and this inequality can be interpreted as the length of a side of a triangle is never longer than the sum of the lengths of the two other sides.

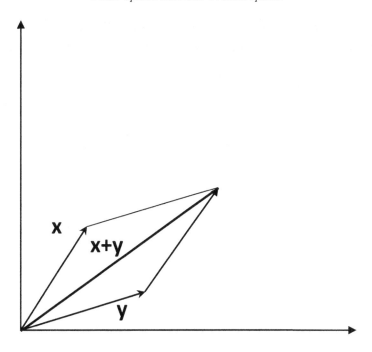

Figure 2.2. Illustration of the triangle inequality

Proof. Simple calculation and the Cauchy-Schwarz inequality imply that

$$
\begin{aligned}
\left|\mathbf{x}+\mathbf{y}\right|^2 &= (\mathbf{x}+\mathbf{y},\mathbf{x}+\mathbf{y}) = (\mathbf{x},\mathbf{x}+\mathbf{y}) + \left(\mathbf{y},\mathbf{x}+\mathbf{y}\right) \\
&= (\mathbf{x},\mathbf{x}) + (\mathbf{x},\mathbf{y}) + (\mathbf{y},\mathbf{x}) + (\mathbf{y},\mathbf{y}) = (\mathbf{x},\mathbf{x}) + 2(\mathbf{x},\mathbf{y}) + (\mathbf{y},\mathbf{y}) \\
&= \left|\mathbf{x}\right|^2 + 2(\mathbf{x},\mathbf{y}) + \left|\mathbf{y}\right|^2 \le \left|\mathbf{x}\right|^2 + 2\left|(\mathbf{x},\mathbf{y})\right| + \left|\mathbf{y}\right|^2 \\
&\le \left|\mathbf{x}\right|^2 + 2\left|\mathbf{x}\right|\cdot\left|\mathbf{y}\right| + \left|\mathbf{y}\right|^2 = \left(\left|\mathbf{x}\right|+\left|\mathbf{y}\right|\right)^2 .
\end{aligned}
$$

♣

Definition 2.14. Let $\mathbf{x}$ and $\mathbf{y}$ be two nonzero elements of an inner-product space. The angle φ between these elements is defined as the unique $\varphi \in [0, \pi]$ such that

$$
\cos\varphi = \frac{(\mathbf{x},\mathbf{y})}{\left|\mathbf{x}\right|\cdot\left|\mathbf{y}\right|} . \tag{2.4}
$$

The right-hand side of this equation is always between -1 and $+1$, therefore φ exists and is unique. Notice that this definition is equivalent to the definition of inner-products

$$(\mathbf{x},\mathbf{y}) = |\mathbf{x}| \cdot |\mathbf{y}| \cos\varphi$$

in the case of two or three-dimensional vectors, which is known from analytic geometry.

Example 2.25. Let

$$\mathbf{x} = \begin{pmatrix} 1 \\ 2 \\ 1 \end{pmatrix} \text{ and } \mathbf{y} = \begin{pmatrix} -4 \\ 1 \\ 2 \end{pmatrix}.$$

Then

$$(\mathbf{x},\mathbf{y}) = 1 \cdot (-4) + 2 \cdot 1 + 1 \cdot 2 = 0,$$

$$|\mathbf{x}| = \sqrt{1^2 + 2^2 + 1^2} = \sqrt{6} \text{ and } |\mathbf{y}| = \sqrt{(-4)^2 + 1^2 + 2^2} = \sqrt{21},$$

therefore $\cos\varphi = 0$, consequently $\varphi = 90°$. The two vectors are orthogonal.

♦

Theorem 2.11. Let $\{\mathbf{x}_1, \mathbf{x}_2, \ldots, \mathbf{x}_k\}$ be an orthogonal system of nonzero elements in an inner product space. Then vectors $\mathbf{x}_1, \mathbf{x}_2, \ldots, \mathbf{x}_k$ are linearly independent.

Proof. Assume that $\mathbf{x}_1, \mathbf{x}_2, \ldots, \mathbf{x}_k$ are linearly dependent. Then with some scalars,

$$a_1\mathbf{x}_1 + \ldots + a_i\mathbf{x}_i + \ldots + a_k\mathbf{x}_k = \mathbf{0},$$

where at least one coefficient is nonzero. Say, $a_i \neq 0$. Multiply both sides by $\mathbf{x}_i$ to see that

$$a_1(\mathbf{x}_1, \mathbf{x}_i) + \ldots + a_i(\mathbf{x}_i, \mathbf{x}_i) + \ldots + a_k(\mathbf{x}_k, \mathbf{x}_i) = (\mathbf{0}, \mathbf{x}_i). \tag{2.5}$$

The right-hand side is zero, and the orthogonality of the vectors $\mathbf{x}_1, \ldots, \mathbf{x}_k$ implies that

$$(\mathbf{x}_1, \mathbf{x}_i) = \ldots = (\mathbf{x}_{i-1}, \mathbf{x}_i) = (\mathbf{x}_{i+1}, \mathbf{x}_i) = \ldots = (\mathbf{x}_k, \mathbf{x}_i) = 0,$$

therefore equation (2.5) can be rewritten as

$$a_i(\mathbf{x}_i, \mathbf{x}_i) = 0.$$

This is a contradiction, since $a_i \neq 0$ and $(\mathbf{x}_i, \mathbf{x}_i) > 0$.

♣

We have seen earlier that vector space R^n has an orthonormal basis consisting of the vectors $\mathbf{e}_1, \mathbf{e}_2, \ldots, \mathbf{e}_n$. In the next theorem we will show that this holds for any finite dimensional inner-product space.

Theorem 2.12. Any finite dimensional inner-product space V has an orthonormal basis.

Proof. Assume that dim $(V) = k$, and let $\{\mathbf{x}_1, \mathbf{x}_2, \ldots, \mathbf{x}_k\}$ denote a basis of V. Based on these elements we introduce an inductive procedure for constructing an orthonormal basis. The method is the following. Select

$$\mathbf{z}_1 = \frac{\mathbf{y}_1}{|\mathbf{y}_1|}, \tag{2.6}$$

where $\mathbf{y}_1 = \mathbf{x}_1$, and for $i = 1, 2, 3, \ldots, k-1$, let

$$\mathbf{z}_{i+1} = \frac{\mathbf{y}_{i+1}}{|\mathbf{y}_{i+1}|}, \tag{2.7}$$

with

$$\mathbf{y}_{i+1} = \mathbf{x}_{i+1} - \sum_{j=1}^{i} \left(\mathbf{x}_{i+1}, \mathbf{z}_j\right)\mathbf{z}_j. \tag{2.8}$$

In order to prove that $\{z_1, z_2, ..., z_k\}$ is an orthonormal basis we have to verify that

      (a) $|z_i| = 1$ for $i = 1, 2, ..., k$
      (b) $(z_i, z_j) = 0$ for all $i \neq j$
      (c) set $\{z_1, z_2, ..., z_k\}$ is a generating system of V.

Property (a) follows from (2.7), since

$$|z_i|^2 = (z_i, z_i) = \left(\frac{y_i}{|y_i|}, \frac{y_i}{|y_i|}\right) = \frac{1}{|y_i|^2}(y_i, y_i) = \frac{1}{|y_i|^2}|y_i|^2 = 1.$$

The orthogonality of the elements $\{z_1, z_2, ..., z_k\}$ is proved by finite induction. Notice first that

$$(z_2, z_1) = \left(\frac{y_2}{|y_2|}, \frac{y_1}{|y_1|}\right) = \frac{1}{|y_1| \cdot |y_2|}(y_2, y_1),$$

where

$$(y_2, y_1) = (x_2 - (x_2, z_1)z_1, y_1) = (x_2, y_1) - (x_2, z_1) \cdot (z_1, y_1)$$

$$= (x_2, y_1) - \left(x_2, \frac{y_1}{|y_1|}\right) \cdot \left(\frac{y_1}{|y_1|}, y_1\right) = (x_2, y_1) - \frac{1}{|y_1|}(x_2, y_1) \cdot \frac{1}{|y_1|}(y_1, y_1)$$

$$= (x_2, y_1) - \frac{1}{|y_1|^2}(x_2, y_1) \cdot |y_1|^2 = 0.$$

Assume next that for some i, z_i is orthogonal to $z_1, ..., z_{i-1}$. Then we will prove that z_{i+1} is orthogonal to all elements $z_1, z_2, ..., z_i$. Let l ($1 \leq l \leq i$) be arbitrary. Then

$$(z_{i+1}, z_l) = \left(\frac{y_{i+1}}{|y_{i+1}|}, \frac{y_l}{|y_l|}\right) = \frac{1}{|y_{i+1}| \cdot |y_l|}(y_{i+1}, y_l),$$

where

$$(\mathbf{y}_{i+1},\mathbf{y}_l)=\left(\mathbf{x}_{i+1}-\sum_{j=1}^{i}(\mathbf{x}_{i+1},\mathbf{z}_j)\mathbf{z}_j,\mathbf{y}_l\right)=(\mathbf{x}_{i+1},\mathbf{y}_l)-\sum_{j=1}^{i}(\mathbf{x}_{i+1},\mathbf{z}_j)\cdot(\mathbf{z}_j,\mathbf{y}_l)$$

$$=(\mathbf{x}_{i+1},\mathbf{y}_l)-\sum_{j=1}^{i}(\mathbf{x}_{i+1},\mathbf{z}_j)\cdot(\mathbf{z}_j,|\mathbf{y}_l|\cdot\mathbf{z}_l)=(\mathbf{x}_{i+1},\mathbf{y}_l)-\sum_{j=1}^{i}(\mathbf{x}_{i+1},\mathbf{z}_j)\cdot|\mathbf{y}_l|\cdot(\mathbf{z}_j,\mathbf{z}_l).$$

Notice that $(\mathbf{z}_j,\mathbf{z}_l)=0$ if $j\neq 1$ and $(\mathbf{z}_l,\mathbf{z}_l)=1$, therefore

$$(\mathbf{y}_{i+1},\mathbf{y}_l)=(\mathbf{x}_{i+1},\mathbf{y}_l)-(\mathbf{x}_{i+1},\mathbf{z}_l)\cdot|\mathbf{y}_l|\cdot(\mathbf{z}_l,\mathbf{z}_l)$$
$$=(\mathbf{x}_{i+1},\mathbf{y}_l)-\left(\mathbf{x}_{i+1},|\mathbf{y}_l|\cdot\mathbf{z}_l\right)=0,$$

since $|\mathbf{y}_l|\cdot\mathbf{z}_l=\mathbf{y}_l$. Hence, property (b) is verified.

In order to prove that $\{\mathbf{z}_1,\mathbf{z}_2,\dots,\mathbf{z}_k\}$ is a basis in V it is sufficient to mention that dim $(V)=k$ and system $\{\mathbf{z}_1,\mathbf{z}_2,\dots,\mathbf{z}_k\}$ consists of k linearly independent elements.

♣

Remark. Procedure (2.6), (2.7), (2.8) is known as the *Gram-Schmidt orthogonalization process*.
Assume now that $\{\mathbf{z}_1,\mathbf{z}_2,\dots,\mathbf{z}_k\}$ is an orthonormal basis in V. Then arbitrary vectors $\mathbf{x},\mathbf{y}\in V$ can be uniquely expressed as the linear combinations of the basis elements:

$$\mathbf{x}=x_1\mathbf{z}_1+x_2\mathbf{z}_2+\dots+x_k\mathbf{z}_k$$

and

$$\mathbf{y}=y_1\mathbf{z}_1+y_2\mathbf{z}_2+\dots+y_k\mathbf{z}_k,$$

where x_i and y_i are real scalars for $i=1,2,\dots,k$. Then the inner product of $\mathbf{x}$ and $\mathbf{y}$ has the simple form:

$$(\mathbf{x},\mathbf{y})=\left(\sum_{i=1}^{k}x_i\,\mathbf{z}_i,\sum_{j=1}^{k}y_j\,\mathbf{z}_j\right)=\sum_{i=1}^{k}\sum_{j=1}^{k}x_i\,y_j\left(\mathbf{z}_i,\mathbf{z}_j\right)=\sum_{i=1}^{k}x_i\,y_i,$$

since for $i \neq j$, $(\mathbf{z}_i, \mathbf{z}_j) = 0$, and for all i, $(\mathbf{z}_i, \mathbf{z}_i) = 1$. Notice that this simple derivation shows that the inner product can be easily obtained as the product of the row vector $\mathbf{x}^T = (x_1, x_2, ..., x_k)$ by the column vector

$$\mathbf{y} = \begin{pmatrix} y_1 \\ y_2 \\ \vdots \\ y_k \end{pmatrix}.$$

The Gram-Schmidt orthogonalization procedure is illustrated by the following numerical examples.

Example 2.26. Let $V = R^3$ and consider the vectors

$$\mathbf{x}_1 = \begin{pmatrix} 1 \\ 0 \\ 0 \end{pmatrix}, \quad \mathbf{x}_2 = \begin{pmatrix} 1 \\ 1 \\ 0 \end{pmatrix}, \quad \text{and} \quad \mathbf{x}_3 = \begin{pmatrix} 1 \\ 1 \\ 1 \end{pmatrix}.$$

It is easy to show that they are linearly independent, therefore they form a basis in V. The application of the Gram-Schmidt process consists of the following elementary steps.

For $i = 1$,

$$\mathbf{y}_1 = \mathbf{x}_1 = \begin{pmatrix} 1 \\ 0 \\ 0 \end{pmatrix}, \quad |\mathbf{y}_1| = \sqrt{1^2 + 0^2 + 0^2} = 1, \quad \mathbf{z}_1 = \frac{\mathbf{y}_1}{|\mathbf{y}_1|} = \begin{pmatrix} 1 \\ 0 \\ 0 \end{pmatrix}.$$

For $i = 2$,

$$\mathbf{y}_2 = \mathbf{x}_2 - (\mathbf{x}_2, \mathbf{z}_1)\mathbf{z}_1 = \begin{pmatrix} 1 \\ 1 \\ 0 \end{pmatrix} - 1 \cdot \begin{pmatrix} 1 \\ 0 \\ 0 \end{pmatrix} = \begin{pmatrix} 0 \\ 1 \\ 0 \end{pmatrix}, \quad |\mathbf{y}_2| = \sqrt{0^2 + 1^2 + 0^2} = 1,$$

and so,

$$\mathbf{z}_2 = \frac{\mathbf{y}_2}{|\mathbf{y}_2|} = \begin{pmatrix} 0 \\ 1 \\ 0 \end{pmatrix}.$$

For $i = 3$

$$\mathbf{y}_3 = \mathbf{x}_3 - (\mathbf{x}_3, \mathbf{z}_1)\mathbf{z}_1 - (\mathbf{x}_3, \mathbf{z}_2)\mathbf{z}_2 = \begin{pmatrix} 1 \\ 1 \\ 1 \end{pmatrix} - 1 \cdot \begin{pmatrix} 1 \\ 0 \\ 0 \end{pmatrix} - 1 \cdot \begin{pmatrix} 0 \\ 1 \\ 0 \end{pmatrix} = \begin{pmatrix} 0 \\ 0 \\ 1 \end{pmatrix},$$

and since $|\mathbf{y}_3| = \sqrt{0^2 + 0^2 + 1^2} = 1,$

$$\mathbf{z}_3 = \frac{\mathbf{y}_3}{|\mathbf{y}_3|} = \begin{pmatrix} 0 \\ 0 \\ 1 \end{pmatrix}.$$

♦

Example 2.27. Consider next the set of all real polynomials of degree at most 2 in the interval $[-1, 1]$. We know that

$$p_0(x) = 1, p_1(x) = x, p_2(x) = x^2$$

is a basis. In this case the Gram-Schmidt orthogonalization process has the following steps, where for the sake of using usual notation, q and r will be used instead of y and z.

For $i = 0$,

$$q_0(x) = p_0(x) = 1, \ |q_0| = \sqrt{\int_{-1}^{1} 1 dx} = \sqrt{2},$$

$$r_0(x) = \frac{q_0(x)}{|q_0|} = \frac{1}{\sqrt{2}}.$$

For $i = 1$,

$$q_1(x) = p_1(x) - (p_1, r_0) r_0(x) = x$$

since

$$(p_1, r_0) = \int_{-1}^{1} x \cdot \frac{1}{\sqrt{2}} dx = \left[\frac{x^2}{2\sqrt{2}} \right]_{-1}^{1} = 0.$$

Notice that

$$|q_1| = \sqrt{\int_{-1}^{1} x^2 dx} = \sqrt{\left[\frac{x^3}{3} \right]_{-1}^{1}} = \sqrt{\frac{2}{3}},$$

therefore

$$r_1(x) = \frac{q_1(x)}{|q_1|} = \frac{x}{\sqrt{\frac{2}{3}}} = \sqrt{\frac{3}{2}} \cdot x.$$

For $i = 2$,

$$q_2(x) = p_2(x) - (p_2, r_0) r_0(x) - (p_2, r_1) r_1(x).$$

Simple calculation shows that

$$(p_2, r_0) = \int_{-1}^{1} x^2 \frac{1}{\sqrt{2}} dx = \left[\frac{x^3}{3\sqrt{2}} \right]_{-1}^{1} = \frac{2}{3\sqrt{2}} = \frac{\sqrt{2}}{3},$$

$$(p_2, r_1) = \int_{-1}^{1} x^2 \sqrt{\frac{3}{2}} x dx = \left[\frac{x^4 \sqrt{3}}{4\sqrt{2}} \right]_{-1}^{1} = 0,$$

which imply that

$$q_2(x) = x^2 - \frac{\sqrt{2}}{3} \cdot \frac{1}{\sqrt{2}} - 0 \cdot \sqrt{\frac{3}{2}} \cdot x = x^2 - \frac{1}{3}.$$

Since

$$|q_2| = \sqrt{\int_{-1}^{1}\left(x^2 - \frac{1}{3}\right)^2 dx} = \sqrt{\int_{-1}^{1}\left(x^4 - \frac{2}{3}x^2 + \frac{1}{9}\right)dx} = \sqrt{\left[\frac{x^5}{5} - \frac{2x^3}{9} + \frac{x}{9}\right]_{-1}^{1}}$$

$$= \sqrt{\frac{2}{5} - \frac{4}{9} + \frac{2}{9}} = \sqrt{\frac{8}{45}},$$

$$r_2(x) = \frac{x^2 - \frac{1}{3}}{\sqrt{\frac{8}{45}}} = \sqrt{\frac{45}{8}}x^2 - \sqrt{\frac{5}{8}}.$$

Hence

$$r_0(x) = \frac{1}{\sqrt{2}}, \quad r_1(x) = \sqrt{\frac{3}{2}} \cdot x, \quad r_2(x) = \sqrt{\frac{45}{8}}x^2 - \sqrt{\frac{5}{8}}$$

is an orthonormal basis.

♦

In this section inner products were defined only in real vector spaces, and the properties of only real inner product spaces have been examined. Similar to the real case, inner products in complex vector spaces as well as complex inner product spaces can be introduced. If V is a complex vector space, then it is assumed that the inner product is a bivariable function that assigns to each pair $\mathbf{x}, \mathbf{y}$ of elements of V a complex number denoted by $(\mathbf{x}, \mathbf{y})$, such that

(i) $(\mathbf{x}, \mathbf{x}) \geq 0$ for all $\mathbf{x} \in V$, and $(\mathbf{x}, \mathbf{x}) = 0$ if and only if $\mathbf{x} = 0$;

(ii) $(\mathbf{x}, \mathbf{y}) = \overline{(\mathbf{y}, \mathbf{x})}$ for all $\mathbf{x}, \mathbf{y} \in V$, where overbar denotes complex conjugate;

(iii) $(\mathbf{x}_1 + \mathbf{x}_2, \mathbf{y}) = (\mathbf{x}_1, \mathbf{y}) + (\mathbf{x}_2, \mathbf{y})$ for all $\mathbf{x}_1, \mathbf{x}_2, \mathbf{y} \in V$;

(iv) $(c\mathbf{x}, \mathbf{y}) = c(\mathbf{x}, \mathbf{y})$ for all $\mathbf{x}, \mathbf{y} \in V$ and complex number c.

Notice that properties (ii) and (iv) imply that for all $\mathbf{x}$, $\mathbf{y} \in V$ and complex number c,

$$\left(\mathbf{x}, c\,\mathbf{y}\right) = \overline{(c\mathbf{y},\mathbf{x})} = \overline{c} \cdot \overline{(\mathbf{y},\mathbf{x})} = \overline{c}(\mathbf{x},\mathbf{y})$$

Orthogonality in complex inner product spaces can be defined in the same way as it has been done for the real case. With obvious modifications all previous results and the results of the next section can be extended to the complex case. For example, assume that $(\mathbf{z}_1, \mathbf{z}_2, ..., \mathbf{z}_k)$ is an orthonormal basis in a complex inner product space V, then any arbitrary elements $\mathbf{x}$, $\mathbf{y} \in V$ can be uniquely represented as

$$\mathbf{x} = x_1\mathbf{z}_1 + x_2\mathbf{z}_2 + ... + x_k\mathbf{z}_k$$

and

$$\mathbf{y} = y_1\mathbf{z}_1 + y_2\mathbf{z}_2 + ... + y_k\mathbf{z}_k$$

where x_i, y_i are complex scalars for $i = 1, 2, ..., k$. Then the inner product of $\mathbf{x}$ and $\mathbf{y}$ can be given as

$$\left(\mathbf{x},\mathbf{y}\right) = \left(\sum_{i=1}^{k} x_i\,\mathbf{z}_i, \sum_{j=1}^{k} y_j\,\mathbf{z}_j \right) = \sum_{i=1}^{k}\sum_{j=1}^{k} x_i\,\overline{y}_j\left(\mathbf{z}_i,\mathbf{z}_j\right) = \sum_{i=1}^{k} x_i\,\overline{y}_i,$$

since for $i \neq j$, $(\mathbf{z}_i, \mathbf{z}_j) = 0$ and for all i, $(\mathbf{z}_i, \mathbf{z}_i) = 1$. Notice that in the above equation $\overline{y}_i$ denotes the complex conjugate of y_i. Hence the inner product of $\mathbf{x}$ and $\mathbf{y}$ can be easily obtained as the product of the row vector $\mathbf{x}^T = \left(x_1, x_2, ..., x_k\right)$ by the column vector

$$\overline{\mathbf{y}} = \begin{pmatrix} \overline{y}_1 \\ \overline{y}_2 \\ \vdots \\ \overline{y}_k \end{pmatrix}.$$

Finally, we note that in the linear algebra literature real and complex inner-product spaces are called *Euclidean* and *Unitary spaces*, respectively.

2.5 Direct Sums and Orthogonal Complementary Subspaces

Let V be a vector-space, and let V_1 and V_2 be two subspaces of V. We do not assume first that there is an inner product in V. From Theorem 2.3 we know that the intersection $V_1 \cap V_2$ is also a subspace of V. However, the union $V_1 \cup V_2$ is not necessarily a subspace as the following example illustrates.

Example 2.28. Assume that $V = R^2$, and let $V_1 = \left\{ a \cdot \begin{pmatrix} 1 \\ 0 \end{pmatrix} \middle| a \text{ is real} \right\}$, and

$V_2 = \left\{ a \cdot \begin{pmatrix} 0 \\ 1 \end{pmatrix} \middle| a \text{ is real} \right\}$. The vectors $\begin{pmatrix} 1 \\ 0 \end{pmatrix}$ and $\begin{pmatrix} 0 \\ 1 \end{pmatrix}$ are in V_1 and V_2,

respectively. However their sum, the vector $\mathbf{x} = \begin{pmatrix} 1 \\ 1 \end{pmatrix}$ does not belong to

$V_1 \cup V_2$, since $\mathbf{x} \notin V_1$ and $\mathbf{x} \notin V_2$.

♦

The subspace generated by the union $V_1 \cup V_2$ is the smallest subspace containing both V_1 and V_2. By using the notation introduced in Section 2.2, we might use the symbols $V(V_1 \cup V_2)$ or $V(V_1, V_2)$ to denote this subspace. The construction of $V(V_1 \cup V_2)$ is given in the following theorem.

Theorem 2.13. Let V_1 and V_2 be subspaces of a vector space V. Then

$$V(V_1 \cup V_2) = \left\{ \mathbf{x} + \mathbf{y} \mid \mathbf{x} \in V_1 \text{ and } \mathbf{y} \in V_2 \right\} \qquad (2.9)$$

Proof. Let U denote the right-hand side of (2.9). First we show that any element of $V(V_1 \cup V_2)$ can be written as the sum of an element of V_1 and an element of V_2. Theorem 2.4 implies that any arbitrary element $\mathbf{z} \in V(V_1 \cup V_2)$ can be written as a linear combination of finitely many elements of $V_1 \cup V_2$, that is,

$$\mathbf{z} = a_1 \mathbf{x}_1 + \ldots + a_k \mathbf{x}_k + b_1 \mathbf{y}_1 + \ldots + b_l \mathbf{y}_l,$$

where $\mathbf{x}_1$, $\mathbf{x}_2$, ..., $\mathbf{x}_k \in V_1$ and $\mathbf{y}_1$, $\mathbf{y}_2$, ..., $\mathbf{y}_l \in V_2$. Here we used the fact that each element of the union $V_1 \cup V_2$ belongs either to V_1 or to V_2. Since V_1 and V_2 are subspaces,

$$\mathbf{x} = a_1\mathbf{x}_1 + \cdots + a_k\mathbf{x}_k \in V_1 \quad \text{and} \quad \mathbf{y} = b_1\mathbf{y}_1 + \cdots + b_l\mathbf{y}_l \in V_2,$$

and therefore $\mathbf{z} = \mathbf{x} + \mathbf{y}$. Hence $V(V_1 \cup V_2) \subseteq U$.

In order to complete the proof, we have to show that $U \subseteq V(V_1 \cup V_2)$ by verifying that if $\mathbf{x} \in V_1$ and $\mathbf{y} \in V_2$ are arbitrary elements, then $\mathbf{x} + \mathbf{y} \in V(V_1 \cup V_2)$. Since V_1 and V_2 are subsets of $V(V_1 \cup V_2)$, both $\mathbf{x}$ and $\mathbf{y}$ belong to $V(V_1 \cup V_2)$, and the fact that $V(V_1 \cup V_2)$ is a subspace implies that $\mathbf{x} + \mathbf{y} \in V(V_1 \cup V_2)$. Therefore $U \subseteq V(V_1 \cup V_2)$, which completes the proof.

♣

Remark. The above theorem guarantees that every element of the generated subspace $V(V_1 \cup V_2)$ can be written as the sum of an element of V_1 and an element of V_2. In most cases this decomposition is not unique, as it is shown in the following example.

Example 2.29. Select the vector space $V = R^3$, and the vectors

$$\mathbf{e}_1 = \begin{pmatrix} 1 \\ 0 \\ 0 \end{pmatrix}, \; \mathbf{e}_2 = \begin{pmatrix} 0 \\ 1 \\ 0 \end{pmatrix}, \; \text{and} \; \mathbf{e}_3 = \begin{pmatrix} 0 \\ 0 \\ 1 \end{pmatrix}.$$

Let V_1 be the subspace generated by $\mathbf{e}_1$ and $\mathbf{e}_2$ and let V_2 be generated by $\mathbf{e}_2$ and $\mathbf{e}_3$. Then

$$V_1 = \left\{ a_1 \begin{pmatrix} 1 \\ 0 \\ 0 \end{pmatrix} + a_2 \begin{pmatrix} 0 \\ 1 \\ 0 \end{pmatrix} = \begin{pmatrix} a_1 \\ a_2 \\ 0 \end{pmatrix} \; \middle| \; a_1 \text{ and } a_2 \text{ are real} \right\}$$

and

$$V_2 = \left\{ b_1 \begin{pmatrix} 0 \\ 1 \\ 0 \end{pmatrix} + b_2 \begin{pmatrix} 0 \\ 0 \\ 1 \end{pmatrix} = \begin{pmatrix} 0 \\ b_1 \\ b_2 \end{pmatrix} \middle| \; b_1 \text{ and } b_2 \text{ are real} \right\}.$$

In Example 2.13 we have seen that e_1, e_2 and e_3 generate the entire vector space R^3, therefore $V(V_1 \cup V_2) = R^3$, since all of the vectors e_1, e_2 and e_3 belong to $V_1 \cup V_2$. Consider next a three-element vector with entries x_1, x_2 and x_3. Therefore, it can be rewritten as the sum of a vector of V_1 and a vector of V_2. That is,

$$\begin{pmatrix} x_1 \\ x_2 \\ x_3 \end{pmatrix} = \begin{pmatrix} a_1 \\ a_2 \\ 0 \end{pmatrix} + \begin{pmatrix} 0 \\ b_1 \\ b_2 \end{pmatrix}. \tag{2.10}$$

Comparing the corresponding elements of the left-hand side and right-hand side vectors we see that this equation is equivalent to the following system of equations:

$$x_1 = a_1$$
$$x_2 = a_2 + b_1$$
$$x_3 = b_2.$$

If vector $\mathbf{x}$ is given, then the values of x_1, x_2 and x_3 are also given. The values of a_1 and b_2 are unique, however x_2 can be rewritten as the sum of two real numbers in infinitely many different ways, therefore infinitely many decompositions (2.10) exist.

♦

The following result gives a sufficient and necessary condition for the uniqueness of the decomposition of the elements of $V(V_1 \cup V_2)$ as the sum of an element of V_1 and an element of V_2.

Theorem 2.14. Let V_1 and V_2 be two subspaces of a vector space V. Each element of $V(V_1 \cup V_2)$ can be uniquely written as $\mathbf{x} + \mathbf{y}$ with $\mathbf{x} \in V_1$ and $\mathbf{y} \in V_2$ if and only if the intersection $V_1 \cap V_2$ consists of only the zero vector.

Proof. Assume first, that $V_1 \cap V_2 = \{\mathbf{0}\}$, and assume in addition, that for a $\mathbf{z}$, two such decompositions hold:

$$\mathbf{z} = \mathbf{x}_1 + \mathbf{y}_1 = \mathbf{x}_2 + \mathbf{y}_2,$$

where $\mathbf{x}_1, \mathbf{x}_2 \in V_1$ and $\mathbf{y}_1, \mathbf{y}_2 \in V_2$. We can rewrite this equality as

$$\mathbf{x}_1 - \mathbf{x}_2 = \mathbf{y}_2 - \mathbf{y}_1.$$

The left-hand side is in V_1 and the right-hand side is in V_2, therefore both sides are in $V_1 \cap V_2$, which consists of only the zero element. So,

$$\mathbf{x}_1 - \mathbf{x}_2 = \mathbf{y}_2 - \mathbf{y}_1 = \mathbf{0},$$

that is, $\mathbf{x}_1 = \mathbf{x}_2$ and $\mathbf{y}_1 = \mathbf{y}_2$ proving the uniqueness of the decomposition. Assume next, that $V_1 \cap V_2$ contains at least one nonzero element which can be denoted by $\mathbf{w}$. If $\mathbf{z} = \mathbf{x} + \mathbf{y}$ with some $\mathbf{x} \in V_1$ and $\mathbf{y} \in V_2$ then $\mathbf{z} = (\mathbf{x} + \mathbf{w}) + (\mathbf{y} - \mathbf{w})$, where $\mathbf{x} \neq \mathbf{x} + \mathbf{w} \in V_1$ and $\mathbf{y} \neq \mathbf{y} - \mathbf{w} \in V_2$ showing that the decomposition is not unique.

♣

Definition 2.15. Let V_1 and V_2 be subspaces of a vector space V such that $V_1 \cap V_2 = \{\mathbf{0}\}$, then the subspace $V(V_1 \cup V_2)$ is called the *direct sum* of V_1 and V_2, and is denoted by $V_1 \oplus V_2$.

From Theorem 2.14 we know that any element of $V_1 \oplus V_2$ can be uniquely represented as $\mathbf{x} + \mathbf{y}$ with some $\mathbf{x} \in V_1$ and $\mathbf{y} \in V_2$. This fact has the following consequence. Assume that V_1 and V_2 are finitely generated, and let $(\mathbf{x}_1, ..., \mathbf{x}_k)$ and $(\mathbf{y}_1, ..., \mathbf{y}_l)$ be a basis of V_1 and V_2, respectively. Then $\{\mathbf{x}_1, ..., \mathbf{x}_k, \mathbf{y}_1, ..., \mathbf{y}_l\}$ is a basis of $V_1 \oplus V_2$. In order to prove this assertion, we have to show that
(a) these elements are linearly independent;
(b) they generate the entire subspace $V_1 \oplus V_2$.

In order to verify (a), assume that a linear combination of the elements $\mathbf{x}_1, ..., \mathbf{x}_k, \mathbf{y}_1, ..., \mathbf{y}_l$ is zero. Then with some scalars a_i and b_j ($1 \leq i \leq k$, $1 \leq j \leq l$),

$$a_1 \mathbf{x}_1 + \ldots + a_k \mathbf{x}_k + b_1 \mathbf{y}_1 + \ldots + b_l \mathbf{y}_l = \mathbf{0}$$

which implies that

$$a_1 \mathbf{x}_1 + \ldots + a_k \mathbf{x}_k = (-b_1)\mathbf{y}_1 + \ldots + (-b_l \mathbf{y}_l).$$

The left-hand side is in V_1, the right hand side is in V_2, therefore both are in $V_1 \cap V_2$, which contains only the zero element. Therefore,

$$a_1 \mathbf{x}_1 + \ldots + a_k \mathbf{x}_k = \mathbf{0} \qquad \text{and} \qquad (-b_1)\mathbf{y}_1 + \ldots + (-b_l)\mathbf{y}_l = \mathbf{0}.$$

Since $\mathbf{x}_1, \ldots, \mathbf{x}_k$ as well as $\mathbf{y}_1, \ldots, \mathbf{y}_l$ are linearly independent, $a_1 = \ldots = a_k = 0$ and $b_1 = \ldots = b_l = 0$ showing that the set $\{\mathbf{x}_1, \ldots, \mathbf{x}_k, \mathbf{y}_1, \ldots, \mathbf{y}_l\}$ consists of only linearly independent elements. In order to prove statement (b), consider an arbitrary element $\mathbf{z}$ of $V_1 \oplus V_2$. Then with some $\mathbf{x} \in V_1$ and $\mathbf{y} \in V_2$, $\mathbf{z} = \mathbf{x} + \mathbf{y}$. Since $\{\mathbf{x}_1, \ldots, \mathbf{x}_k\}$ is a basis of V_1, $\mathbf{x} = a_1\mathbf{x}_1 + \ldots + a_k\mathbf{x}_k$ with some scalars $a_1, \ldots, a_k$, and since $\{\mathbf{y}_1, \ldots, \mathbf{y}_l\}$ is a basis in V_2, $\mathbf{y} = b_1\mathbf{y}_1 + \ldots + b_l\mathbf{y}_1$. Therefore

$$\mathbf{z} = \mathbf{x} + \mathbf{y} = a_1\mathbf{x}_1 + \ldots + a_k\mathbf{x}_k + b_1\mathbf{y}_1 + \ldots + b_l\mathbf{y}_l,$$

that is, $\mathbf{z}$ is a linear combination of vectors $\mathbf{x}_1, \ldots, \mathbf{x}_k, \mathbf{y}_1, \ldots, \mathbf{y}_l$. Hence, they generate the entire direct sum $V_1 \oplus V_2$.

This statement has the following important consequence:

$$\dim(V_1 \oplus V_2) = \dim(V_1) + \dim(V_2), \tag{2.11}$$

since the dimension of finitely generated subspaces equals the number of basis elements.

Let's drop next the assumption that $V_1 \cap V_2 = \{\mathbf{0}\}$. Then $V_1 \oplus V_2$ is not defined, but the subspace $V(V_1 \cup V_2)$ generated by the union of V_1 and V_2 is defined. We will next show that a straightforward generalization of relation (2.11) holds in this more general case.

Theorem 2.15. Let V_1 and V_2 be finitely generated subspaces. Then

$$\dim\left(V\left(V_1 \cup V_2\right)\right) + \dim\left(V_1 \cap V_2\right) = \dim\left(V_1\right) + \dim\left(V_2\right).$$

Proof. If $V_1 \cap V_2 = \{0\}$ then relation (2.11) is equivalent to the assertion, since $\dim(\{0\}) = 0$. Assume next that $V_1 \cap V_2 \neq \{0\}$. Let $\{x_1, \ldots, x_k\}$ be a basis of $V_1 \cap V_2$. Corollary 2 of Theorem 2.9 implies the existence of elements $y_{k+1}, \ldots, y_m$ and $z_{k+1}, \ldots, z_n$ such that $\{x_1, \ldots, x_k, y_{k+1}, \ldots, y_m\}$ is a basis of V_1 and $\{x_1, \ldots, x_k, z_{k+1}, \ldots, z_n\}$ is a basis of V_2. By using the method which was applied in proving relation (2.11) one may easily verify that $\{x_1, \ldots, x_k, y_{k+1}, \ldots, y_m, z_{k+1}, \ldots, z_n\}$ is a basis of $V(V_1 \cup V_2)$. That is, $\dim\left(V(V_1 \cup V_2)\right) = m + n - k$. The assertion then becomes clear, since $\dim\left(V_1 \cap V_2\right) = k$, $\dim\left(V_1\right) = m$, and $\dim\left(V_2\right) = n$.

♣

Direct sums of more than two subspaces can be defined in the following way.

Definition 2.16. Let $V_1, V_2, \ldots, V_k$ be subspaces of V. Then $V(V_1, V_2, \ldots, V_k)$ is said to be the *direct sum* of $V_1, \ldots, V_k$, if each element $v \in V(V_1, V_2, \ldots, V_k)$ can be uniquely expressed as a sum

$$v = v_1 + \ldots + v_k, \tag{2.12}$$

where for $i = 1, 2, \ldots, k$, $v_i \in V_i$. In this case we use the notation

$$V\left(V_1, V_2, \ldots, V_k\right) = V_1 \oplus V_2 \oplus \ldots \oplus V_k.$$

Notice first that for $k = 2$, this definition is equivalent to Definition 2.15. To show that a vector space (or a subspace) V is the direct sum of given subspaces $V_1, V_2, \ldots, V_k$, we have to verify two things. First, we have to prove that each element can be written as the sum (2.12), and second, that this decomposition is unique for all $v \in V$. In practical cases, however we do not need to check uniqueness for all elements of v as it is given in the following result.

Theorem 2.16. Let V be a vector space (or a subspace) and $V_1, V_2, \ldots, V_k$ be subspaces of V. Then $V = V_1 \oplus V_2 \oplus \ldots \oplus V_k$ if and only if

a) for all $\mathbf{v} \in V$, $\mathbf{v}$ can be expressed as

$$\mathbf{v} = \mathbf{v}_1 + \mathbf{v}_2 + \ldots + \mathbf{v}_k,$$

where $\mathbf{v}_i \in V_i$ $(i = 1, 2, \ldots, k)$;
b) if $\mathbf{v}_i \in V_i$ $(i = 1, 2, \ldots, k)$ are elements such that

$$\mathbf{v}_1 + \ldots + \mathbf{v}_k = \mathbf{0}, \tag{2.13}$$

then $\mathbf{v}_1 = \mathbf{v}_2 = \ldots = \mathbf{v}_k = \mathbf{0}$.

Proof. If $V = V_1 \oplus V_2 \oplus \ldots \oplus V_k$, then (a) is obviously satisfied. Since the selection $\mathbf{v}_1 = \mathbf{v}_2 = \ldots = \mathbf{v}_k = \mathbf{0}$ satisfies equation (2.13), the uniqueness of such decomposition implies that necessarily $\mathbf{v}_i = \mathbf{0}$ for all i.

Assume next that conditions (a) and (b) are satisfied. From (a) we know that all elements $\mathbf{v}$ of V can be written as the sum (2.12). In order to show that $V = V_1 \oplus V_2 \oplus \ldots \oplus V_k$ it is sufficient to prove that decomposition (2.12) is unique. Assume that for some $\mathbf{v} \in V$, we have two such decompositions:

$$\mathbf{v} = \mathbf{v}_1 + \mathbf{v}_2 + \ldots + \mathbf{v}_k = \mathbf{v}_1' + \mathbf{v}_2' + \ldots + \mathbf{v}_k',$$

where $\mathbf{v}_i, \mathbf{v}_i' \in V_i$ for all i. Then

$$\left(\mathbf{v}_1 - \mathbf{v}_1'\right) + \left(\mathbf{v}_2 - \mathbf{v}_2'\right) + \ldots + \left(\mathbf{v}_k - \mathbf{v}_k'\right) = \mathbf{0},$$

and since $\mathbf{v}_i, \mathbf{v}_i' \in V_i$, condition (b) implies that for all i, $\mathbf{v}_i - \mathbf{v}_i' = \mathbf{0}$, that is, $\mathbf{v}_i = \mathbf{v}_i'$ proving the required uniqueness.

♣

As the conclusion of this section a particular case of direct sums of subspaces of inner-product spaces is introduced, which will play important roles in later chapters of this book.

Definition 2.17. Let V_1 and V_2 be two subspaces of an inner-product space V. We say that V_2 is the *orthogonal complementary subspace* of V_1 in V, if

(i) $V_1 \cap V_2 = \{0\}$;

(ll) $V_1 \oplus V_2 = V$;

(iii) for all $\mathbf{x} \in V_1$ and $\mathbf{x} \in V_2$, $(\mathbf{x}, \mathbf{y}) = 0$. That is, all elements of V_1 are orthogonal to all elements of V_2.

If V_2 is the orthogonal complementary subspace of V_1 then V_1 is also the orthogonal complementary subspace of V_2 in the same inner-product space.

Example 2.30. Select $V = R^2$, and let $V_1 = V(\mathbf{e}_1)$ and $V_2 = V(\mathbf{e}_2)$. That is, V_1 and V_2 are the subspaces generated by vectors $\mathbf{e}_1$ and $\mathbf{e}_2$, respectively. We will show that V_1 and V_2 are orthogonal complementary subspaces in V. Notice first, that

$$V_1 = \left\{ a \begin{pmatrix} 1 \\ 0 \end{pmatrix} = \begin{pmatrix} a \\ 0 \end{pmatrix} \,\middle|\, a \text{ is real} \right\}$$

and

$$V_2 = \left\{ b \begin{pmatrix} 0 \\ 1 \end{pmatrix} = \begin{pmatrix} 0 \\ b \end{pmatrix} \,\middle|\, b \text{ is real} \right\}.$$

Property (i) is obvious, since for any vector $\mathbf{x} = (x_i)$ such that $\mathbf{x} \in V_1 \cap V_2$, $x_2 = 0$ (since $\mathbf{x} \in V_1$) and $x_1 = 0$ (since $\mathbf{x} \in V_2$), that is $\mathbf{x} = \mathbf{0}$. In order to prove condition (ii) assume that $\mathbf{x} = (x_i)$ is an arbitrary vector in R^2. Then $\mathbf{x} = x_1 \mathbf{e}_1 + x_2 \mathbf{e}_2$, where the first term is in V_1 and the second term is in V_2, and this is the unique such decomposition.

Property (iii) is obvious, since with arbitrary real numbers a and b, the inner product of vectors $\begin{pmatrix} a \\ 0 \end{pmatrix}$ and $\begin{pmatrix} 0 \\ b \end{pmatrix}$ is zero, since

$$\left(\begin{pmatrix} a \\ 0 \end{pmatrix}, \begin{pmatrix} 0 \\ b \end{pmatrix} \right) = (a,0) \begin{pmatrix} 0 \\ b \end{pmatrix} = a \cdot 0 + 0 \cdot b = 0.$$

♦

Consider next a subspace $V_1 \neq \{0\}$ in a finitely generated inner-product space V, and let V_2 be the orthogonal complementary subspace of V_1 in V. The following characterization results are useful in constructing V_2.

Theorem 2.17. The orthogonal complementary subspace consists of all elements of V which are orthogonal to all elements of V_1, that is,

$$V_2 = \left\{ \mathbf{y} \mid \mathbf{y} \in V, \text{ such that } (\mathbf{y}, \mathbf{x}) = 0 \text{ for all } \mathbf{x} \in V_1 \right\}.$$

Proof. We have to verify that all conditions of Definition 2.17 are satisfied.
(i) If $\mathbf{x} \in V_1 \cap V_2$, then $\mathbf{x} \in V_1$ and $\mathbf{x} \in V_2$, therefore $\mathbf{x}$ is orthogonal to itself. That is, $(\mathbf{x}, \mathbf{x}) = 0$, which implies that $\mathbf{x} = \mathbf{0}$;
(ii) Let $\{\mathbf{x}_1, \ldots, \mathbf{x}_k\}$ be a basis in V_1. Corollary 2 of Theorem 2.9 implies that there are elements $\mathbf{y}_{k+1}, \ldots, \mathbf{y}_n \in V$ such that $\{\mathbf{x}_1, \ldots, \mathbf{x}_k, \mathbf{y}_{k+1}, \ldots, \mathbf{y}_n\}$ is a basis of V. Apply the Gram-Schmidt orthogonalization process for these vectors, then an orthonormal basis $\{\mathbf{z}_1, \ldots, \mathbf{z}_k, \mathbf{z}_{k+1}, \ldots, \mathbf{z}_n\}$ of V is obtained.
From the inductive nature of the procedure it follows that $\{\mathbf{z}_1, \ldots, \mathbf{z}_k\}$, is a basis of V_1. Denote the subspace generated by the other basis elements $\mathbf{z}_{k+1}, \ldots, \mathbf{z}_n$ by W. We will verify that $W = V_2$. First, we show that $W \subseteq V_2$. Let $\mathbf{z} \in W$ be arbitrary, then

$$\mathbf{z} = a_{k+1}\mathbf{z}_{k+1} + \ldots + a_n\mathbf{z}_n$$

and if $\mathbf{x} \in V_1$, then

$$\mathbf{x} = b_1\mathbf{z}_1 + \ldots + b_k\mathbf{z}_k.$$

Obviously, $\mathbf{z}$ is orthogonal to $\mathbf{x}$, since

$$(\mathbf{z},\mathbf{x}) = \left(\sum_{i=k+1}^{n} a_i \mathbf{z}_i, \sum_{j=1}^{k} b_j \mathbf{z}_j \right) = \sum_{i=k+1}^{n} \sum_{j=1}^{k} a_i b_j \left(\mathbf{z}_i, \mathbf{z}_j \right) = 0.$$

That is, all elements of W are orthogonal to all elements of V_1 implying that $W \subseteq V_2$. Next, we show that $V_2 \subseteq W$. Let $\mathbf{y} \in V_2$ be an arbitrary element. Since $V_2 \subseteq V$, necessarily $\mathbf{y} \in V$, therefore

$$\mathbf{y} = a_1 \mathbf{z}_1 + \ldots + a_k \mathbf{z}_k + a_{k+1} \mathbf{z}_{k+1} + \ldots + a_n \mathbf{z}_n$$

with some scalars $a_1, \ldots, a_n$. In order to verify that $\mathbf{y} \in W$ we have to prove that $a_1 = \ldots = a_k = 0$. Multiply both sides of the above equation by $\mathbf{z}_i$ to see that

$$(\mathbf{z}_i, \mathbf{y}) = a_1 \left(\mathbf{z}_i, \mathbf{z}_1 \right) + \ldots + a_i \left(\mathbf{z}_i, \mathbf{z}_i \right) + \ldots + a_n \left(\mathbf{z}_i, \mathbf{z}_n \right).$$

Notice that $\mathbf{z}_i \in V_1$ and $\mathbf{y} \in V_2$, therefore the left-hand side is zero. If $i \neq j$, $(\mathbf{z}_i, \mathbf{z}_j) = 0$, which implies that the right hand side equals $a_i(\mathbf{z}_i, \mathbf{z}_i)$. Since $\mathbf{z}_i \neq 0$ (otherwise vectors $\mathbf{z}_1, \ldots, \mathbf{z}_k$ would be linearly dependent), $(\mathbf{z}_i, \mathbf{z}_i) > 0$, therefore $a_i = 0$ for $i = 1, 2, \ldots, k$ implying that $\mathbf{y} \in W$. Condition (iii) follows from the definition of V_2, which completes the proof.

♣

Example 2.31. Consider the inner-product space R^2, and let

$$V_1 = \left\{ \begin{pmatrix} x \\ x \end{pmatrix} \middle| \ x \text{ is real} \right\}.$$

We first show that V_1 is a subspace by applying Theorem 2.2. Let a and b be two scalars, and $\mathbf{x}$ and $\mathbf{y}$ two elements from V_1. Then with some real values x and y,

$$\mathbf{x} = \begin{pmatrix} x \\ x \end{pmatrix} \text{ and } \mathbf{y} = \begin{pmatrix} y \\ y \end{pmatrix},$$

furthermore

$$a\mathbf{x} + b\mathbf{y} = \begin{pmatrix} ax \\ ax \end{pmatrix} + \begin{pmatrix} by \\ by \end{pmatrix} = \begin{pmatrix} ax + by \\ ax + by \end{pmatrix} \in V_1.$$

The orthogonal complementary subspace of V_1 in R^2 consists of all vectors which are orthogonal to all elements of V_1. Let now $\mathbf{z} = (z_i)$ denote such a vector. Then for all $\mathbf{x} \in V_1$,

$$(\mathbf{z}, \mathbf{x}) = (z_1, z_2) \begin{pmatrix} x \\ x \end{pmatrix} = z_1 x + z_2 x = x(z_1 + z_2) = 0,$$

which holds for all real x if and only if $z_2 = -z_1$.
Hence

$$V_2 = \left\{ \begin{pmatrix} z \\ -z \end{pmatrix} \mid z \text{ is real} \right\}.$$

♦

2.6 Applications

In this final section, two applications will be briefly outlined, and then simple examples will be presented.

1. First a simple algorithm is presented to select a maximum number of linearly independent vectors from a set of finitely many vectors. Consider vectors $\mathbf{x}_1, \mathbf{x}_2, \ldots, \mathbf{x}_m \in R^N$. By using Theorem 2.6, the following algorithm can be suggested:
Step 1. Select $S = \{\mathbf{x}_1, \mathbf{x}_2, \ldots, \mathbf{x}_m\}$ and $k = 2$.
Step 2. Check if $\mathbf{x}_k$ is the linear combination of vectors $\mathbf{x}_1, \mathbf{x}_2, \ldots \mathbf{x}_{k-1}$. If not, then let $k: = k + 1$ and go to Step 3. Otherwise let $m = m-1$, and for $i = k, k + 1, \ldots, m-1$, set $\mathbf{x}_i = \mathbf{x}_{i+1}$, and go to Step 3.
Step 3. If $k \leq m$, then go to Step 2, otherwise stop. The remaining vectors in S form the requested maximum number of linearly independent vectors.

Example 2.32. Select $m = 4$ and

$$S = \left\{ \begin{pmatrix} 1 \\ 1 \\ 0 \end{pmatrix}, \begin{pmatrix} 1 \\ 1 \\ 1 \end{pmatrix}, \begin{pmatrix} 2 \\ 2 \\ 1 \end{pmatrix}, \begin{pmatrix} 3 \\ 3 \\ 2 \end{pmatrix} \right\}.$$

We first select $k = 2$. Since x_2 is not a constant multiple of x_1, it is not a linear combination of x_1. Therefore, we let $k = 3$, and go to Step 3. Since $k < m$, we have to go back to Step 2, where we have to check if x_3 is a linear combination of x_1 and x_2. It is easy to see that $x_3 = x_1 + x_2$,

therefore we let $m = 3$, and $x_3 = \begin{pmatrix} 3 \\ 3 \\ 2 \end{pmatrix}$.

Therefore, system S is modified as $\left\{ \begin{pmatrix} 1 \\ 1 \\ 0 \end{pmatrix}, \begin{pmatrix} 1 \\ 1 \\ 1 \end{pmatrix}, \begin{pmatrix} 3 \\ 3 \\ 2 \end{pmatrix} \right\}.$

In Step 3 we see that $k = m$, so we have to go back to Step 2, where we see that $x_3 = x_1 + 2x_2$. Therefore, we let $m = 2$, and the final set becomes

$$S = \left\{ \begin{pmatrix} 1 \\ 1 \\ 0 \end{pmatrix}, \begin{pmatrix} 1 \\ 1 \\ 1 \end{pmatrix} \right\}.$$

♦

2. Assume next that system $S = \{x_1, x_2, \ldots, x_k\}$ is orthonormal, and a vector x does not belong to the subspace generated by the elements of S. That is, x cannot be expressed as the linear combination of vectors $x_1, x_2, \ldots, x_k$. We will now find the linear combination of these vectors that has the minimal distance from x. This is a particular case of the least squares

problem known from statistics. Mathematically this problem can be formulated as an unconstrained optimization problem:

$$\text{Minimize} \quad \left| c_1 \mathbf{x}_1 + c_2 \mathbf{x}_2 + ... + c_k \mathbf{x}_k - \mathbf{x} \right|.$$

In order to find the optimal values of c_1, c_2, ..., c_k, consider the square of the objective function:

$$\left| \sum_{i=1}^{k} c_i \mathbf{x}_i - \mathbf{x} \right|^2 = \left(\sum_{i=1}^{k} c_i \mathbf{x}_i - \mathbf{x}, \sum_{i=1}^{k} c_i \mathbf{x}_i - \mathbf{x} \right) = \sum_{i=1}^{k} c_i^2 - 2 \sum_{i=1}^{k} c_i \left(\mathbf{x}_i, \mathbf{x} \right) + \left(\mathbf{x}, \mathbf{x} \right),$$

where we used the fact that the elements of S form an orthonormal system. It is easy to see that this objective function can be rewritten as

$$\sum_{i=1}^{k} \left(c_i - \left(\mathbf{x}_i, \mathbf{x} \right) \right)^2 + \left(\mathbf{x}, \mathbf{x} \right) - \sum_{i=1}^{k} \left(\mathbf{x}_i, \mathbf{x} \right)^2.$$

It is clear that this function is minimal if we select

$$c_i = \left(\mathbf{x}_i, \mathbf{x} \right) \tag{2.14}$$

for all $i = 1, 2, ..., k$. If the elements of S are not orthonormal, then Theorem 2.12 implies that the subspace generated by the elements of S has an orthonormal basis $\mathbf{z}_1$, $\mathbf{z}_2$, ..., $\mathbf{z}_l$. Then we have to use the above procedure with the new system $\overline{S} = \{ \mathbf{z}_1, \mathbf{z}_2, ..., \mathbf{z}_l \}$.

Example 2.33. Assume that

$$S = \left\{ \begin{pmatrix} 1 \\ 0 \\ 1 \end{pmatrix}, \begin{pmatrix} 1 \\ 2 \\ 2 \end{pmatrix} \right\} \text{ and } \mathbf{x} = \begin{pmatrix} 2 \\ 2 \\ 2 \end{pmatrix}.$$

First, we show that $\mathbf{x}$ cannot be expressed as the linear combination of the two vectors of S. The equation

$$a \begin{pmatrix} 1 \\ 0 \\ 1 \end{pmatrix} + b \begin{pmatrix} 1 \\ 2 \\ 2 \end{pmatrix} = \begin{pmatrix} 2 \\ 2 \\ 2 \end{pmatrix}$$

can be rewritten as the system

$$a + b = 2$$
$$2b = 2$$
$$a + 2b = 2.$$

From the second equation we see that $b = 1$, and from the first equation we obtain that $a = 1$. However, with these values the third equation is not satisfied. Notice that the inner product of the two given vectors of S equals

$$(1, 0, 1) \begin{pmatrix} 1 \\ 2 \\ 2 \end{pmatrix} = 1 + 0 + 2 = 3 \neq 0,$$

therefore, they are not orthonormal. In order to use the above algorithm, an orthonormal basis of the subspace generated by these two vectors has to be first constructed. The Gram-Schmidt orthogonalization process will be used. In our case $k = 2$,

$$\mathbf{x}_1 = \begin{pmatrix} 1 \\ 0 \\ 1 \end{pmatrix} \quad \text{and} \quad \mathbf{x}_2 = \begin{pmatrix} 1 \\ 2 \\ 2 \end{pmatrix}.$$

We will follow the method introduced in the proof of Theorem 2.12. From equation (2.6) we have

$$\mathbf{y}_1 = \mathbf{x}_1 = \begin{pmatrix} 1 \\ 0 \\ 1 \end{pmatrix}, \quad |\mathbf{y}_1| = \sqrt{1^2 + 0^2 + 1^2} = \sqrt{2},$$

$$\mathbf{z}_1 = \frac{\mathbf{y}_1}{|\mathbf{y}_1|} = \begin{pmatrix} \dfrac{1}{\sqrt{2}} \\ 0 \\ \dfrac{1}{\sqrt{2}} \end{pmatrix}.$$

Equation (2.8) implies that

$$\mathbf{y}_2 = \mathbf{x}_2 - (\mathbf{x}_2, \mathbf{z}_1)\mathbf{z}_1 = \begin{pmatrix} 1 \\ 2 \\ 2 \end{pmatrix} - \frac{3}{\sqrt{2}} \begin{pmatrix} \dfrac{1}{\sqrt{2}} \\ 0 \\ \dfrac{1}{\sqrt{2}} \end{pmatrix} = \begin{pmatrix} -\dfrac{1}{2} \\ 2 \\ \dfrac{1}{2} \end{pmatrix},$$

$$|\mathbf{y}_2| = \sqrt{\left(-\frac{1}{2}\right)^2 + 2^2 + \left(\frac{1}{2}\right)^2} = \sqrt{\frac{18}{4}} = \frac{3}{\sqrt{2}},$$

$$\mathbf{z}_2 = \frac{\mathbf{y}_2}{|\mathbf{y}_2|} = \begin{pmatrix} -\dfrac{\sqrt{2}}{6} \\ \dfrac{2\sqrt{2}}{3} \\ \dfrac{\sqrt{2}}{6} \end{pmatrix}.$$

And finally, from equation (2.14) we conclude that

$$c_1 = (\mathbf{z}_1, \mathbf{x}) = \left(\frac{\sqrt{2}}{2}, 0, \frac{\sqrt{2}}{2} \right) \begin{pmatrix} 2 \\ 2 \\ 2 \end{pmatrix} = 2\sqrt{2},$$

$$c_2 = (\mathbf{z}_2, \mathbf{x}) = \left(-\frac{\sqrt{2}}{6}, \frac{2\sqrt{2}}{3}, \frac{\sqrt{2}}{6} \right) \begin{pmatrix} 2 \\ 2 \\ 2 \end{pmatrix} = \frac{4}{3}\sqrt{2},$$

and so, vector

$$c_1\mathbf{z}_1 + c_2\mathbf{z}_2 = 2\sqrt{2} \begin{pmatrix} \dfrac{\sqrt{2}}{2} \\ 0 \\ \dfrac{\sqrt{2}}{2} \end{pmatrix} + \frac{4}{3}\sqrt{2} \begin{pmatrix} -\dfrac{\sqrt{2}}{6} \\ \dfrac{2\sqrt{2}}{3} \\ \dfrac{\sqrt{2}}{6} \end{pmatrix} = \begin{pmatrix} \dfrac{14}{9} \\ \dfrac{16}{9} \\ \dfrac{22}{9} \end{pmatrix}$$

belonging to S has the smallest distance from $\mathbf{x}$.

♦

The least squares method will be discussed later in Section 3.8, where a different solution method will be introduced.

3. Assume now that $\mathbf{x}_1, \mathbf{x}_2, \ldots, \mathbf{x}_n$ form an orthonormal basis in R^n, and construct the matrix

$$\mathbf{A} = (\mathbf{x}_1, \mathbf{x}_2, \ldots, \mathbf{x}_n)$$

with columns $\mathbf{x}_1, \mathbf{x}_2, \ldots, \mathbf{x}_n$. The transpose of this matrix can be written as

$$\mathbf{A}^T = \begin{pmatrix} \mathbf{x}_1^T \\ \mathbf{x}_2^T \\ \ldots \\ \mathbf{x}_n^T \end{pmatrix},$$

and since $\mathbf{x}_i^T \mathbf{x}_i = 1$ and $\mathbf{x}_i^T \mathbf{x}_j = 0$ ($i \neq j$), $\mathbf{A}^T\mathbf{A} = \mathbf{I}$ showing that in this special case,

$$\mathbf{A}^{-1} = \mathbf{A}^T.$$

In Chapter 7 we will examine special matrices including the ones satisfying this relation.

4. If $\mathbf{x}_1, \mathbf{x}_2, \ldots, \mathbf{x}_n$ form an orthogonal basis, then $\mathbf{x}_i^T \mathbf{x}_j = 0$ for all $i \neq j$. Denote the products $\mathbf{x}_i^T \mathbf{x}_j$ by d_i. Then similarly to the previous application we have

$$\mathbf{A}^T \mathbf{A} = \begin{pmatrix} d_1 & & & \\ & d_2 & & \\ & & \ddots & \\ & & & d_n \end{pmatrix}.$$

5. Many problems of operational research can be formulated by using inner-products of finite dimensional vectors.

The most simple example is the objective function of linear programming problem, which are usually written a $c_1 x_1 + c_2 x_2 + \ldots + c_n x_n$, where $c_1, c_2, \ldots, c_n$ are given coefficients. Notice that this objective function can be rewritten in the form of $\mathbf{c}^T \mathbf{x} = (\mathbf{c}, \mathbf{x})$, where the inner-product introduced earlier in Example 2.21 is used. Here c_i is the ith entry of the column vector $\mathbf{c}$, and x_i is the ith entry of $\mathbf{x}$.

Nonlinear complementarity problems can be formulated in the following way. Let function $\mathbf{f} \colon R^n \to R^n$ be defined for all nonnegative vectors $\mathbf{x}$. Then the corresponding nonlinear complementarity problem has the form:

$$\mathbf{x}^T \mathbf{f}(\mathbf{x}) = 0$$
$$\mathbf{x} \geq \mathbf{0}$$
$$\mathbf{f}(\mathbf{x}) \geq \mathbf{0}.$$

Using again the inner-product of Example 2.21, this problem can be reformulated as

$$\left(\mathbf{x}, \mathbf{f}\left(\mathbf{x}\right)\right) = 0$$
$$\mathbf{x} \geq \mathbf{0}$$
$$\mathbf{f}\left(\mathbf{x}\right) \geq \mathbf{0}.$$

Notice that for all i, either x_i or $f_i(\mathbf{x})$ has to be zero, where $f_i(\mathbf{x})$ denotes the ith entry of vector $\mathbf{f}(\mathbf{x})$. In the special case, when $\mathbf{f}(\mathbf{x}) = \mathbf{Mx} + \mathbf{b}$ with some $n \times n$ matrix $\mathbf{M}$ and $\mathbf{b} \in R^n$, the problem simplifies to the following:

$$\mathbf{x}^T \left(\mathbf{Mx} + \mathbf{b}\right) = 0$$
$$\mathbf{x} \geq \mathbf{0}$$
$$\mathbf{Mx} + \mathbf{b} \geq \mathbf{0},$$

which is known as a linear complementarity problem.

Variational inequalities have the usual form

$$\left(\mathbf{x} - \mathbf{x}^*\right)^T \mathbf{f}\left(\mathbf{x}^*\right) \leq 0, \quad \left(\text{all } \mathbf{x} \in R^n\right)$$

where $\mathbf{f}: R^n \rightarrow R^n$ is a given function, and $\mathbf{x}^* \in R^n$ is a given vector from the domain of $\mathbf{f}$. Notice again that the left hand side can be expressed as an inner-product of Example 2.21, the problem can be rewritten as

$$\left(\mathbf{x} - \mathbf{x}^*, \mathbf{f}\left(\mathbf{x}^*\right)\right) \leq 0 \quad \left(\text{all } \mathbf{x} \in R^n\right).$$

2.7 Exercises

1. Show that the real numbers form a real vector space with the usual addition and multiplication. Is the set of rational numbers a subspace of this vector space?

2. Let V and W be two real vector spaces. Define $V \times W$ as the set of all ordered pairs (v, w) such that $v \in V$ and $w \in W$. Define addition and multiplication by scalars element-wise, that is,

$(v_1, w_1) + (v_2, w_2) = (v_1 + v_2, w_1 + w_2)$ and $a(v,w) = (av, aw)$. Is $V \times W$ a real vector space?

3. Determine which of the following subsets of R^n are subspaces. The set of

a) all vectors (x_i) such that $x_1 = 0$;

b) all vectors (x_i) such that $x_1 \neq 0$;

c) all vectors (x_i) such that $x_1 = 0$;

d) all vectors (x_i) such that $x_1 + x_2 + \ldots + x_n = 0$;

e) all vectors (x_i) such that $x_1 + x_2 + \ldots + x_n \neq 0$;

f) all vectors (x_i) such that $x_1 + x_2 + \ldots + x_n = 1$;

g) all vectors (x_i) such that $a_1 x_1 + a_2 x_2 + \ldots + a_n x_n = 0$ with fixed real numbers $a_1, \ldots, a_n$;

h) all vectors (x_i) such that $a_1 x_1 + a_2 x_2 + \ldots + a_n x_n \neq 0$ with fixed real numbers $a_1, \ldots, a_n$;

i) all vectors (x_i) such that $a_1 x_1 + a_2 x_2 + \ldots + a_n x_n = 1$ with fixed real numbers $a_1, \ldots, a_n$;

4. Let V be the set of all continuous functions on $[0, 1]$. Determine which of the following subsets are subspaces. Set of
a) all polynomials;

b) all functions such that $f(0) = 0$;

c) all functions such that $f(0) \neq 0$;

d) all functions such that $f(0) = 1$;

e) all functions such that $\int_0^1 f(x)\,dx = 0$;

f) all functions such that $\int_0^1 f(x)\,dx \neq 0$;

g) all functions such that $\int_0^1 f(x)\,dx = 1$;

h) all functions such that $f(0) + f(1) = 0$.

5. Do vectors $\begin{pmatrix} 1 \\ 2 \end{pmatrix}$ and $\begin{pmatrix} 1 \\ 1 \end{pmatrix}$ span the vector space R^2?

6. Show that vector $\begin{pmatrix} 1 \\ 1 \end{pmatrix}$ cannot be expressed as the linear combination

   of $\begin{pmatrix} 1 \\ 2 \end{pmatrix}$ and $\begin{pmatrix} 3 \\ 6 \end{pmatrix}$.

7. Verify that vectors $\begin{pmatrix} 1 \\ 2 \end{pmatrix}$ and $\begin{pmatrix} 1 \\ 1 \end{pmatrix}$ form a basis in R^2.

8. A vector space is spanned by four vectors. What can be said about the dimension of this vector space? How your answer has to be modified if the four vectors are linearly independent?

9. Assume that $x \in R^n$ is a linear combination of vectors $x_1, x_2, \dots, x_k$, and for all $i = 1, 2, \dots, k$, x_i is the linear combination of vectors $a_1, a_2, \dots, a_l$. Prove that x is a linear combination of $a_1, a_2, \dots, a_l$.

10. Assume that $V = V(x_1, x_2, \dots, x_k)$, where for $i = 1, 2, \dots, k$, x_i is the linear combination of the linearly independent vectors $a_1, a_2, \dots, a_l$. What can you say about $\dim(V)$?

11. Assume that vectors $\mathbf{a}_1, \mathbf{a}_2, \ldots, \mathbf{a}_k$ are linearly independent. Examine the linear independence of vectors

   a) $\mathbf{a}_1, \mathbf{a}_1 + \mathbf{a}_2, \mathbf{a}_3, \ldots, \mathbf{a}_k$;

   b) $\mathbf{a}_1, \mathbf{a}_1, \mathbf{a}_2, \mathbf{a}_3, \ldots, \mathbf{a}_k$;

   c) $-\mathbf{a}_1, \mathbf{a}_2, \mathbf{a}_3, \ldots, \mathbf{a}_k$;

   d) $\mathbf{a}_1, 2\mathbf{a}_2, 3\mathbf{a}_3, \ldots, k\mathbf{a}_k$.

12. Are the following polynomials linearly independent?

   a) $1, x, x^2, \ldots, x^k$;

   b) $1, 1+x, 1+x+x^2, 1+2x+x^2, 1+x+x^2+x^3$;

   c) $1, 1+x, 1+x+x^2, 1+x+x^2+x^3, \ldots, 1+x+x^2+\ldots+x^k$.

13. Determine whether the row vector $(1, 2, 3, 4)$ belongs to the subspace spanned by vectors $(1, 1, 1, 1)$, $(0, 1, 1, 1)$, and $(0, 0, 1, 1)$.

14. Are the following vectors linearly independent?

   a) $\begin{pmatrix} 1 \\ 1 \end{pmatrix}, \begin{pmatrix} 1 \\ 2 \end{pmatrix}$;

   b) $\begin{pmatrix} 1 \\ 1 \\ 1 \end{pmatrix}, \begin{pmatrix} 1 \\ 2 \\ 2 \end{pmatrix}, \begin{pmatrix} 1 \\ 3 \\ 3 \end{pmatrix}$;

   c) $\begin{pmatrix} 1 \\ 1 \\ 1 \end{pmatrix}, \begin{pmatrix} 1 \\ 2 \\ 3 \end{pmatrix}, \begin{pmatrix} 1 \\ 2 \\ 5 \end{pmatrix}$.

15. Prove that if V_1 is a subspace of V_2, and V_2 is a subspace of V_3, then V_1 is a subspace of V_3.

16. Let V be a real vector space, and $A \subseteq B \subseteq V$. Prove that $V(A) \subseteq V(B)$.

17. Let $V = \left\{ \mathbf{x} \mid \mathbf{x} = (x_i) \in R^n, \ x_1 + x_2 + \ldots + x_n = 0 \right\}$. Find a basis in V, and determine the dimension of V.

18. Let $\mathbf{A}$ be an invertible $n \times n$ matrix (that is, $\mathbf{A}^{-1}$ exists), $\mathbf{x}_1, \mathbf{x}_2, \ldots, \mathbf{x}_k$ ($k \leq n$) linearly independent n-element vectors. Prove that vectors $\mathbf{Ax}_1, \mathbf{Ax}_2, \ldots, \mathbf{Ax}_k$ are also linearly independent.

19. Is the assertion of the previous Exercise valid if $\mathbf{A}$ is not invertible, or vectors $\mathbf{x}_1, \mathbf{x}_2, \ldots, \mathbf{x}_k$ are linearly dependent?

20. Illustrate Exercise 18 with matrix $\mathbf{A} = \begin{pmatrix} 1 & 2 \\ 1 & 1 \end{pmatrix}$ and vectors

$$\mathbf{x}_1 = \begin{pmatrix} 1 \\ 0 \end{pmatrix} \text{ and } \mathbf{x}_2 = \begin{pmatrix} 0 \\ 1 \end{pmatrix}.$$

21. Repeat Example 2.25 with vectors

$$\mathbf{x} = \begin{pmatrix} 1 \\ 1 \\ 1 \end{pmatrix} \text{ and } \mathbf{y} = \begin{pmatrix} 0 \\ 1 \\ 1 \end{pmatrix}.$$

22. Prove that if a vector $\mathbf{u}$ has zero length, then $\mathbf{u} = \mathbf{0}$.

23. Prove that for any two vectors of the same size,

$$\left| \mathbf{u} - \mathbf{v} \right|^2 + \left| \mathbf{u} + \mathbf{v} \right|^2 = 2 \left(\left| \mathbf{u} \right|^2 + \left| \mathbf{v} \right|^2 \right).$$

24. Prove that if $\mathbf{u}$ and $\mathbf{v}$ are real vectors and $\left| \mathbf{u} + \mathbf{v} \right| = \left| \mathbf{u} \right| + \left| \mathbf{v} \right|$ then $\mathbf{u}$ is a scalar multiple of $\mathbf{v}$.

25. Repeat Example 2.26 with vectors

$$\mathbf{x}_1 = \begin{pmatrix} 1 \\ 1 \\ 1 \\ 1 \end{pmatrix}, \quad \mathbf{x}_2 = \begin{pmatrix} 0 \\ 1 \\ 1 \\ 1 \end{pmatrix}, \quad \mathbf{x}_3 = \begin{pmatrix} 0 \\ 0 \\ 1 \\ 1 \end{pmatrix}, \quad \mathbf{x}_4 = \begin{pmatrix} 0 \\ 0 \\ 0 \\ 1 \end{pmatrix}.$$

26. Repeat Example 2.27 with functions

$$p_0(x) = 1, \ p_1(x) = 1 + x, \ p_2(x) = 1 + x + x^2, \ p_3(x) = 1 + x + x^2 + x^3.$$

27. Select a maximum number of linearly independent vectors from the set

$$\left\{ \begin{pmatrix} 1 \\ 1 \\ 1 \\ 1 \end{pmatrix}, \begin{pmatrix} 1 \\ 2 \\ 2 \\ 2 \end{pmatrix}, \begin{pmatrix} 2 \\ 3 \\ 3 \\ 3 \end{pmatrix}, \begin{pmatrix} 0 \\ 1 \\ 1 \\ 1 \end{pmatrix}, \begin{pmatrix} 1 \\ 2 \\ 3 \\ 4 \end{pmatrix} \right\}.$$

28. Repeat Example 2.33 with vectors

$$\mathbf{x}_1 = \begin{pmatrix} 1 \\ 1 \\ 1 \end{pmatrix}, \quad \mathbf{x}_2 = \begin{pmatrix} 0 \\ 1 \\ 1 \end{pmatrix}, \quad \text{and } \mathbf{x} = \begin{pmatrix} 1 \\ 2 \\ 3 \end{pmatrix}.$$

Chapter 3

Systems of Linear Equations, and Inverses of Matrices

3.1 Introduction

Before presenting a general definition for systems of linear equations consider the following problem. Let $\mathbf{a}_1$, $\mathbf{a}_2$, ..., $\mathbf{a}_n$ and $\mathbf{b}$ be given m-element real (or complex) vectors. Assume that we wish to check if $\mathbf{b}$ is in the subspace generated by the given vectors $\mathbf{a}_1$, $\mathbf{a}_2$, ..., $\mathbf{a}_n$. From Theorem 2.4 we know that $\mathbf{b} \in V(\mathbf{a}_1, \mathbf{a}_2, ..., \mathbf{a}_n)$ if and only if there exist scalars x_1, x_2, ..., x_n such that

$$x_1\mathbf{a}_1 + x_2\mathbf{a}_2 + ... + x_n\mathbf{a}_n = \mathbf{b} \tag{3.1}$$

To obtain an equivalent formulation for this equation, introduce the notation

$$\mathbf{a}_1 = \begin{pmatrix} a_{11} \\ a_{21} \\ \vdots \\ a_{m1} \end{pmatrix}, \ \mathbf{a}_2 = \begin{pmatrix} a_{12} \\ a_{22} \\ \vdots \\ a_{m2} \end{pmatrix}, \ ..., \mathbf{a}_n = \begin{pmatrix} a_{1n} \\ a_{2n} \\ \vdots \\ a_{mn} \end{pmatrix}, \ \mathbf{b} = \begin{pmatrix} b_1 \\ b_2 \\ \vdots \\ b_m \end{pmatrix}.$$

Here a_{ij} denotes the ith element of vector $\mathbf{a}_j$. That is, the second subscript refers to the vector, and the first subscript indicates the position of the element. By using this notation, equation (3.1) can be rewritten as

$$x_1 \begin{pmatrix} a_{11} \\ a_{21} \\ \vdots \\ a_{m1} \end{pmatrix} + x_2 \begin{pmatrix} a_{12} \\ a_{22} \\ \vdots \\ a_{m2} \end{pmatrix} + \ldots + x_n \begin{pmatrix} a_{1n} \\ a_{2n} \\ \vdots \\ a_{mn} \end{pmatrix} = \begin{pmatrix} b_1 \\ b_2 \\ \vdots \\ b_m \end{pmatrix},$$

and by comparing the like elements of the two sides we obtain the following set of equations:

$$
\begin{aligned}
a_{11}x_1 + a_{12}x_2 + \ldots + a_{1n}x_n &= b_1 \\
a_{21}x_1 + a_{22}x_2 + \ldots + a_{2n}x_n &= b_2 \\
\vdots \quad \vdots \quad \vdots \quad \vdots \\
a_{m1}x_1 + a_{m2}x_2 + \ldots + a_{mn}x_n &= b_m.
\end{aligned}
\tag{3.2}
$$

Definition 3.1. If the coefficients a_{ij} and right-hand side scalars b_i are given real (or complex) numbers for all $i = 1, 2, \ldots, m$ and $j = 1, 2, \ldots, n$, then equations (3.2) are called a *system of linear equations* for the unknowns $x_1, \ldots, x_n$.

Here n is the number of unknowns, m is the number of equations, and the system of linear equations is called an $m \times n$ system.

Introduce the $m \times n$ matrix **A** with (i, j) element a_{ij}, and let **x** be the n-vector with j^{th} element x_j. Then equations (3.2) can be summarized as

$$\mathbf{Ax} = \mathbf{b}. \tag{3.3}$$

This compact representation can be verified by noticing that for all $i = 1, 2, \ldots, m$, the ith element of the left-hand side is the product of the ith row of matrix **A** by the column vector **x**:

$$\left(a_{i1}, a_{i2}, \ldots, a_{in} \right) \begin{pmatrix} x_1 \\ x_2 \\ \vdots \\ x_n \end{pmatrix} = a_{i1}x_1 + a_{i2}x_2 + \ldots + a_{in}x_n,$$

which is the left-hand side of the ith equation of system (3.2). The ith element of the right-hand side of equation (3.3) is b_i, which is the right-hand side of the ith equation of system (3.2). Hence equations (3.1), (3.2), and (3.3) are equivalent to each other. The construction of matrix **A** implies that its columns are the vectors $\mathbf{a}_1$, $\mathbf{a}_2$, ..., $\mathbf{a}_n$.

Definition 3.2. If $\mathbf{b} = \mathbf{0}$, then equation (3.3) is called *homogeneous*, otherwise the equation is called *inhomogeneous*.
For example,

$$2x_1 + x_2 = 0$$
$$x_1 - x_2 = 0$$

is a 2×2 homogeneous system of linear equations, and the system

$$2x_1 + x_2 = 1$$
$$x_1 - x_2 = 0$$

is inhomogeneous, since there is at least one nonzero right-hand side number.

3.2 Existence and Uniqueness of a Solution

In this section we will find necessary and sufficient conditions for the existence of solutions for systems of linear equations. In the case of the existence of a solution we will also find conditions for the uniqueness of the solution. In the cases of unique and multiple solutions practical algorithms will be introduced to find the solution or to characterize all solutions.

Our first example shows that a system of linear equations might not have any solution at all.

Example 3.1. Consider the 2×2 system

$$x_1 + x_2 = 0$$
$$x_1 + x_2 = 1,$$

which must not have solution, since $x_1 + x_2$ must not have two different values in the same time.

♦

The following example shows a case, when infinitely many solutions exist.

Example 3.2. Consider system

$$x_1 + x_2 = 0$$
$$x_1 + x_2 = 0,$$

which has infinitely many solutions: x_1 is arbitrary, and $x_2 = -x_1$.

♦

In certain cases, a unique solution might exist, as it is illustrated next.

Example 3.3. Consider now the system

$$x_1 + x_2 = 2$$
$$x_1 - x_2 = 0,$$

which has a unique solution: $x_2 - x_1 = 1$ which can be obtained by adding and subtracting the two equations.

♦

We will first show that the above examples have covered all possibilities concerning the number of solutions of linear equations by verifying that in the case of multiple solutions infinitely many solutions exist. Consider the linear equations represented in the compact form (3.3), and assume that $\mathbf{x}_1$ and $\mathbf{x}_2$ ($\neq \mathbf{x}_1$) are two solutions. For arbitrary real (or complex) number t, consider the vector

$$\mathbf{x} = t \cdot \mathbf{x}_1 + (1-t) \cdot \mathbf{x}_2 = \mathbf{x}_2 + t \cdot (\mathbf{x}_1 - \mathbf{x}_2).$$

Since $\mathbf{x}_1 - \mathbf{x}_2 \neq \mathbf{0}$ different values of t give different $\mathbf{x}$ vectors. Substitute $\mathbf{x}$ into equation (3.3) to see that

$$\mathbf{Ax} = \mathbf{A}\left(t \cdot \mathbf{x}_1 + \left(1-t\right) \cdot \mathbf{x}_2\right) = t \cdot \mathbf{Ax}_1 + \left(1-t\right) \cdot \mathbf{Ax}_2 = t \cdot \mathbf{b} + \left(1-t\right) \cdot \mathbf{b} = \mathbf{b}.$$

That is, for all t, $\mathbf{x}$ solves the equation. Hence infinitely many solutions exist.

From equation (3.1) and Theorem 2.4 we immediately obtain the following important result.

Theorem 3.1. A system of linear equations has at least one solution if and only if the right hand side vector belongs to the subspace generated by the columns of the coefficient matrix.

The assertion of the theorem is illustrated in the following example.

Example 3.4. Consider the system of linear equations

$$x_1 + 2x_2 + x_3 = 2$$
$$2x_1 + 4x_2 + 2x_3 = 5.$$

The columns of the coefficient matrix are

$$\mathbf{a}_1 = \begin{pmatrix} 1 \\ 2 \end{pmatrix}, \quad \mathbf{a}_2 = \begin{pmatrix} 2 \\ 4 \end{pmatrix}, \quad \text{and} \quad \mathbf{a}_3 = \begin{pmatrix} 1 \\ 2 \end{pmatrix}.$$

Notice that $\mathbf{a}_3 = \mathbf{a}_1$ and $\mathbf{a}_2 = 2\mathbf{a}_1$, therefore $\{\mathbf{a}_1\}$ is a basis of the subspace generated by the columns. Since it is generated by only one vector, it consists of all scalar multiples of $\mathbf{a}_1$. For any scalar c,

$$c\mathbf{a}_1 = c\begin{pmatrix} 1 \\ 2 \end{pmatrix} = \begin{pmatrix} c \\ 2c \end{pmatrix},$$

where the second element is twice the first element. Since the right-hand side vector $\begin{pmatrix} 2 \\ 5 \end{pmatrix}$ does not have this property, it does not belong to the subspace generated by $\mathbf{a}_1$. Therefore, the above system of linear equations have no solution.

♦

Using the notation $V(\mathbf{a}_1, \ldots, \mathbf{a}_n)$ for the subspace generated by the columns of matrix $\mathbf{A}$, the condition of Theorem 3.1 can be formulated as

$$\mathbf{b} \in V\left(\mathbf{a}_1, \ldots, \mathbf{a}_n\right).$$

For easier reference $V(\mathbf{a}_1, \ldots, \mathbf{a}_n)$ is often called the column space of matrix $\mathbf{A}$. (Similarly, the subspace generated by the rows of a matrix $\mathbf{A}$ is called the row space of $\mathbf{A}$.) Combining Theorems 3.1 and 2.7 leads to the following condition for the uniqueness of a solution.

Theorem 3.2. A system of linear equations has a unique solution if and only if the right-hand side vector belongs to the subspace generated by the columns of the coefficient matrix, and the columns are linearly independent.

The statements of the above two theorems will be restated in a more convenient way by using the following concepts:

Definition 3.3. Let $\mathbf{A}$ be an $m \times n$ matrix with column vectors $\mathbf{a}_1, \ldots, \mathbf{a}_n$. The rank of matrix $\mathbf{A}$ is defined as the dimension of the column space $V(\mathbf{a}_1, \ldots, \mathbf{a}_n)$.

Definition 3.4. The *augmented matrix* of the system of linear equations (3.3) is the $m \times (n+1)$ matrix with columns $\mathbf{a}_1, \ldots, \mathbf{a}_n, \mathbf{b}$.

The next result follows immediately from these definitions and Theorems 3.1 and 3.2.

Theorem 3.3. A system of linear equations has a solution if and only if the rank of its coefficient matrix equals the rank of the augmented matrix. The solution is unique if and only if in addition, the common rank is n, where n is the number of unknowns.

The statement of the theorem is illustrated next.

Example 3.5. Consider again the system of linear equations

$$x_1 + 2x_2 + x_3 = 2$$
$$2x_1 + 4x_2 + 2x_3 = 5,$$

which was the subject of our previous example, where we have seen that the column space of the coefficient matrix is generated by only one vector, that is, the rank of the coefficient matrix equals 1. The columns of the augmented matrix are the vectors

$$\mathbf{a}_1 = \begin{pmatrix} 1 \\ 2 \end{pmatrix}, \quad \mathbf{a}_2 = \begin{pmatrix} 2 \\ 4 \end{pmatrix}, \quad \mathbf{a}_3 = \begin{pmatrix} 1 \\ 2 \end{pmatrix}, \quad \text{and} \quad \mathbf{b} = \begin{pmatrix} 2 \\ 5 \end{pmatrix}.$$

We will next prove that the rank of the augmented matrix is 2, which differs from the rank of the coefficient matrix implying that no solution exists. Let

$$\mathbf{z} = \begin{pmatrix} z_1 \\ z_2 \end{pmatrix}$$

be an arbitrary vector in R^2.

We will verify that $\mathbf{z}$ can be expressed as a linear combination of the columns of the augmented matrix showing that its rank equals 2. We will prove that $\mathbf{z}$ is a linear combination of $\mathbf{a}_1$ and $\mathbf{b}$ which means that with some scalars

$$\begin{pmatrix} z_1 \\ z_2 \end{pmatrix} = c_1 \begin{pmatrix} 1 \\ 2 \end{pmatrix} + c_2 \begin{pmatrix} 2 \\ 5 \end{pmatrix}.$$

Comparing the corresponding elements of the left-hand and right-hand sides we get the equations

$$z_1 = c_1 + 2c_2$$
$$z_2 = 2c_1 + 5c_2.$$

From the first equation we have

$$c_1 = z_1 - 2c_2,$$

and substitution of this relation into the second equation gives equation

$$z_2 = 2z_1 - 4c_2 + 5c_2,$$

which implies that

$$c_2 = z_2 - 2z_1,$$

and hence

$$c_1 = z_1 - 2c_2 = 5z_1 - 2z_2.$$

Hence, for arbitrary vector z, there is a unique pair of scalars c_1 and c_2.

♦

In the case of multiple solution it is very important to characterize all solutions. The following theorem provides such a characterization.

Theorem 3.4. Assume that the inhomogeneous system of linear equations $A \cdot x = b$ has a solution x_0. Then for all solutions x^* of the corresponding homogeneous system $A \cdot x = 0$, $x_0 + x^*$ is a solution of the inhomogeneous system, and all solutions of the inhomogeneous system can be represented in this form.

Proof. Assume first that x_0 solves the inhomogeneous system and x^* is a solution of the corresponding homogeneous system. Then $z = x_0 + x^*$ solves the inhomogeneous system, since

$$Az = A(x_0 + x^*) = Ax_0 + Ax^* = b + 0 = b.$$

Assume next that z and x_0 are solutions of the inhomogeneous system. Then $x^* = z - x_0$ is a solution of the homogeneous system, since

$$Ax^* = A(z - x_0) = Az - Ax_0 = b - b = 0.$$

♣

Corollary. An inhomogeneous system of linear equations has a unique solution if and only if $\mathbf{x} = \mathbf{0}$ is the only solution of the corresponding homogeneous system.

The assertion of Theorem 3.4 can be reformulated by saying that the general solution of an inhomogeneous system can be obtained as the sum of a particular solution of the inhomogeneous system and the general solution of the corresponding homogeneous system. Therefore, the complete characterization of all solutions of homogeneous systems is the first step before the same for inhomogeneous systems can be examined.

3.3 Systems of Homogeneous Linear Equations

In this section the set of all solutions of systems of homogeneous equations will be characterized. Our first result gives a characterization based on the concept of orthogonal complementary subspaces (which was introduced earlier in Definition 2.17).

Theorem 3.5. The set of all solutions of an $m \times n$ homogeneous system $\mathbf{Ax} = \mathbf{0}$ is a subspace of R^n, which coincides with the orthogonal complementary subspace of the row-space of matrix $\mathbf{A}$.

Proof. Let $\mathbf{r}_1^T$, $\mathbf{r}_2^T$, ..., $\mathbf{r}_m^T$ denote the rows of $\mathbf{A}$, then the homogeneous system $\mathbf{Ax} = \mathbf{0}$ is equivalent to the set of equations:

$$
\begin{aligned}
\mathbf{r}_1^T \mathbf{x} &= 0 \\
\mathbf{r}_2^T \mathbf{x} &= 0 \\
&\vdots \\
\mathbf{r}_m^T \mathbf{x} &= 0.
\end{aligned}
\tag{3.4}
$$

Assume first that a vector $\mathbf{z} \in R^n$ is orthogonal to all vectors of the subspace generated by the rows of matrix $\mathbf{A}$. Then it is orthogonal to all rows of $\mathbf{A}$, since these rows belong to the row-space of the matrix.

Therefore, vector $\mathbf{z}$ satisfies equations (3.4), and so, it is a solution of the homogeneous equation $\mathbf{Ax} = \mathbf{0}$.

Assume next that $\mathbf{z}$ solves the homogeneous equation, then it satisfies equations (3.4). We will next prove that $\mathbf{z}$ is orthogonal to all elements of the row-space of matrix $\mathbf{A}$. Let $\mathbf{r}^T$ be an arbitrary element of the row-space, then with some scalars c_i ($1 \le i \le m$),

$$\mathbf{r}^T = c_1\mathbf{r}_1^T + c_2\mathbf{r}_2^T + \dots + c_m\mathbf{r}_m^T,$$

and therefore,

$$\mathbf{r}^T\mathbf{z} = c_1\mathbf{r}_1^T\mathbf{z} + c_2\mathbf{r}_2^T\mathbf{z} + \dots + c_m\mathbf{r}_m^T\mathbf{z} = 0,$$

since each term is zero. Hence, $\mathbf{z}$ is orthogonal to $\mathbf{r}$, which completes the proof.

♣

Our second result on characterizing the solutions of homogeneous systems of linear equations provides a basis for the subspace of all solutions. Before formulating the theorem, some notations are introduced. Let $\mathbf{a}_1, \dots, \mathbf{a}_n$ denote the columns of the coefficient matrix of the $m \times n$ homogeneous system of linear equations $\mathbf{Ax} = \mathbf{0}$. Assume that the columns (as well as the unknowns) are ordered so that $\{\mathbf{a}_1, \dots, \mathbf{a}_r\}$ is a basis of the column space. Then vectors $\mathbf{a}_{r+1}, \dots, \mathbf{a}_n$ are linear combinations of the basis elements. That is, for $k = r+1, \dots, n$,

$$\mathbf{a}_k = c_{k1}\mathbf{a}_1 + \dots + c_{kr}\mathbf{a}_r, \tag{3.5}$$

with some scalars $c_{k1}, \dots, c_{kr}$. For each $k = r+1, \dots, n$, introduce the vectors

$$\mathbf{z}_k = \begin{pmatrix} c_{k1} \\ \vdots \\ c_{kr} \\ 0 \\ \vdots \\ 0 \\ -1 \\ 0 \\ \vdots \\ 0 \end{pmatrix}, \quad \leftarrow k^{\text{th}} \text{ element}$$

where the first r elements are the coefficients of the linear combination (3.5), the kth element is -1, and all other elements are equal to zero.

Theorem 3.6. Set $\{\mathbf{z}_{r+1}, \ldots, \mathbf{z}_n\}$ is a basis of the subspace of all solutions of the homogeneous system of linear equations $\mathbf{Ax} = \mathbf{0}$.

Proof. First, we prove that for all $k = r+1, \ldots, n$, $\mathbf{z}_k$ is a solution of the homogeneous system. Simple substitution and equation (3.5) imply that

$$\mathbf{Az}_k = c_{k1}\mathbf{a}_1 + \ldots + c_{kr}\mathbf{a}_r + (-1)\mathbf{a}_k = \mathbf{0}.$$

Next, we prove that vectors $\mathbf{z}_{r+1}, \ldots, \mathbf{z}_n$ are linearly independent. Consider a zero valued linear combination of these vectors:

$$
b_{r+1}\begin{pmatrix} c_{r+1,1} \\ \vdots \\ c_{r+1,r} \\ -1 \\ 0 \\ 0 \\ \vdots \\ 0 \\ 0 \\ 0 \end{pmatrix} + b_{r+2}\begin{pmatrix} c_{r+2,1} \\ \vdots \\ c_{r+z,r} \\ 0 \\ -1 \\ 0 \\ \vdots \\ 0 \\ 0 \\ 0 \end{pmatrix} + \ldots + b_{n}\begin{pmatrix} c_{n1} \\ \vdots \\ c_{nr} \\ 0 \\ 0 \\ 0 \\ \vdots \\ 0 \\ 0 \\ -1 \end{pmatrix} = \begin{pmatrix} 0 \\ \vdots \\ 0 \\ 0 \\ 0 \\ 0 \\ \vdots \\ 0 \\ 0 \\ 0 \end{pmatrix}.
$$

Comparing the $(r+1)^{\text{st}}$, $(r+2)^{\text{nd}}$, ..., $(r+n)^{\text{th}}$ elements we see that

$$
-b_{r+1} = -b_{r+2} = \ldots = -b_{n} = 0,
$$

that is, all coefficients are necessarily equal to zero, proving the linear independence of these vectors.

We will finally prove that vectors z_{r+1}, ..., z_n generate the entire solution set. Let $x = (x_i)$ be a solution. The definition of vectors z_{r+1}, ..., z_n implies that vector

$$
\mathbf{x}^* = \mathbf{x} + x_{r+1}\mathbf{z}_{r+1} + \ldots + x_n\mathbf{z}_n \tag{3.6}
$$

satisfies the following properties:
(a) $\mathbf{x}^*$ solves the homogeneous system, since it is a linear combination of solutions;
(b) for $k = r+1$, ..., n, the kth element of $\mathbf{x}^*$ equals

$$
x_k + x_{r+1} \cdot 0 + \ldots + x_k(-1) + \ldots + x_n \cdot d = 0.
$$

Therefore

$$\mathbf{x}^* = \begin{pmatrix} x_1^* \\ \vdots \\ x_r^* \\ 0 \\ \vdots \\ 0 \end{pmatrix}$$

with some elements $x_1^*, \ldots, x_r^*$. Property (a) implies that

$$x_1^* \mathbf{a}_1 + \ldots + x_r^* \mathbf{a}_r = \mathbf{0},$$

and since vectors $\mathbf{a}_1, \ldots, \mathbf{a}_r$ are linearly independent, necessarily, $x_1^* = \ldots = x_r^* = 0$. That is, $\mathbf{x}^* = \mathbf{0}$, and from equation (3.6) we conclude that

$$\mathbf{x} = -x_{r+1} \mathbf{z}_{r+1} - \ldots - x_n \mathbf{z}_n.$$

Hence $\mathbf{x}$ can be expressed as a linear combination of vectors $\mathbf{z}_{r+1}, \ldots, \mathbf{z}_n$. Thus, the proof is complete.

♣

Example 3.6. Consider the homogeneous system

$$x_1 + 2x_2 + 4x_3 = 0$$
$$x_1 - x_2 + x_3 = 0.$$

The columns of the coefficient matrix are

$$\mathbf{a}_1 = \begin{pmatrix} 1 \\ 1 \end{pmatrix}, \mathbf{a}_2 = \begin{pmatrix} 2 \\ -1 \end{pmatrix}, \quad \text{and} \quad \mathbf{a}_3 = \begin{pmatrix} 4 \\ 1 \end{pmatrix}.$$

Vectors $\mathbf{a}_1$ and $\mathbf{a}_2$ are linearly independent and $\mathbf{a}_3 = 2\mathbf{a}_1 + \mathbf{a}_2$, therefore $\{\mathbf{a}_1, \mathbf{a}_2\}$ is a basis for the column space of the coefficient matrix. In our case, $n = 3$ and $r = 2$. Since $\mathbf{a}_3 = 2\mathbf{a}_1 + \mathbf{a}_2$, from equation (3.5) we have

$$c_{31} = 2 \quad \text{and} \quad c_{32} = 1,$$

and therefore, the basis of the solution space consists of only one element:

$$\mathbf{z}_3 = \begin{pmatrix} 2 \\ 1 \\ -1 \end{pmatrix}.$$

Hence the general solution has the form:

$$\mathbf{x} = \begin{pmatrix} 2c \\ c \\ -c \end{pmatrix}$$

with arbitrary scalar c.

♦

3.4 Systems of Inhomogeneous Linear Equations

In this section the solutions of systems of inhomogeneous linear equations of the form

$$\mathbf{Ax} = \mathbf{b}$$

will be discussed, where $\mathbf{A}$ is an $m \times n$ real (or complex) matrix, and $\mathbf{b}$ is a real (or complex) n-vector such that $\mathbf{b} \neq \mathbf{0}$. In Theorem 3.4 we have proved that the solution set of the inhomogeneous system is given as

$$S = \left\{ \mathbf{x}_0 + \mathbf{x}^* \;\middle|\; \begin{array}{l} \mathbf{x}_0 \text{ is a particular solution} \\ \text{of the inhomogeneous system and } \mathbf{x}^* \in V_1 \end{array} \right\}$$

where V_1 is the set of all solutions of the corresponding homogeneous system. We have also shown in Theorem 3.4 that V_1 is a subspace. That is, the solutions can be obtained as the sum of a fixed vector and all elements of a subspace. This structure can be generally defined as follows.

Definition 3.5. Let V be a vector space, let V_1 be a subspace of V, and assume that a given element $x_0 \in V$ is not in V_1. The set

$$S = \left\{ x_0 + x \mid x \in V_1 \right\} \tag{3.7}$$

is called a *linear manifold*. Subspace V_1 is called the *directing space* of S. The dimension of S is defined as the dimension of V_1, and is denoted by $\dim(S)$.

We know from Theorem 3.5 that the set of all solutions of an inhomogeneous system of linear equation form a linear manifold. We can also prove that for any linear manifold $S \subseteq R^n$, there exists a certain system of linear equations such that its solution set coincides with S.

Theorem 3.7. Let S be an r-dimensional linear manifold in R^n. Then there exists an $(n - r) \times n$ system of linear equations such that the rank of the coefficient matrix is r, and its solution set coincides with S.

Proof. Assume that

$$S = \left\{ x_0 + z \mid z \in V_1 \right\},$$

where $\dim(V_1) = r$. Let $a_1, \ldots, a_{n-r}$ be a basis of the orthogonal complementary subspace of V_1. (From Theorem 2.15 we know that the dimension of the orthogonal complementary subspace is $n - r$.) Define matrix A as the matrix with rows $a_1^T, \ldots, a_{n-r}^T$. Let $b = Ax_0$. We will next prove that the system $Ax = b$ satisfies the assertion, that is, its solution set coincides with S.

Assume first, that $x \in S$. Then $x = x_0 + z$ with some $z \in V_1$. Then

$$Ax = Ax_0 + Az = b,$$

since each row of A is orthogonal to z, that is, $Az = 0$. Therefore, x is a solution of system $Ax = b$.

Assume next that x is a solution of the system $Ax = b$. We will prove that $x = x_0 + z$ with some $z \in V_1$. Consider the vector $z = x - x_0$. Then

$$\mathbf{Az} = \mathbf{Ax} - \mathbf{Ax}_0 = \mathbf{b} - \mathbf{b} = \mathbf{0}.$$

Consequently, for all $k = 1, 2, \ldots, n-r$, $\mathbf{a}_k^T\mathbf{z} = 0$. Hence $\mathbf{z}$ is orthogonal to all basis vectors of V_1, consequently it is orthogonal to all elements of V_1.

♣

Theorem 3.8. Let a linear manifold be defined as given in (3.7). Then

$$V_1 = \left\{ \mathbf{y} - \mathbf{z} \mid \mathbf{y} \text{ and } \mathbf{z} \in S \right\}. \tag{3.8}$$

Proof. Let W denote the right-hand side of relation (3.8). We will prove that $W \subseteq V_1$ and $V_1 \subseteq W$.

Assume first that $\mathbf{u} \in W$ is an arbitrary element. Then $\mathbf{u} = \mathbf{y} - \mathbf{z}$ with some $\mathbf{y}$ and $\mathbf{z} \in S$. Therefore $\mathbf{y} = \mathbf{x}_0 + \mathbf{x}$ and $\mathbf{z} = \mathbf{x}_0 + \mathbf{x}^*$, where $\mathbf{x}$ and $\mathbf{x}^*$ are from V_1. Then

$$\mathbf{u} = \mathbf{y} - \mathbf{z} = \left(\mathbf{x}_0 + \mathbf{x}\right) - \left(\mathbf{x}_0 + \mathbf{x}^*\right) = \mathbf{x} - \mathbf{x}^* \in V_1.$$

Assume next that $\mathbf{x} \in V_1$ is an arbitrary element. Then both $\mathbf{y} = \mathbf{x}_0 + 2\mathbf{x}$ and $\mathbf{z} = \mathbf{x}_0 + \mathbf{x}$ belong to S (since $\mathbf{x} \in V_1$ implies that $2\mathbf{x} \in V_1$), therefore

$$\mathbf{x} = \left(\mathbf{x}_0 + 2\mathbf{x}\right) - \left(\mathbf{x}_0 + \mathbf{x}\right) = \mathbf{y} - \mathbf{z} \in W.$$

In summary, $V_1 = W$, which completes the proof.

♣

Notice that the assertion of the theorem implies that the directing subspace V_1 does not depend on the selection of vector $\mathbf{x}_0$.

Example 3.7. Assume that the vector space is R^2, and V_1 is generated by a nonzero 2-element vector $\mathbf{a}$. Then all elements of V_1 can be expressed as $c \cdot \mathbf{a}$ with some scalar c. As it is illustrated in Figure 3.1, the vectors of V_1 form a straight line passing through the origin, and the direction of the line is the same as the direction of vector $\mathbf{a}$. Adding $\mathbf{x}_0$ to all vectors of this line geometrically means that the line is shifted by the vector $\mathbf{x}_0$. Hence, the linear manifold is parallel to the one-dimensional subspace V_1.

◆

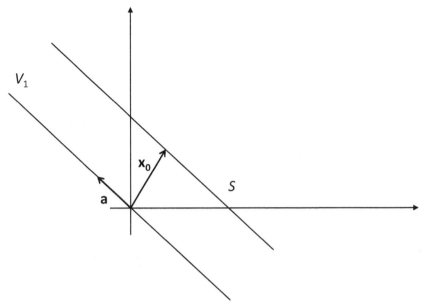

Figure 3.1. Illustration of a one-dimensional manifold

In analyzing particular linear structures a new concept is often applied, which is a common generalization of straight lines of two-dimensional spaces and planes of three-dimensional spaces.

Definition 3.6. Let V be a finite dimensional inner-product space and let $\mathbf{a} \in V$ be a given nonzero element. Assume that b is a given scalar. The set

$$H = \left\{ \mathbf{x} \mid \mathbf{x} \in V, \; (\mathbf{a}, \mathbf{x}) = b \right\} \tag{3.9}$$

is called a *hyperplane* in V.

In order to show that every hyperplane is a linear manifold, let V_1 denote the orthogonal complementary subspace of the subspace generated by vector $\mathbf{a}$. That is,

$$V_1 = \left\{ \mathbf{x} \in V \,\middle|\, (\mathbf{a}, \mathbf{x}) = 0 \right\},$$

and $\dim(V_1) = n - 1$, where n is the dimension of V. Select the vector

$$\mathbf{x}_0 = \frac{b}{(\mathbf{a}, \mathbf{a})} \cdot \mathbf{a},$$

and consider the linear manifold

$$S = \left\{ \mathbf{x}_0 + \mathbf{x} \,\middle|\, \mathbf{x} \in V_1 \right\}.$$

Theorem 3.9. The hyperplane H coincides with the linear manifold S.

Proof. We will prove that $H \subseteq S$ and $S \subseteq H$. Assume first that $\mathbf{z} \in H$. We will show that necessarily $\mathbf{z} \in S$ by verifying that $\mathbf{x} = \mathbf{z} - \mathbf{x}_0 \in V_1$. Simple calculation shows that

$$(\mathbf{a}, \mathbf{x}) = (\mathbf{a}, \mathbf{z} - \mathbf{x}_0) = (\mathbf{a}, \mathbf{z}) - (\mathbf{a}, \mathbf{x}_0) = b - \left(\mathbf{a}, \frac{b}{(\mathbf{a}, \mathbf{a})} \mathbf{a} \right) = b - \frac{b}{(\mathbf{a}, \mathbf{a})} (\mathbf{a}, \mathbf{a}) = 0,$$

which implies that $\mathbf{x} \in V_1$, and therefore $\mathbf{z} \in S$.

Assume next that $\mathbf{z} \in S$. Then $\mathbf{z} - \mathbf{x}_0 \in V_1$, which means that $(\mathbf{a}, \mathbf{z} - \mathbf{x}_0) = 0$. Since

$$(\mathbf{a}, \mathbf{z} - \mathbf{x}_0) = (\mathbf{a}, \mathbf{z}) - \left(\mathbf{a}, \frac{b}{(\mathbf{a}, \mathbf{a})} \mathbf{a} \right) = (\mathbf{a}, \mathbf{z}) - \frac{b}{(\mathbf{a}, \mathbf{a})} (\mathbf{a}, \mathbf{a}) = (\mathbf{a}, \mathbf{z}) - b,$$

necessarily $(\mathbf{a}, \mathbf{z}) = b$ which implies that $\mathbf{z} \in H$.
Thus, the proof is complete.

♣

In the special case when $V = R^n$, the assertion of the theorem is obvious, since H is the solution set of the single linear equation

$$\mathbf{a}^T \mathbf{x} = b,$$

and hence, from Theorem 3.4 we know that the solution set is a linear manifold. Its directing space is the solution set of the homogeneous equation

$$\mathbf{a}^T \mathbf{x} = 0.$$

3.5 Rank of Matrices

In Section 3.2 we have already introduced the concept of the rank of an $m \times n$ matrix as the dimension of the subspace generated by its columns. We have also seen that conditions for the existence of solutions of systems of linear equations could be conveniently formulated by using this concept. In this brief section the fundamental properties of matrix ranks will be introduced. These results can be summarized as follows.

Theorem 3.10. The following relations are true:

(i) $\operatorname{rank}(\mathbf{A}) = \operatorname{rank}(\mathbf{A}^T)$;

(ii) $\operatorname{rank}(\mathbf{A} + \mathbf{B}) \leq \operatorname{rank}(\mathbf{A}) + \operatorname{rank}(\mathbf{B})$;

(iii) $\operatorname{rank}(\mathbf{AB}) \leq \min\{\operatorname{rank}(\mathbf{A}); \operatorname{rank}(\mathbf{B})\}$

where we assume that $\mathbf{A}$ and $\mathbf{B}$ are arbitrary matrices such that the indicated operations are defined.

Proof. (i) Assume that matrix $\mathbf{A}$ is $m \times n$. Consider the homogeneous equation $\mathbf{Ax} = \mathbf{0}$. From Theorem 3.5 we know that the set of all solutions of this equation is the orthogonal complementary subspace of the subspace generated by the rows of $\mathbf{A}$. The rows of $\mathbf{A}$ are the columns of $\mathbf{A}^T$, the dimension of the solution set is therefore $n - \operatorname{rank}(\mathbf{A}^T)$. Theorem 3.6 implies that the dimension of the solution set is $n - \operatorname{rank}(\mathbf{A})$. Hence,

$$n - \operatorname{rank}(\mathbf{A}^T) = n - \operatorname{rank}(\mathbf{A}),$$

from which the assertion follows.

(ii) Let $\mathbf{u}_1, \ldots, \mathbf{u}_r$ and $\mathbf{v}_1, \ldots, \mathbf{v}_s$ be bases of the column spaces of $\mathbf{A}$ and $\mathbf{B}$, respectively. It means, that all columns of $\mathbf{A}$ can be expressed as a linear combination of vectors $\mathbf{u}_1, \ldots, \mathbf{u}_r$:

$$\mathbf{a}_k = c_{k1}\mathbf{u}_1 + \ldots + c_{kr}\mathbf{u}_r.$$

Similarly, all columns of $\mathbf{B}$ can also be expressed as

$$\mathbf{b}_k = d_{k1}\mathbf{v}_1 + \ldots + d_{ks}\mathbf{v}_s.$$

Since

$$\mathbf{a}_k + \mathbf{b}_k = c_{k1}\mathbf{u}_1 + \ldots + c_{kr}\mathbf{u}_r + d_{k1}\mathbf{v}_1 + \ldots + d_{ks}\mathbf{v}_s,$$

all columns of $\mathbf{A} + \mathbf{B}$ are in the subspace generated by the set $\{\mathbf{u}_1, \ldots, \mathbf{u}_r, \mathbf{v}_1, \ldots, \mathbf{v}_s\}$. Therefore, the dimension of the column-space of matrix $\mathbf{A} + \mathbf{B}$ is not greater than the number $r + s$ of vectors in the generating system. Notice that $r = \text{rank}(\mathbf{A})$, $s = \text{rank}(\mathbf{B})$, from which the assertion follows.

(iii) We will first prove that rank $(\mathbf{AB}) \leq$ rank $(\mathbf{A})$ by verifying that each column of $\mathbf{AB}$ belongs to the column-space of $\mathbf{A}$. The kth column of $\mathbf{AB}$ equals the product of matrix $\mathbf{A}$ and the kth column $\mathbf{b}_k$ of $\mathbf{B}$. Let $b_{1k}, b_{2k}, \ldots, b_{nk}$ denote the elements of this column, then

$$\mathbf{Ab}_k = b_{1k}\mathbf{a}_1 + b_{2k}\mathbf{a}_2 + \ldots + b_{nk}\mathbf{a}_n,$$

where $\mathbf{a}_1, \ldots, \mathbf{a}_n$ are the columns of $\mathbf{A}$. This equation shows that $\mathbf{Ab}_k$ is a linear combination of the columns of $\mathbf{A}$, consequently it belongs to the column-space of $\mathbf{A}$.

Next, we verify that rank $(\mathbf{AB}) \leq$ rank $(\mathbf{B})$. By using properties (i) and (ii) we have

$$\text{rank}\left(\mathbf{AB}\right) = \text{rank}\left(\left(\mathbf{AB}\right)^T\right) = \text{rank}\left(\mathbf{B}^T\mathbf{A}^T\right) \leq \text{rank}\left(\mathbf{B}^T\right) = \text{rank}\left(\mathbf{B}\right).$$

Thus, the proof is complete.

♣

Let $\mathbf{A}$ and $\mathbf{B}$ be $n \times n$ and $n \times m$ matrices where $\mathbf{A}$ is invertible. Then

$$\text{rank}(\mathbf{AB}) \le \text{rank}(\mathbf{B})$$

and

$$\text{rank}(\mathbf{B}) = \text{rank}\left(\mathbf{A}^{-1}(\mathbf{AB})\right) \le \text{rank}(\mathbf{AB}).$$

That is, rank($\mathbf{AB}$) = rank($\mathbf{B}$).
Assume next that $\mathbf{A}$ and $\mathbf{B}$ are $n \times n$ and $m \times n$ matrices where $\mathbf{A}$ is invertible. Then

$$\text{rank}(\mathbf{BA}) \le \text{rank}(\mathbf{B})$$

and

$$\text{rank}(\mathbf{B}) = \text{rank}\left((\mathbf{BA})\mathbf{A}^{-1}\right) \le \text{rank}(\mathbf{BA}).$$

That is rank($\mathbf{BA}$) = rank($\mathbf{B}$).

These relations show that the rank of a matrix does not change if it is multiplied by an invertible matrix.

3.6 Matrix Equations and Inverses of Matrices

In many applications, matrix equations of the form

$$\mathbf{AX} = \mathbf{B} \tag{3.10}$$

have to be solved, where $\mathbf{A}$ is a given $m \times n$ matrix, and $\mathbf{B}$ is an $m \times p$ given matrix. The solution of equation (3.10) has to be $n \times p$, otherwise either the product $\mathbf{AX}$ is not defined, or the product does not have the type $m \times p$. Let $\mathbf{x}_1, \mathbf{x}_2, \ldots, \mathbf{x}_p$ denote the unknown columns of $\mathbf{X}$, and assume that $\mathbf{b}_1, \mathbf{b}_2, \ldots, \mathbf{b}_p$ are the columns of $\mathbf{B}$. We may therefore use the symbol

$$\mathbf{X} = \left(\mathbf{x}_1, \mathbf{x}_2, \ldots, \mathbf{x}_p\right) \text{ and } \mathbf{B} = \left(\mathbf{b}_1, \mathbf{b}_2, \ldots, \mathbf{b}_p\right),$$

where we wish to show the columns of these matrices. Equation (3.10) can be rewritten as

$$A\left(x_1, x_2, ..., x_p\right) = \left(b_1, b_2, ..., b_p\right),$$

and comparing the corresponding columns of the left-hand and right-hand side matrices we obtain the following set of equations:

$$Ax_1 = b_1$$
$$Ax_2 = b_2$$
$$...$$
$$Ax_p = b_p.$$
$$\tag{3.11}$$

Hence the matrix equation (3.10) is equivalent to p systems of linear equations. Equation (3.10) has at least one solution if and only if all systems (3.11) of linear equations have solutions. The solution of equation (3.10) is unique if and only if all systems (3.11) have unique solutions. Based on Theorems 3.1, 3.2 and 3.3 we immediately obtain the following results.

Theorem 3.11. The following conditions are equivalent to each other:
(a) Matrix equation (3.10) has at least one solution;
(b) All vectors b_1, b_2, ..., b_p belong to the subspace generated by the columns of matrix A;
(c) rank(A) = rank([A, B]), where the first n columns of matrix [A, B] are those of A and the next p columns of [A, B] are the columns of B.

Theorem 3.12. Equation (3.10) has a unique solution if and only if condition (b) or (c) of Theorem 3.11 holds and the columns of matrix A are linearly independent.

Example 3.8. Consider the matrix equation

$$\begin{pmatrix} 1 & 1 & 1 \\ 1 & 2 & 1 \\ 1 & 1 & 2 \end{pmatrix} \begin{pmatrix} x_{11} & x_{12} \\ x_{21} & x_{22} \\ x_{31} & x_{32} \end{pmatrix} = \begin{pmatrix} 3 & 4 \\ 4 & 5 \\ 4 & 6 \end{pmatrix}.$$

Here **A** is 3×3, **X** and **B** are 3×2. Equations (3.11) have now the forms:

$$\begin{pmatrix} 1 & 1 & 1 \\ 1 & 2 & 1 \\ 1 & 1 & 2 \end{pmatrix} \begin{pmatrix} x_{11} \\ x_{21} \\ x_{31} \end{pmatrix} = \begin{pmatrix} 3 \\ 4 \\ 4 \end{pmatrix}$$

and

$$\begin{pmatrix} 1 & 1 & 1 \\ 1 & 2 & 1 \\ 1 & 1 & 2 \end{pmatrix} \begin{pmatrix} x_{12} \\ x_{22} \\ x_{32} \end{pmatrix} = \begin{pmatrix} 4 \\ 5 \\ 6 \end{pmatrix}.$$

Simple calculation shows that the solutions are:

$$x_{11} = x_{21} = x_{31} = 1 \quad \text{and} \quad x_{12} = x_{22} = 1, \ x_{32} = 2.$$

Hence the solution of the matrix equation is as follows:

$$\mathbf{X} = \begin{pmatrix} 1 & 1 \\ 1 & 1 \\ 1 & 2 \end{pmatrix}.$$

$\blacklozenge$

In Section 1.4 we have already introduced the concept of the inverse of a matrix. If **A** is an $n \times n$ square matrix, then the inverse $\mathbf{A}^{-1}$ of **A** is an $n \times n$ matrix which satisfies equations

$$\mathbf{A} \cdot \mathbf{A}^{-1} = \mathbf{A}^{-1} \cdot \mathbf{A} = \mathbf{I}_n,$$

where $\mathbf{I}_n$ is the $n \times n$ identity matrix. If $\mathbf{A}^{-1}$ exists, then matrix **A** is called *invertible or nonsingular*, otherwise it is called *singular*. We may speak about the inverse of a matrix **A**, since, as we have verified in Section 1.4, the inverse of any matrix, if exists, is necessarily unique.

Our first result gives a sufficient and necessary condition for the existence of the inverse of an $n \times n$ matrix.

Theorem 3.13. Let **A** be a given $n \times n$ matrix. The following conditions are equivalent to each other:

(a) $\mathbf{A}^{-1}$ exists;

(b) The columns of **A** are linearly independent;

(c) rank(**A**) = n.

Proof. Using Theorem 3.11 we will prove that condition (b) is equivalent to the existence of at least one solution of matrix equation

$$\mathbf{A} \cdot \mathbf{X} = \mathbf{I}_n.$$

Notice first that the columns of $\mathbf{I}_n$ are the natural basis vectors e_1, e_2, ..., e_n, and therefore these vectors are in the column-space of **A** if and only if the columns of **A** generate the entire space R^n (or C^n). Since **A** has n columns, this is the case if and only if condition (b) holds. The definition of matrix ranks implies that conditions (b) and (c) are equivalent. Notice in addition that condition (c) implies that the solution of the matrix equation $\mathbf{A} \cdot \mathbf{X} = \mathbf{I}_n$ is unique. In order to complete the proof, we have to show that this unique solution also satisfies equation $\mathbf{X} \cdot \mathbf{A} = \mathbf{I}_n$. Observe first that this equation also has a unique solution, since it can be rewritten as

$$\mathbf{A}^T \mathbf{X}^T = \mathbf{I}_n$$

with rank($\mathbf{A}^T$) = n, therefore condition (c) holds. Let **Y** be the unique matrix such that $\mathbf{Y} \cdot \mathbf{A} = \mathbf{I}_n$. Then

$$\mathbf{X} = \mathbf{I}_n \mathbf{X} = (\mathbf{YA})\mathbf{X} = \mathbf{Y}(\mathbf{AX}) = \mathbf{YI}_n = \mathbf{Y},$$

that is, $\mathbf{X} = \mathbf{Y}$, which completes the proof.

♣

Corollary. Let $\mathbf{A}$ be an $n \times n$ matrix. Then $\mathbf{X} = \mathbf{A}^{-1}$ if and only if $\mathbf{A} \cdot \mathbf{X} = \mathbf{I}_n$. Since the columns of $\mathbf{I}_n$ are the natural basis vectors $\mathbf{e}_1, \ldots, \mathbf{e}_n$, the columns of $\mathbf{X}$ are the solutions of the following systems of linear equations:

$$\mathbf{A}\mathbf{x}_1 = \mathbf{e}_1$$
$$\mathbf{A}\mathbf{x}_2 = \mathbf{e}_2$$
$$\cdots \qquad\qquad (3.12)$$
$$\mathbf{A}\mathbf{x}_n = \mathbf{e}_n.$$

Example 3.9. Let

$$\mathbf{A} = \begin{pmatrix} 2 & 1 & 3 \\ 4 & 4 & 7 \\ 2 & 5 & 9 \end{pmatrix}.$$

If $\mathbf{x}_1$, $\mathbf{x}_2$ and $\mathbf{x}_3$ denote the columns of $\mathbf{A}^{-1}$, then they are the solutions of equations

$$\begin{pmatrix} 2 & 1 & 3 \\ 4 & 4 & 7 \\ 2 & 5 & 9 \end{pmatrix}\begin{pmatrix} x_{11} \\ x_{21} \\ x_{31} \end{pmatrix} = \begin{pmatrix} 1 \\ 0 \\ 0 \end{pmatrix},$$

$$\begin{pmatrix} 2 & 1 & 3 \\ 4 & 4 & 7 \\ 2 & 5 & 9 \end{pmatrix}\begin{pmatrix} x_{12} \\ x_{22} \\ x_{32} \end{pmatrix} = \begin{pmatrix} 0 \\ 1 \\ 0 \end{pmatrix},$$

and

$$\begin{pmatrix} 2 & 1 & 3 \\ 4 & 4 & 7 \\ 2 & 5 & 9 \end{pmatrix}\begin{pmatrix} x_{13} \\ x_{23} \\ x_{33} \end{pmatrix} = \begin{pmatrix} 0 \\ 0 \\ 1 \end{pmatrix},$$

where x_{ik} denotes the ith element of $\mathbf{x}_k$. By substitution one may easily check that the solutions are:

$$\mathbf{x}_1 = \begin{pmatrix} x_{11} \\ x_{21} \\ x_{31} \end{pmatrix} = \begin{pmatrix} \dfrac{1}{16} \\ -\dfrac{11}{8} \\ \dfrac{3}{4} \end{pmatrix}, \quad \mathbf{x}_2 = \begin{pmatrix} x_{12} \\ x_{22} \\ x_{32} \end{pmatrix} = \begin{pmatrix} \dfrac{3}{8} \\ \dfrac{3}{4} \\ -\dfrac{1}{2} \end{pmatrix}, \quad \mathbf{x}_3 = \begin{pmatrix} x_{13} \\ x_{23} \\ x_{33} \end{pmatrix} = \begin{pmatrix} -\dfrac{5}{16} \\ -\dfrac{1}{8} \\ \dfrac{1}{4} \end{pmatrix},$$

therefore

$$\mathbf{A}^{-1} = \left(\mathbf{x}_1, \mathbf{x}_2, \mathbf{x}_3 \right) = \begin{pmatrix} \dfrac{1}{16} & \dfrac{3}{8} & -\dfrac{5}{16} \\ -\dfrac{11}{8} & \dfrac{3}{4} & -\dfrac{1}{8} \\ \dfrac{3}{4} & -\dfrac{1}{2} & \dfrac{1}{4} \end{pmatrix}.$$

$\blacklozenge$

Equations (3.12) show a strong relation between matrix inversion and the solution of systems of linear equations. If one has a solver for systems of linear equations, then the repeated application of the solver provides matrix inverses. We will next show that the reverse is also true: if one has an algorithm for inverting matrices, then the matrix inverses can be used to get the solutions of systems of linear equations with nonsingular coefficient matrices. Consider the system of linear equations

$$\mathbf{A}\mathbf{x} = \mathbf{b}$$

where $\mathbf{A}$ is an $n \times n$ nonsingular matrix. Multiply both sides from the left by $\mathbf{A}^{-1}$ to see that

$$\mathbf{A}^{-1}\left(\mathbf{A}\mathbf{x} \right) = \mathbf{A}^{-1}\mathbf{b}.$$

Notice that

$$\mathbf{A}^{-1}\left(\mathbf{A}\mathbf{x} \right) = \left(\mathbf{A}^{-1}\mathbf{A} \right)\mathbf{x} = \mathbf{I}_n \mathbf{x} = \mathbf{x},$$

that is, the unique solution is obtained as $\mathbf{x} = \mathbf{A}^{-1}\mathbf{b}$.

Example 3.10. Consider the system of linear equations

$$2x_1 + x_2 + 3x_3 = 3$$
$$4x_1 + 4x_2 + 7x_3 = 8$$
$$2x_1 + 5x_2 + 9x_3 = 7,$$

where the coefficient matrix and right-hand side vector are as follows:

$$\mathbf{A} = \begin{pmatrix} 2 & 1 & 3 \\ 4 & 4 & 7 \\ 2 & 5 & 9 \end{pmatrix} \quad \text{and} \quad \mathbf{b} = \begin{pmatrix} 3 \\ 8 \\ 7 \end{pmatrix}.$$

In the previous example we have seen that

$$\mathbf{A}^{-1} = \begin{pmatrix} \dfrac{1}{16} & \dfrac{3}{8} & -\dfrac{5}{16} \\[2mm] -\dfrac{11}{8} & \dfrac{3}{4} & -\dfrac{1}{8} \\[2mm] \dfrac{3}{4} & -\dfrac{1}{2} & \dfrac{1}{4} \end{pmatrix},$$

therefore, the solution can be obtained as

$$\mathbf{x} = \mathbf{A}^{-1}\mathbf{b} = \begin{pmatrix} \dfrac{1}{16} & \dfrac{3}{8} & -\dfrac{5}{16} \\[2mm] -\dfrac{11}{8} & \dfrac{3}{4} & -\dfrac{1}{8} \\[2mm] \dfrac{3}{4} & -\dfrac{1}{2} & \dfrac{1}{4} \end{pmatrix}\begin{pmatrix} 3 \\ 8 \\ 7 \end{pmatrix} = \begin{pmatrix} \dfrac{3}{16} & +3 & -\dfrac{35}{16} \\[2mm] -\dfrac{33}{8} & +6 & -\dfrac{7}{8} \\[2mm] \dfrac{9}{4} & -4 & +\dfrac{7}{4} \end{pmatrix} = \begin{pmatrix} 1 \\ 1 \\ 0 \end{pmatrix}.$$

That is, $x_1 = x_2 = 1$ and $x_3 = 0$.

♦

3.7 The Elimination Method

In this section the most frequently used method for solving systems of linear equations, matrix equations, and for finding matrix inverses will be introduced. By "elimination" we refer to the important property of the method that by adding constant multiples of an equation to the other equations we can make the coefficients of a given variable in the other equations equal to zero. This procedure is repeated until the resulting system of equations has a diagonal coefficient matrix. Before giving the mathematical formulation of the method a small example is presented.

Example 3.11. Consider the system of three linear equations

$$x_1 + x_2 + x_3 = 3$$
$$2x_1 + 3x_2 - x_3 = 4$$
$$3x_1 + 5x_2 + x_3 = 9.$$

The term x_1 of the first equation is used to eliminate x_1 from the other two equations. Multiply the first equation by 2, and subtract the resulting equation from the original second equation:

$$(2x_1 + 3x_2 - x_3) - (2x_1 + 2x_2 + 2x_3) = 4 - 6,$$

that is,

$$x_2 - 3x_3 = -2.$$

Notice that this new equation does not contain the unknown x_1. Multiply next the first equation by 3 and subtract the resulting equation from the original third equation:

$$(3x_1 + 5x_2 + x_3) - (3x_1 + 3x_2 + 3x_3) = 9 - 9,$$

which can be simplified as

$$2x_2 - 2x_3 = 0.$$

Notice that x_1 is not contained in this new equation. By replacing the original second and third equations by these "new" equations we get the so called *first derived system*:

$$x_1 + x_2 + x_3 = 3$$
$$x_2 - 3x_3 = -2 \qquad (3.13)$$
$$2x_2 - 2x_3 = 0.$$

Select next the first term x_2 of the new second equation to eliminate x_2 from the first and third equations. Subtract the second equation from the first one to have

$$\left(x_1 + x_2 + x_3\right) - \left(x_2 - 3x_3\right) = 3 - \left(-2\right),$$

that is,

$$x_1 + 4x_3 = 5.$$

To eliminate x_2 from the third equation, subtract the 2-multiple of the second equation from the third equation:

$$\left(2x_2 - 2x_3\right) - \left(2x_2 - 6x_3\right) = 0 - \left(-4\right),$$

which simplifies as

$$4x_3 = 4.$$

By replacing the first and third equations of the first derived system by these new versions, we have

$$x_1 \qquad\qquad + 4x_3 = 5$$
$$x_2 \qquad 3x_3 = -2 \qquad (3.14)$$
$$4x_3 = 4,$$

which is called the *second derived system*. As the last stage of the method, we eliminate x_3 from the first and second equations. Divide the last equation by 4 then the coefficient of x_3 becomes 1. This small step makes the elimination of x_3 from the other two equations easy, since we have to multiply this new third equation

$$x_3 = 1 \tag{3.15}$$

by the coefficients of x_3 of the other equations. Subtract the 4-multiple of equation (3.15) from the first equation of (3.14) to have

$$x_1 + 4x_3 - \left(4x_3\right) = 5 - 4,$$

that is,

$$x_1 = 1.$$

Multiply equation (3.15) by (–3) and subtract the resulting equation from the second equation of (3.14):

$$x_2 - 3x_3 - \left(-3x_3\right) = -2 - \left(-3\right),$$

which can be simplified as

$$x_2 = 1.$$

The resulted third derived system becomes:

$$x_1 = 1$$
$$x_2 = 1$$
$$x_3 = 1,$$

which provides the solution of the original system.

$$\blacklozenge$$

Following the procedure of the above example some general observations can be made:

1. If we wish to eliminate a variable from all but one equations, then a nonzero coefficient of that variable is first selected. This coefficient is called the *pivot*, and it must be nonzero. If we divide the equation of the pivot coefficient by the pivot, then this coefficient becomes 1. (This was done by dividing the third equation of (3.14) by 4.)

2. Pivot element can be selected only once from an equation, otherwise we bring back already eliminated variables into the equations. To illustrate this point, assume that in system (3.13) the term x_2 of the first equation is selected as the pivot. That is, pivot is selected from the first equation for the second time. Subtract the first equation from the second one, and subtract the 2-multiple of the first equation from the third equation to have

$$(x_2 - 3x_3) - (x_1 + x_2 + x_3) = -2 - 3$$

and

$$(2x_2 - 2x_3) - (2x_1 + 2x_2 + 2x_3) = 0 - 6,$$

that is,

$$- x_1 - 4x_3 = -5$$
$$-2x_1 - 4x_3 = -6.$$

As we see, variable x_1 has "come back" into the second and third equations.

3. In the above example always diagonal elements were selected as pivots. It does not need to be in this way all the times. In case of the selection of a non-diagonal pivot element, the pivot can be placed into the diagonal by interchanging equations and variables, if necessary. To illustrate this point, assume that in the original system of the previous example the second term $5x_2$ of the last equation is selected as the first pivot. We can bring this element to the first diagonal position by interchanging the first and third equations, and then by interchanging variables x_1 and x_2:

$$5x_2 + 3x_1 + x_3 = 9$$
$$3x_2 + 2x_1 - x_3 = 4$$
$$x_2 + x_1 + x_3 = 3.$$

4. Notice that at each step of the elimination process we subtract a constant multiple of an equation from another equation. It is done by

subtracting the constant multiples of each coefficient of an equation from the corresponding coefficients of another equation. This is the same as subtracting a constant multiple of a row from another row. This observation makes the calculation faster, since there is no need to copy entire equations with all the variables. It is sufficient to copy only the coefficients and the right-hand side numbers into the elimination table and manipulating with the entire rows of the table. It is very important to indicate the variables at the top of the table, since in cases when variables are interchanged, we have to indicate these changes there.

In Table 3.1 we show the entire calculation of solving the system of the previous example. In each derived system the new pivot row is first constructed by dividing the row containing the pivot element by the pivot coefficient. The new pivot element (which is circled in the table) is therefore always equal to one. Elimination of the pivot variable from the other equations is the next step by subtracting constant multiples of the pivot row from the other rows.

Table 3.1. Elimination process

	x_1	x_2	x_3		
	1	1	1	3	
	2	3	-1	4	Original system
	3	5	1	9	
pivot row	①	1	1	3	1st derived system
(row 2)-2×(pivot row)		1	-3	-2	
(row 3)-3×(pivot row)		2	-2	0	
(row 1)-1×(pivot row)	1		4	5	
pivot row		①	-3	-2	2nd derived system
(row 3)-2×(pivot row)			4	4	
(row 1)-4×(pivot row)	1			1	
(row 2)-(-3)×(pivot row)		1		1	3rd derived system
pivot row			①	1	

5. Each derived system is equivalent to the original system of linear equations, which can be proved as follows. Notice first that by adding back the constant multiples of the pivot row to the other rows in any derived system, the corresponding equations of the previous derived system are obtained, and by multiplying back the new pivot row by the pivot coefficient, the original pivot row of the previous derived system is obtained. Therefore, the solutions of any derived system necessarily satisfy all previous derived systems, hence they satisfy the original system. The solutions of the original system also satisfy all derived systems since all derived systems are the consequences of the original system of equations. In the above example the last derived system had the identify matrix as coefficient matrix, and therefore the unique

solution is given by the corresponding right hand side vector. This is not always the case, since solution might not exist and in many cases, there are infinitely many solutions. We will discuss these cases later in this section.

The elimination process illustrated in the above example can be generally formulated in the following way. Assume that the coefficient matrix $\mathbf{A} = (a_{ij})$ is $m \times n$, that is, there are m equations for n unknowns. For the sake of simplicity the right hand side numbers are denoted by $a_{1,n+1}$, $a_{2,n+1}$, ..., $a_{m,n+1}$ (that is, they form the $(n + 1)^{st}$ column of $\mathbf{A}$).

Step 1. Set $k = 1$.

Step 2. Consider all elements a_{ij} for $i = k, k+1, ..., m$ and $j = k, k + 1, ..., n$. If they are all zeros, then elimination terminates. Otherwise, select an $a_{ij} \neq 0$ as the pivot coefficient. If $i \neq k$, then interchange rows k and i. If $j \neq k$, then interchange variables j and k by interchanging columns j and k in the elimination table as well as interchanging the symbols of the jth and k^{th} variables on the top of the table. The pivot element is now located in the kth diagonal position.

Step 3. Divide all coefficients of the kth row by a_{kk}:

$$a_{kj} \leftarrow \frac{a_{kj}}{a_{kk}} \qquad (j = k + 1, ..., n + 1)$$

$$a_{kk} \leftarrow 1,$$

and then eliminate the kth variable from all other equations:

$$a_{ij} \leftarrow a_{ij} - a_{ik} \cdot a_{kj} \quad (i \neq k, \; i = 1, ..., m; \; j = k + 1, ..., n + 1),$$

$$a_{ik} \leftarrow 0 \qquad (i \neq k, \; i = 1, ..., m).$$

After this elimination step, set $k \leftarrow k + 1$ and go back to Step 2.

After the process terminates, the resulted final derived system has one of the following structures:

Case 1. The coefficient matrix is the identity matrix. It occurs only if $m = n$, and the final table is as follows:

x_{i_1}	x_{i_2}	$\ldots$	x_{i_n}	
1				$a_{1,n+1}$
	1			$a_{2,n+1}$
		$\ddots$		$\vdots$
			1	$a_{n,n+1}$

Here $a_{i,n+1}$ is the right hand side number after elimination is performed, $i = 1, 2, \ldots, n$.

In this case a unique solution exists:

$$x_{i_1} = a_{1,n+1}, \; x_{i_2} = a_{2,n+1}, \ldots, \; x_{i_n} = a_{n,n+1}.$$

Using matrix notation, the table and the solution can be given as

$\mathbf{x}^T$	
$\mathbf{I}_n$	$\mathbf{a}_{n+1}$

and

$$\mathbf{x} = \mathbf{a}_{n+1}. \tag{3.16}$$

We have to mention here, that the order of the elements of the solution vector $\mathbf{x}$ might differ from the original order, if we had to interchange columns during the elimination process.

Case 2. All variables have been eliminated, but there are some additional equations. This case may occur only if $m > n$. The final table has the form:

x_{i_1}	x_{i_2}	...	x_{i_n}	
1				$a_{1,n+1}$
	1			$a_{2,n+1}$
		$\ddots$		$\vdots$
			1	$a_{n,n+1}$
				$a_{n+1,n+1}$
				$\vdots$
				$a_{m,n+1}$

Since all variables are eliminated, all other elements than the right-hand side numbers and the indicated 1's are equal to zero. Therefore, no more nonzero elements are in the table. If all of the "extra" right hand side numbers $a_{n+1,n+1}, \ldots, a_{m,n+1}$ are equal to zero, then a unique solution exists:

$$x_{i_1} = a_{1,n+1}, \; x_{i_2} = a_{2,n+1}, \ldots, \; x_{i_n} = a_{n,n+1}.$$

Otherwise, there is at least one nonzero value $a_{k,n+1}$ among the numbers $a_{n+1,n+1}, \ldots, a_{m,n+1}.$ Then the corresponding equation

$$0 \cdot x_{i_1} + 0 \cdot x_{i_2} + \ldots + 0 \cdot x_{i_n} = a_{k,n+1}$$

is a contradiction, since the right-hand side is nonzero, and the left-hand side is zero. Hence, no solution exists. Using matrix notation, the final table can be written as

$\mathbf{x}^T$	
$\mathbf{I}_n$	$\mathbf{a}_{n+1}^{(1)}$ $\mathbf{a}_{n+1}^{(2)}$

where $\mathbf{a}_{n+1}^{(1)}$ and $\mathbf{a}_{n+1}^{(2)}$ denote the vectors containing the first n and last $(m-n)$ elements of $\mathbf{a}_{n+1}$. If $\mathbf{a}_{n+1}^{(2)} \neq \mathbf{0}$, then no solution exists, and if $\mathbf{a}_{n+1}^{(2)} = \mathbf{0}$, then the unique solution is

$$\mathbf{x} = \mathbf{a}_{n+1}^{(1)}. \tag{3.17}$$

Case 3. We could not eliminate all variables, because there are no more equations to select a new pivot element from. This case may occur only if $m < n$. In this case the final table is the following:

x_{i_1}	x_{i_2}	...	x_{i_m}	$x_{i_{m+1}}$	...	x_{i_n}	
1				$a_{1,m+1}$	...	$a_{1,n}$	$a_{1,n+1}$
	1			$a_{2,m+1}$	...	$a_{2,n}$	$a_{2,n+1}$
		$\ddots$		$\vdots$		$\vdots$	$\vdots$
			1	$a_{m,m+1}$	...	$a_{m,n}$	$a_{m,n+1}$

In this case infinitely many solutions exist. Variables $x_{i_{m+1}}$, ..., x_{i_n} can have arbitrary values, therefore they are called the free variables. The values of x_{i_1}, x_{i_2}, ..., x_{i_m} depend on the particular selection of the values of the free variables. Solve the first equation for x_{i_1}, the second equation for x_{i_2}, and so on, to get the general solution:

$$x_{i_1} = a_{1,n+1} - \sum_{j=m+1}^{n} a_{1j} x_{i_j}$$

$$x_{i_2} = a_{2,n+1} - \sum_{j=m+1}^{n} a_{2j} x_{i_j}$$

...

$$x_{i_m} = a_{m,n+1} - \sum_{j=m+1}^{n} a_{mj} x_{i_j}.$$

Using matrix notation, the final table can be given as

$\mathbf{x}^{(1)T} \ \mathbf{x}^{(2)T}$	
$\mathbf{I}_m \ \mathbf{A}^{(2)}$	$\mathbf{a}_{n+1}$

and the general solution has the form:

$$\mathbf{x}^{(1)} = \mathbf{a}_{n+1} - \mathbf{A}^{(2)} \mathbf{x}^{(2)}, \tag{3.18}$$

where vector $\mathbf{x}^{(2)}$ is arbitrary.

Case 4. We could not eliminate all variables, because all coefficients equal zero in the additional equations. The final table is now the following:

x_{i_1}	x_{i_2}	...	x_{i_k}	$x_{i_{k+1}}$	...	x_{i_n}	
1				$a_{1,k+1}$	...	$a_{1,n}$	$a_{1,n+1}$
	1			$a_{2,k+1}$	...	$a_{2,n}$	$a_{2,n+1}$
		$\ddots$		$\vdots$		$\vdots$	$\vdots$
			1	$a_{k,k+1}$	...	$a_{k,n}$	$a_{k,n+1}$
							$a_{k+1,n+1}$
							$\vdots$
							$a_{m,n+1}$

Notice that this table is a combination of Cases 2 and 3. If at last one of the "extra" right hand side numbers $a_{k+1,n+1}$, ..., $a_{m,n+1}$ has nonzero value, then no solution exists, since the corresponding equation is a contradiction. If all of these numbers equal zero, then simply ignore the additional equations and the resulting table coincides with the final table of Case 3 with $m = k$. Therefore, infinitely many solutions exist, and the general solution is obtained as it was shown in Case 3. In matrix notation this case results in a table of the form:

$\mathbf{x}^{(1)T}$ $\mathbf{x}^{(2)T}$	
$\mathbf{I}_k$ $\mathbf{A}^{(2)}$	$\mathbf{a}^{(1)}_{n+1}$ $\mathbf{a}^{(2)}_{n+1}$

If $\mathbf{a}^{(2)}_{n+1} \neq \mathbf{0}$, then no solution exists. Otherwise, there are infinitely many solutions, and the general solution has the form:

$$x^{(1)} = \mathbf{a}^{(1)}_{n+1} - \mathbf{A}^{(2)}\mathbf{x}^{(2)}, \qquad (3.19)$$

where vector $\mathbf{x}^{(2)}$ is arbitrary.

The following examples illustrate the above four cases and the ways how the solutions are obtained.

Example 3.12. We will first solve the system

$$x_1 - x_2 + x_3 = 1$$
$$2x_1 + x_2 - x_3 = -1$$
$$x_1 + 2x_2 + x_3 = 4.$$

The elimination procedure is given in Table 3.2.

Table 3.2. Elimination of Example 3.12

	x_1	x_2	x_3		
	1	-1	1	1	
	2	1	-1	-1	original system
	1	2	1	4	
pivot row	①	-1	1	1	
(row 2)– 2×(pivot row)		3	-3	-3	1st derived system
(row 3) – 1×(pivot row)		3	0	3	
(row 1) – (–1)×(pivot row)	1		0	0	
pivot row		①	-1	-1	2nd derived system
(row 3) – 3×(pivot row)			3	6	
row 1	1			0	
(row 2) + 1×(pivot row)		1		1	3rd derived system
pivot row			①	2	

Notice that in the first equation of the second derived system the coefficient of x_3 is zero. We simply copied this equation into the third derived system, since zero coefficient does not need to be eliminated. The pivot elements (after dividing the pivot terms by the pivot coefficients) are circled in all derived systems. The last system gives the unique solution:

$$x_1 = 0, \ x_2 = 1, \ x_3 = 2.$$

♦

Example 3.13. Consider next the system

$$\begin{aligned}
x_1 - x_2 + x_3 &= 1 \\
2x_1 + x_2 - x_3 &= -1 \\
x_1 + 2x_2 + x_3 &= 4 \\
3x_1 + 2x_2 - x_3 &= 0 \\
x_1 + x_2 - 2x_2 &= -5.
\end{aligned}$$

The elimination process is shown in Table 3.3. The last two equations have all zero coefficients, and in the case of the last equation the right-hand side is nonzero. Therefore, no solution exists. Modify the original last equation as

$$x_1 + x_2 - 2x_2 = -3,$$

and repeat the elimination process. The calculations are also shown in Table 3.3, where the new right-hand side numbers of the last equations are given next to their original values.

Table 3.3. Elimination of Example 3.13

	x_1	x_2	x_3		
	1	-1	1	1	
	2	1	-1	-1	
	1	2	1	4	original system
	3	2	-1	0	
	1	1	-2	-5	-3
pivot row	①	-1	1	1	
(row 2)-2×(pivot row)		3	-3	-3	
(row 3)-1×(pivot row)		3	0	3	1st derived system
(row 4)-3×(pivot row)		5	-4	-3	
(row 5)-1×(pivot row)		2	-3	-6	-4
(row 1)+1×(pivot row)	1		0	0	
pivot row		①	-1	-1	
(row 3)-3×(pivot row)			3	6	2nd derived system
(row 4)-5×(pivot row)			1	2	
(row 5)-2×(pivot row)			-1	-4	-2
row 1	1			0	
(row 2)+1×(pivot row)		1		1	
pivot row			①	2	3rd derived system
(row 4)-1×(pivot row)				0	
(row 5)-(-1)×(pivot row)				-2	0

Notice that only the last right-hand side numbers become different, since we did not select pivot from the last equation. The right-hand side numbers of the last two equations become zero showing that a unique solution exists:

$$x_1 = 0, \ x_2 = 1, \ x_3 = 2.$$

♦

Example 3.14. We solve now the system

$$x_1 + x_2 + x_3 + x_4 = 2$$
$$x_1 + x_2 + 2x_3 + 2x_4 = 2$$
$$x_1 + x_2 + 2x_3 + 4x_4 = 2,$$

which has only three equations for four unknowns. The elimination process is shown in Table 3.4.

Table 3.4. Elimination of Example 3.14

	X_1	X_2	X_3	X_4		
	X_1	X_3	X_2	X_4		
	X_1	X_3	X_4	X_2		
	1	1	1	1	2	
	1	1	2	2	2	original system
	1	1	2	4	2	
pivot row	①	1	1	1	2	
(row 2)-1×(pivot row)		0	1	1	0	1st derived system
(row 3)-1×(pivot row)		0	1	3	0	
	1	1	1	1	2	columns 2 and 3 are
		1	0	1	0	interchanged in the
		1	0	3	0	first derived system
(row 1)-1×(pivot row)	1		1	0	2	
pivot row		①	0	1	0	2nd derived system
(row 3)-1×(pivot row)			0	2	0	
	1		0	1	2	columns 3 and 4 are
		1	1	0	0	interchanged in the
			2	0	0	2nd derived system
row 1	1			1	2	
(row 2)-1×(pivot row)		1		0	0	3rd derived system
pivot row			①	0	0	

Since in the first derived system, the coefficient of x_2 is zero in the second and third equations, no pivot element could be selected from the second column. We had therefore selected the x_3 term from the second equation as the pivot term. We had to interchange the second and third columns in order to bring the pivot element into the diagonal position. At the same time we had to interchange x_2 and x_3 on the top of the table. Since the third coefficient in the third equation of the second derived system is also zero, the last pivot element had to be selected as the fourth coefficient of this equation. We had to interchange the third and fourth columns in order to bring the pivot element into the diagonal position again. In addition, we had to interchange x_2 and x_4 on the top of the table. From the third derived system we see that x_2 is the only free variable, and

$$x_1 = 2 - x_2, \ x_3 = 0, \ x_4 = 0$$

is the general solution.

$\blacklozenge$

Example 3.15. Consider next the system

$$
\begin{aligned}
x_1 &+ x_2 &+ x_3 &+ x_4 &= 2 \\
2x_1 &+ x_2 &+ 2x_3 &+ 2x_4 &= 3 \\
3x_1 &+ 2x_2 &+ 3x_3 &+ 3x_4 &= 5 \\
x_1 & &+ x_3 &+ x_4 &= 1.
\end{aligned}
$$

Table 3.5 shows the elimination process. From the last table we see that infinitely many solutions exist.

Table 3.5. Elimination of Example 3.15

	x_1	x_2	x_3	x_4		
	1	1	1	1	2	
	2	1	2	2	3	
	3	2	3	3	5	original system
	1	0	1	1	1	
pivot row	①	1	1	1	2	
(row 2)-2×(pivot row)		-1	0	0	-1	1st derived system
(row 3)-3×(pivot row)		-1	0	0	-1	
(row 4)-1×(pivot row)		-1	0	0	-1	
(row 1)-1×(pivot row)	1		1	1	1	2nd derived system
pivot row		①	0	0	1	last two equations
(row3)-(-1)×(pivot row)						should be deleted
(row4)-(-1)×(pivot row)						as meaningless with all zeros

The free variables are x_3 and x_4, and the general solution is:

$$x_1 = 1 - x_3 - x_4, \; x_2 = 1.$$

♦

The solution of matrix equations and inverting matrices are based on the repeated application of the elimination method. Since the coefficient matrix is common in the systems to be solved, the elimination process has to be performed only once, but several right-hand side vectors should be considered simultaneously. This idea is illustrated in the next two examples.

Example 3.16. We will first solve the matrix equation

$$\begin{pmatrix} 1 & 1 & 1 \\ 1 & 2 & 1 \\ 1 & 1 & 2 \end{pmatrix} \cdot \mathbf{X} = \begin{pmatrix} 3 & 4 \\ 4 & 5 \\ 4 & 6 \end{pmatrix},$$

which was the subject earlier in Example 3.8. In the elimination table the entire right-hand side matrix has to replace the right-hand side vector. The process is shown in Table 3.6.

Table 3.6. Elimination of Example 3.16

	x_1	x_2	x_3			
	1	1	1	3	4	
	1	2	1	4	5	original system
	1	1	2	4	6	
pivot row	①	1	1	3	4	
(row 2)-1×(pivot row)		1	0	1	1	1st derived system
(row 3)-1×(pivot row)		0	1	1	2	
(row 1)-1×(pivot row)	1		1	2	3	
pivot row		①	0	1	1	2nd derived system
row 3			1	1	2	
(row 1)-1×(pivot row)	1			1	1	
row 2		1		1	1	3rd derived system
pivot row			①	1	2	

The "right hand side matrix" gives the solution:

$$\mathbf{X} = \begin{pmatrix} 1 & 1 \\ 1 & 1 \\ 1 & 2 \end{pmatrix}.$$

♦

Example 3.17. We will next find the inverse of matrix

$$\mathbf{A} = \begin{pmatrix} 2 & 1 & 3 \\ 4 & 4 & 7 \\ 2 & 5 & 9 \end{pmatrix}.$$

The elimination procedure is given in Table 3.7. The identity matrix is placed as the right-hand matrix. From the last derived system we see that

$$\mathbf{A}^{-1} = \begin{pmatrix} \dfrac{1}{16} & \dfrac{3}{8} & -\dfrac{5}{16} \\ -\dfrac{11}{8} & \dfrac{3}{4} & -\dfrac{1}{8} \\ \dfrac{3}{4} & -\dfrac{1}{2} & \dfrac{1}{4} \end{pmatrix}.$$

Table 3.7. Elimination of Example 3.17

	x_1	x_2	x_3				
	2	1	3	1	0	0	original system
	4	4	7	0	1	0	
	2	5	9	0	0	1	
pivot row	①	1/2	3/2	1/2	0	0	1st derived system
(row 2)-4×(pivot row)		2	1	-2	1	0	
(row 3)-2×(pivot row)		4	6	-1	0	1	
(row1)-1/2×(pivot row)	1		5/4	1	-1/4	0	2nd derived system
pivot row		①	1/2	-1	1/2	0	
(row 3)-4×(pivot row)			4	3	-2	1	
(row1)-5/4×(pivot row)	1			1/16	3/8	-5/16	3rd derived system
(row2)-1/2×(pivot row)		1		-11/8	3/4	-1/8	
pivot row			①	3/4	-1/2	1/4	

♦

We conclude this section with the example of a matrix which has no inverse.

Example 3.18. Consider matrix

$$\mathbf{A} = \begin{pmatrix} 1 & 1 & 1 \\ 2 & 3 & 3 \\ 3 & 4 & 4 \end{pmatrix}.$$

The elimination process is shown in Table 3.8. The process terminates at the second derived system, since the last coefficient of the third equation is zero, so no more pivot element can be selected. The right-hand side numbers have nonzero values in the last equation, which implies that no solution exists. We have seen earlier that a matrix inverse is unique if it exists. Therefore, no such case may occur during matrix inversion that would indicate the existence of multiple solutions.

Table 3.8. Elimination of Example 3.18

	x_1	x_2	x_3				
	1	1	1	1	0	0	original
	2	3	3	0	1	0	system
	3	4	4	0	0	1	
pivot row	①	1	1	1	0	0	1st derived
(row 2)-2×(pivot row)		1	1	-2	1	0	system
(row 3)-3×(pivot row)		1	1	-3	0	1	
(row 1)-1×(pivot row)	1		0	3	-1	0	2nd
pivot row		①	1	-2	1	0	derived
(row 3)-1×(pivot row)			0	-1	-1	1	system

♦

In the discussions on elimination as well as in the above examples we always assumed that the pivot element was selected from the diagonal of the coefficient matrix, otherwise by interchanging rows and columns (which is equivalent to interchanging equations and unknowns) the pivot element was "moved" into a diagonal position. This interchange has certain advantages in simplifying the matrix notation of the solution, however in practical cases it is not necessary. If a non-diagonal pivot element is selected, then the coefficients of the same variable have to be eliminated from all other equations. When the elimination process terminates, the final coefficient matrix will not necessarily contain the identity matrix, but it will always contain a matrix that can be obtained from the identity matrix by column interchanges. The solution of the system can be then obtained in the same way as shown before with the minor difference that the order in which the unknowns are determined might differ from the natural order. This modified procedure is illustrated in the following example.

Example 3.19. We will now solve equations

$$2x_1 + x_2 - x_3 = 3$$
$$x_1 + x_2 + x_3 = 2$$
$$x_1 - x_2 + 2x_3 = 0.$$

The elimination process is shown in Table 3.9.

Introduction to Matrix Theory

Table 3.9. Elimination of Example 3.19

	x_1	x_2	x_3		
	2	1	-1	3	
	1	1	1	2	original system
	1	-1	2	0	
(row 1)-2×(pivot row)	0	-1	-3	-1	
pivot row	①	1	1	2	1st derived system
(row 3)-1×(pivot row)	0	-2	1	-2	
pivot row	0	①	3	1	
(row 2)-1×(pivot row)	1	0	-2	1	2nd derived system
(row3)-(-2)×(pivot row)	0	0	7	0	
(row 1)-3×(pivot row)	0	1	0	1	
(row2)-(-2)×(pivot row)	1	0	0	1	3rd derived system
pivot row	0	0	①	0	

The first equation of the last derived system shows that $x_2 = 1$, the second equation implies that $x_1 = 1$, and from the third equation we obtain $x_3 = 0$.

♦

The elimination method discussed in this section is known as the Gauss-Jordan elimination, where all nonzero elements under and above the pivot elements are eliminated (made zero).

This process has an important application in optimization, since it is the basis of the simplex method to solve linear programming problems.

A less elegant variant of Gauss-Jordan elimination is defined when nonzero elements are eliminated only under the diagonal. Consider the most general final table of Case 4. If no elimination is done above the diagonal, then above the k diagonal unities nonzero values might show up. If at least one of the right hand side values $a_{k+1,n+1},\dots,a_{m,n+1}$ is nonzero, then no solution exists. Otherwise x_{i_k} can be determined from the last nonzero row, maybe in terms of the free variables. Then this solution is substituted into the $(k-1)^{st}$ equation to get $x_{i_{k-1}}$. In the same way the solution can be obtained in reverse order $x_{i_{k-1}},\dots,x_{i_2},\dots,x_{i_2},x_{i1}$.

This method is called the Gauss elimination which consists of two stages. The *elimination* part provides the final table, and the successive determination of the solution components is called *back substitution*.

3.8 Applications

In this section some particular problems will be introduced and solved by using the methodology of this chapter.

1. The first model is known as *Leontief's input-output* system. Consider an economy of n productive sectors where n kinds of goods are produced, traded, and consumed. The goods are labeled by $i = 1, 2, ..., n$. It is assumed that each sector produces a single kind of good, so no joint production prevails. Distinct sectors produce distinct kinds of goods, therefore there is a one-to-one correspondence between goods and the sectors. The sector producing the ith good is also denoted by i. In each sector, production transforms several (possible all) kinds of goods in some quantities into a single kind of good in some amount. It is also assumed that to produce one unit of the jth good, a_{ij} units of the ith good $(i = 1, 2, ..., n)$ are needed as inputs for sector j.

The a_{ij} quantities are called the *input coefficients*. Let x_i denote the output of the ith good per unit of time, say, per annum. This x_i is the gross output, and part of it is consumed as input needed for production activities. The net output y_i of good i can be obtained as the difference of the gross output and the total amount of good i consumed by all sectors as input:

$$y_i = x_i - \sum_{j=1}^{n} a_{ij} x_j \quad (i = 1, 2, ..., n). \tag{3.20}$$

Introduce the following notation:

$$\mathbf{y} = \begin{pmatrix} y_1 \\ y_2 \\ \vdots \\ y_n \end{pmatrix}, \quad \mathbf{x} = \begin{pmatrix} x_1 \\ x_2 \\ \vdots \\ x_n \end{pmatrix}, \quad \mathbf{A} = \begin{pmatrix} a_{11} & a_{12} & \cdots & a_{1n} \\ a_{21} & a_{22} & \cdots & a_{2n} \\ \vdots & \vdots & & \vdots \\ a_{n1} & a_{n2} & \cdots & a_{nn} \end{pmatrix}.$$

Vectors $\mathbf{x}$ and $\mathbf{y}$ are called the gross and net output vectors, respectively, and matrix $\mathbf{A}$ is called the input coefficient matrix. Equations (3.20) can be written in an equivalent form:

$$\mathbf{x} - \mathbf{A}\mathbf{x} = \mathbf{y},$$

that is,

$$(\mathbf{I}_n - \mathbf{A})\mathbf{x} = \mathbf{y}. \tag{3.21}$$

If the gross production vector is known, then the corresponding net production vector is obtained by multiplying matrix $\mathbf{I}_n - \mathbf{A}$ by vector $\mathbf{x}$. It is an important question in economic planning to determine the gross production vector for given net productions. That is, vector $\mathbf{x}$ has to be determined for a given matrix $\mathbf{A}$ and a net production vector $\mathbf{y}$.

Notice that (3.21) is a system of linear equations with coefficient matrix $\mathbf{I}_n - \mathbf{A}$, right hand side vector $\mathbf{y}$, and unknown vector $\mathbf{x}$. Hence, the elimination method introduced in the previous section can be directly used to find vector $\mathbf{x}$.

Example 3.20. As an illustration, consider a three-sector economy with input coefficient matrix

$$\mathbf{A} = \begin{pmatrix} \dfrac{1}{10} & \dfrac{2}{10} & \dfrac{2}{10} \\ \dfrac{2}{10} & \dfrac{1}{10} & \dfrac{2}{10} \\ \dfrac{2}{10} & \dfrac{2}{10} & \dfrac{1}{10} \end{pmatrix}.$$

Assume that an economic plan specifies the net output vector

$$\mathbf{y} = \begin{pmatrix} 310 \\ 200 \\ 90 \end{pmatrix}.$$

Then the gross output vector is the solution of equation (3.21), which has now the form

$$\frac{9}{10}x_1 - \frac{2}{10}x_2 - \frac{2}{10}x_3 = 310$$

$$-\frac{2}{10}x_1 + \frac{9}{10}x_2 - \frac{2}{10}x_3 = 200$$

$$-\frac{2}{10}x_1 - \frac{2}{10}x_2 + \frac{9}{10}x_3 = 90.$$

The elimination process can be applied, and the solution

$$x_1 = 500, \quad x_2 = 400, \quad \text{and} \quad x_3 = 300$$

can be found. The details of elimination are left as an exercise.

$\blacklozenge$

We mention here that a comprehensive analysis of linear input-output models is given in Nikaido (1968).

2. The second model to be discussed here is known as *polynomial interpolation*. Assume that $x_1, x_2, \ldots, x_n$ are given different real numbers, and $f_1, f_2, \ldots, f_n$ are arbitrary real values. The f_i values are not necessarily different. The x_i values can be interpreted as the values of the independent variable where a certain unknown real function is measured, and for $i = 1, 2, \ldots, n$, f_i is the measured function value at x_i. The unknown function is approximated by a polynomial of least degree that is consistent with the measurements. That is, a polynomial p of degree as small as possible is determined such that for $i = 1, 2, \ldots, n$, $p(x_i) = f_i$. We have therefore n equations. In order to have the same number of unknowns as equations, a polynomial of degree $n - 1$ is selected:

$$p(x) = a_0 + a_1 x + a_2 x^2 + \ldots + a_{n-1} x^{n-1}.$$

The unknown coefficients must satisfy the equations:

$$a_0 + a_1 x_i + a_2 x_i^2 + \ldots + a_{n-1} x_i^{n-1} = f_i \quad (i = 1, 2, \ldots, n),$$

which is a system of n linear equations for n unknowns. Using matrix and vector notation, these equations can be summarized as follows:

$$\begin{pmatrix} 1 & x_1 & x_1^2 & \ldots & x_1^{n-1} \\ 1 & x_2 & x_2^2 & \ldots & x_2^{n-1} \\ \vdots & \vdots & \vdots & & \vdots \\ 1 & x_n & x_n^2 & \ldots & x_n^{n-1} \end{pmatrix} \begin{pmatrix} a_0 \\ a_1 \\ \vdots \\ a_{n-1} \end{pmatrix} = \begin{pmatrix} f_1 \\ f_2 \\ \vdots \\ f_n \end{pmatrix} \tag{3.22}$$

It can be proved that this system has a unique solution, if the x_i values are different (that is, when no repeated measurement is used). For details, see for example, Szidarovszky and Yakowitz (1978). The resulting polynomial is known as the *interpolation polynomial*.

Example 3.21. Consider four measurements:

$$x_1 = 0, \quad x_2 = 1, \quad x_3 = 2, \quad x_4 = 3$$

and

$$f_1 = 0, \quad f_2 = 2, \quad f_3 = 10, \quad f_4 = 30.$$

The unique cubic interpolation polynomial can be determined by solving system (3.22), which has now the form:

$$\begin{pmatrix} 1 & 0 & 0 & 0 \\ 1 & 1 & 1 & 1 \\ 1 & 2 & 4 & 8 \\ 1 & 3 & 9 & 27 \end{pmatrix} \begin{pmatrix} a_0 \\ a_1 \\ a_2 \\ a_3 \end{pmatrix} = \begin{pmatrix} 0 \\ 2 \\ 10 \\ 30 \end{pmatrix}.$$

Easy calculation (which is left as an exercise) shows that the solution is the following:

$$a_0 = 0, \quad a_1 = 1, \quad a_2 = 0, \quad \text{and} \quad a_3 = 1.$$

Thus, the interpolation polynomial is the following:

$$p(x) = x + x^3.$$

♦

We mention here that alternative methods to find interpolation polynomials can be found in the numerical analysis literature. See, for example, Szidarovszky and Yakowitz (1978).

3. From the previous case study we know that by increasing the number of measurement points the degree of the interpolation polynomial also increases. In many applications, instead of finding the high degree interpolation polynomial with exact fit, a polynomial of lower degree is determined which is not required to fit the data points exactly, but which fits the data as well as possible. Depending on the selection of the measure of the goodness of fit, several concepts have been developed. The most popular approach is the *least squares method*, which can be described as follows. Assume again that $x_1, x_2, \ldots, x_n$ are different real numbers, and $f_1, f_2, \ldots, f_n$ are arbitrary real values. A polynomial

$$p(x) = a_0 + a_1 x + \ldots + a_m x^m \qquad (3.23)$$

of fixed degree $m \leq n - 1$ has to be determined which minimizes the sum of the squared residuals:

$$Q = \sum_{i=1}^{n} \left(f_i - p(x_i) \right)^2 = \sum_{i=1}^{n} \left(f_i - \sum_{j=0}^{m} a_j x_i^j \right)^2.$$

Since for all i, x_i and f_i are known, this quantity depends on only the $m + 1$ unknown coefficients $a_0, a_1, \ldots, a_m$. It can be proved (see, for example, Ross, 1987) that this function has a unique minimizer, and it can be obtained by solving the system

$$\left(\mathbf{X}^T \mathbf{X} \right) \mathbf{a} = \mathbf{X}^T \mathbf{f} \qquad (3.24)$$

where

$$X = \begin{pmatrix} 1 & x_1 & x_1^2 & \cdots & x_1^m \\ 1 & x_2 & x_2^2 & \cdots & x_2^m \\ \vdots & \vdots & \vdots & & \vdots \\ 1 & x_n & x_n^2 & \cdots & x_n^m \end{pmatrix}, \quad a = \begin{pmatrix} a_0 \\ a_1 \\ \vdots \\ a_m \end{pmatrix}, \quad f = \begin{pmatrix} f_1 \\ f_2 \\ \vdots \\ f_n \end{pmatrix}.$$

The solution of this system provides the values of the unknown coefficients $a_0, a_1, \ldots, a_m$ of the least squares polynomial (3.23).

Example 3.22. As the illustration of the application of system (3.24), we will consider again the data of the previous example, and we will find the quadratic least squares polynomial based on the given data set. Since $m = 2$, in this case,

$$X = \begin{pmatrix} 1 & 0 & 0 \\ 1 & 1 & 1 \\ 1 & 2 & 4 \\ 1 & 3 & 9 \end{pmatrix}, \quad a = \begin{pmatrix} a_0 \\ a_1 \\ a_2 \end{pmatrix}, \quad f = \begin{pmatrix} 0 \\ 2 \\ 10 \\ 30 \end{pmatrix}.$$

Simple calculation shows that

$$X^T X = \begin{pmatrix} 1 & 1 & 1 & 1 \\ 0 & 1 & 2 & 3 \\ 0 & 1 & 4 & 9 \end{pmatrix} \begin{pmatrix} 1 & 0 & 0 \\ 1 & 1 & 1 \\ 1 & 2 & 4 \\ 1 & 3 & 9 \end{pmatrix} = \begin{pmatrix} 4 & 6 & 14 \\ 6 & 14 & 36 \\ 14 & 36 & 98 \end{pmatrix},$$

$$X^T f = \begin{pmatrix} 1 & 1 & 1 & 1 \\ 0 & 1 & 2 & 3 \\ 0 & 1 & 4 & 9 \end{pmatrix} \begin{pmatrix} 0 \\ 2 \\ 10 \\ 30 \end{pmatrix} = \begin{pmatrix} 42 \\ 112 \\ 312 \end{pmatrix},$$

therefore equation (3.24) can be written as

$$\begin{pmatrix} 4 & 6 & 14 \\ 6 & 14 & 36 \\ 14 & 36 & 98 \end{pmatrix} \begin{pmatrix} a_0 \\ a_1 \\ a_2 \end{pmatrix} = \begin{pmatrix} 42 \\ 112 \\ 312 \end{pmatrix}.$$

The application of the elimination process (the details are left as an exercise) shows that

$$a_0 = \frac{3}{10}, \; a_1 = -\frac{37}{10}, \quad \text{and} \quad a_2 = \frac{45}{10}.$$

Hence, the least squares polynomial is given as

$$p(x) = \frac{3}{10} - \frac{37}{10}x + \frac{9}{2}x^2.$$

◆

Finally, we mention that further details on the least squares method (including alternative procedures) can be found in the numerical analysis as well as in the statistics literature. See for example, Szidarovszky and Yakowitz (1978) or Ross (1987).

4. In applying interpolations or least square polynomials in function approximations, the same polynomial is used in the entire domain of the functions. In many applications piecewise polynomial approximations are used, where in each segment the degree of approximation is low. The most popular such approximation is known as cubic splines. Let $x_1, x_2, \ldots, x_n$ be the distinct abcissa values with corresponding function values $f_1, f_2, \ldots, f_n$. In each segment $[x_i, x_{i+1}]$ a cubic approximation is constructed as follows:

$$s_i(x) = a_i + b_i(x - x_i) + c_i(x - x_i)^2 + d_i(x - x_i)^3 \qquad (1 \le i \le n-1),$$

where the unknown coefficients are determined by the following requirements.

a) Continuity and interpolation require that

$$s_i(x_{i+1}) = s_{i+1}(x_{i+1}) = f_i.$$

b) The approxmation has to be twice differentiable:

$$s_i'\left(x_{i+1}\right) = s_{i+1}'\left(x_{i+1}\right) \text{ and } s_i''\left(x_{i+1}\right) = s_{i+1}''\left(x_{i+1}\right).$$

Requirement *a*) can be rewritten as

$$f_{i+1} - a_i + b_i h_i + c_i h_i^2 + d_i h_i^3 = a_{i+1} \text{ with } h_i = x_{i+1} - x_i, \quad (3.25)$$

and assumption b) can be expressed as

$$b_i + 2c_i h_i + 3d_i h_i^2 = b_{i+1} \text{ and } c_i + 3d_i h_i = c_{i+1}, \quad (3.26)$$

Solving the second equation of (3.26) for d_i and substituting it into (3.25) and to the first equation of (3.26) yields

$$a_{i+1} = a_i + b_i h_i + c_i h_i^2 + \frac{c_{i+1} - c_i}{3h_i} h_i^3 = a_i + b_i h_i + \frac{h_i^2}{3}\left(2c_i + c_{i+1}\right) \quad (3.27)$$

and

$$b_{i+1} = b_i + 2c_i h_i + 3\frac{c_{i+1} - c_i}{3h_i} h_i^2 = b_i + h_i\left(c_i + c_{i+1}\right) \quad (3.28)$$

From (3.27) we have

$$b_i = \frac{a_{i+1} - a_i}{h_i} - \frac{h_i}{3}\left(2c_i + c_{i+1}\right)$$

and by replacing *i* by *i*–1 we get

$$b_{i-1} = \frac{a_i - a_{i-1}}{h_{i-1}} - \frac{h_{i-1}}{3}\left(2c_{i-1} + c_i\right)$$

Substituting the last two equations into (3.28) and using the fact that $a_i = f_i$ for $i = 1, 2, \ldots, n - 1$ we obtain

$$h_{i-1}c_{i-1} + 2\left(h_{i-1} + h_i\right)c_i + h_i c_{i+1} = 3\frac{f_{i+1} - f_i}{h_i} - 3\frac{f_i - f_{i-1}}{h_{i-1}} \quad (3.29)$$

f_i for $i = 2, 3, \ldots, n-2$. These equations give $n-3$ relations for the $n-1$ unknown c_i values. In order to have a unique solution we need two more equations. In the natural spline we require that the second derivative vanishes at the first and last points:

$$s_1''(x_1) = 2c_1 + 6d_1(x_1 - x_1) = 2c_1 = 0$$

and

$$s''_{n-1}(x_n) = 2c_{n-1} + 6d_{n-1}h_{n-1} = 0$$

that is, $c_1 = 0$ and $c_{n-1} + 3d_{n-1}h_{n-1} = 0$. After the values of c_i are obtained, then

$$a_i = f_i$$
$$b_i = \frac{f_{i+1} - f_i}{h_i} - \frac{h_i}{3}(2c_i + c_{i+1}) \tag{3.30}$$

and

$$d_i = \frac{c_{i+1} - c_i}{3h_i}$$

The conditions on c_1 and c_{n-1} are easily satisfied if we introduce the (extraneous) parameter $c_n = 0$, then (3.29) holds for $i = 2, 3, \ldots, n-1$. Hence, we have to solve first the system of equations

$$2(h_1 + h_2)c_2 + h_2c_3 = 3\frac{f_3 - f_2}{h_2} - 3\frac{f_2 - f_1}{h_1}$$

$$h_2c_2 + 2(h_2 + h_3)c_3 + h_3c_4 = 3\frac{f_4 - f_3}{h_3} - 3\frac{f_3 - f_2}{h_2}$$

$$\ddots \qquad\qquad \vdots \tag{3.31}$$

$$h_{n-3}c_{n-3} + 2(h_{n-3} + h_{n-2})c_{n-2} + h_{n-2}c_{n-1} = 3\frac{f_{n-1} - f_{n-2}}{h_{n-2}} - 3\frac{f_{n-2} - f_{n-3}}{h_{n-3}}$$

$$h_{n-2}c_{n-2} + 2(h_{n-2} + h_{n-1})c_{n-1} = 3\frac{f_n - f_{n-1}}{h_{n-1}} - 3\frac{f_{n-1} - f_{n-2}}{h_{n-2}}$$

for $c_1, c_2, \ldots, c_{n-1}$ and then the other coefficients are obtained by (3.30).

Example 3.23. Assume

$$x_1 = -1,\ x_2 = 0,\ x_3 = 1 \text{ and } f_1 = 2,\ f_2 = 1, f_3 = 2.$$

In this case $n = 3$, $h_1 = h_2 = 1$, $\dfrac{f_2 - f_1}{h_1} = -1$ and $\dfrac{f_3 - f_2}{h_2} = 1$.

Since $c_1 = c_3 = 0$ we have a single equation for c_2,

$$h_1 c_1 + 2(h_1 + h_2)c_2 + h_2 c_3 = 3\frac{f_3 - f_2}{h_2} - 3\frac{f_2 - f_1}{h_1}$$

or

$$4c_2 = 3 \cdot (1) - 3 \cdot (-1) = 6,$$

so

$$c_2 = \frac{3}{2}.$$

Then $a_1 = 2$, $a_2 = 1$, and

$$b_1 = \frac{f_2 - f_1}{h_1} - \frac{h_1}{3}(2c_1 + c_2) = -1 - \frac{1}{3} \cdot \frac{3}{2} = -\frac{3}{2},$$

$$b_2 = \frac{f_3 - f_2}{h_2} - \frac{h_2}{3}(2c_2 + c_3) = 1 - \frac{1}{3} \cdot 3 = 0,$$

$$d_1 = \frac{c_2 - c_1}{3h_1} = \frac{\frac{3}{2} - 0}{3} = \frac{1}{2}$$

and

$$d_2 = \frac{c_3 - c_2}{3h_2} = \frac{0 - \frac{3}{2}}{3} = -\frac{1}{2}.$$

Hence the cubic spline is

$$s(x) = \begin{cases} 2 - \dfrac{3}{2}(x+1) + \dfrac{1}{2}(x+1)^3 & \text{if } -1 \le x \le 0 \\ 1 + \dfrac{3}{2}x^2 - \dfrac{1}{2}x^3 & \text{if } 0 \le x \le 1 \end{cases}$$

Example 3.24. Consider now the following data

x_i	−1	0	1	2
f_i	−1	0	1	8

which are data from the the $f(x) = x^3$ function. Since it does not satisfy the two conditions on $f''(-1)$ and $f''(2)$, it is not the natural cubic spline.

Table 3.10 shows the data, the values of h_i and $\dfrac{f_{i+1} - f_i}{h_i}$

Table 3.10. Partial results of Example 3.24

i	x_i	f_i	h_i	$\dfrac{f_{i+1}-f_i}{h_i}$
1	−1	−1	1	1
2	0	0	1	1
3	1	1	1	7
4	2	8		

Since $n = 4$ and $c_1 = c_4 = 0$, system (3.31) has the form

$$4c_2 + c_3 = 3 \cdot 1 - 3 \cdot 1 = 0$$
$$c_2 + 4c_3 = 3 \cdot 7 - 3 \cdot 1 = 18$$

implying that

$$c_2 = -\frac{6}{5} \quad \text{and} \quad c_3 = \frac{24}{5}.$$

Therefore

$$b_1 = 1 - \frac{1}{3}\left(2 \cdot 0 - \frac{6}{5}\right) = \frac{7}{5},$$

$$b_2 = 1 - \frac{1}{3}\left(-2 \cdot \frac{6}{5} + \frac{24}{5}\right) = \frac{1}{5},$$

$$b_3 = 7 - \frac{1}{3}\left(2 \cdot \frac{24}{5} + 0\right) = \frac{19}{5},$$

$$d_1 = \frac{-\dfrac{6}{5} - 0}{3} = -\frac{2}{5},$$

$$d_2 = \frac{\dfrac{24}{5} + \dfrac{6}{5}}{3} = 2,$$

$$d_3 = \frac{0 - \dfrac{24}{5}}{3} = -\frac{8}{5},$$

so the cubic spline is as follows

$$s(x) = \begin{cases} -1 + \dfrac{7}{5}(x+1) - \dfrac{2}{5}(x+1)^3 & \text{if } -1 \le x \le 0, \\[2mm] \dfrac{1}{5}x - \dfrac{6}{5}x^2 + 2x^3 & \text{if } 0 \le x \le 1, \\[2mm] 1 + \dfrac{19}{5}(x-1) + \dfrac{24}{5}(x-1)^2 - \dfrac{8}{5}(x-1)^3 & \text{if } 1 \le x \le 2. \end{cases}$$

◆

5. In application 6 of Section 1.5 we have seen that the material cost of the three kinds of biscuits can be obtained from the ingredient prices as

$$c_1 = 0.5p_1 + 0.2p_2 + 0.3p_3$$
$$c_2 = 0.55p_1 + 0.25p_2 + 0.2p_3$$
$$c_3 = 0.6p_1 + 0.15p_2 + 0.25p_3.$$

If the material costs are given, then the prices p_1, p_2, p_3 can be determined by solving the above system of linear equations. As an example, assume that

$$c_1 = 2.3, \ c_2 = 2.2, \text{ and } c_3 = 2.25,$$

then the application of the elimination method shows that

$$p_1 = 2, \ p_2 = 2, \text{ and } p_3 = 3.$$

6. Assume now that at any time period $t = 0, 1, 2, \ldots,$ a system is in one of the states $S_1, S_2, \ldots, S_n$. For example, the weather in a city could be in one of the following possible states: sunny and cold, sunny and warm, cloudy and cold, cloudy and warm, raining and cold, raining and warm. It is also assumed that the transition from one state to another is not deterministic, it can be specified in terms of probabilities that depend on the previous history of the system. In the special case, when the probability of each state depends only on the immidiate history of the system, the process is called a *Markov chain*. The mathematical model is

based on the *transition matrix* **P** which is defined in the following way. The matrix is $n \times n$, and its (i, j)-element P_{ij} gives the probability that if the system is in state j at any time period t, then it will be in state i at the next time period $t + 1$. Clearly,

$$P_{1j} + P_{2j} + ... + P_{nj} = 1,$$

that is, the sum of the elements of each column is one, and all matrix elements are nonnegative. At each time period t, let $x_1(t), x_2(t), ..., x_n(t)$ denote the probability of each state. That is

$$P(\text{state at time period } t = S_k) = x_k(t) \ (1 \le k \le n)$$

Clearly,

$$x_1(t) + x_2(t) + ... + x_n(t) = 1$$

and each number $x_k(t)$ is nonnegative. It is easy to see that for all $t \ge 0$,

$$\mathbf{x}(t+1) = \mathbf{P}\mathbf{x}(t), \tag{3.32}$$

that is, the probabilities of the different states can be recursively computed.

Example 3.25. Assume that

$$\mathbf{P} = \begin{pmatrix} 0.1 & 0.5 & 0.9 \\ 0.4 & 0.2 & 0 \\ 0.5 & 0.3 & 0.1 \end{pmatrix},$$

and at a certain time period $\mathbf{x}(t) = (0.1, 0.5, 0.4)^T$.
Then

$$\mathbf{x}(t+1) = \begin{pmatrix} 0.1 & 0.5 & 0.9 \\ 0.4 & 0.2 & 0 \\ 0.5 & 0.3 & 0.1 \end{pmatrix} \begin{pmatrix} 0.1 \\ 0.5 \\ 0.4 \end{pmatrix} = \begin{pmatrix} 0.62 \\ 0.14 \\ 0.24 \end{pmatrix},$$

that is, at time period $t + 1$, the system will be in S_1 with probability 0.62, in S_2 with probability 0.14, and in state S_3 with probability 0.24. Predicting the probabilities of the different state for the next time period is a simple task: a matrix-vector multiplication has to be performed. We can also raise the question of determining the probabilities of the different states for previous time periods. For example, we may compute $\mathbf{x}(t)$ from $\mathbf{x}(t+1)$ by solving a system of linear equations with coefficient matrix $\mathbf{P}$ and right-hand side vector $\mathbf{x}(t+1)$. Assume now that $\mathbf{x}(4) = (0.62, 0.14, 0.24)^T$, then $\mathbf{x}(3)$ is the solution of linear equations

$$\begin{pmatrix} 0.1 & 0.5 & 0.9 \\ 0.4 & 0.2 & 0 \\ 0.5 & 0.3 & 0.1 \end{pmatrix} \begin{pmatrix} x_1(3) \\ x_2(3) \\ x_3(3) \end{pmatrix} = \begin{pmatrix} 0.62 \\ 0.14 \\ 0.24 \end{pmatrix},$$

which has the usual form:

$$\begin{aligned} 0.1 \cdot x_1(3) + 0.5 \cdot x_2(3) + 0.9 \cdot x_3(3) &= 0.62 \\ 0.4 \cdot x_1(3) + 0.2 \cdot x_2(3) &= 0.14 \\ 0.5 \cdot x_1(3) + 0.3 \cdot x_2(3) + 0.1 \cdot x_3(3) &= 0.24. \end{aligned}$$

Simple calculation shows that the solution is

$$x_1(3) = 0.1, \; x_2(3) = 0.5, \quad \text{and} \quad x_3(3) = 0.4.$$

♦

The repeated application of equation (3.32) shows that for all $t \geq 0$ and $n \geq 1$,

$$\mathbf{x}(t+n) = \mathbf{P}^n \mathbf{x}(t) \tag{3.33}$$

giving a simple way to predict future state probabilities. In addition, by solving equation (3.33) for $\mathbf{x}(t)$, earlier state probabilities can be obtained. Notice also, that an efficient method to compute matrix powers $\mathbf{P}^n$ for large values of n is therefore very useful in examining Markov chains. In Chapter 6 we will introduce an efficient procedure.

The steady state probability vector of a Markov chain is a nonnegative vector $\mathbf{x}^*$ such that the sum of its elements is one, furthermore

$$\mathbf{x}^* = \mathbf{P}\mathbf{x}^*.$$

Hence $\mathbf{x}^*$ can be obtained by solving the linear equations

$$(\mathbf{I} - \mathbf{P})\mathbf{x}^* = \mathbf{0}$$
$$x_1^* + x_2^* + ... + x_n^* = 1. \tag{3.34}$$

Notice that $\mathbf{1}^T(\mathbf{I}-\mathbf{P}) = \mathbf{1}^T \mathbf{I} - \mathbf{1}^T \mathbf{P} = \mathbf{1}^T - \mathbf{1}^T = \mathbf{0}$, so the first n equations of (3.34) are linearly dependent. Therefore, they have at most $n-1$ linearly independent equations for the n unknowns, and at least one nonzero solution. It can be then normalized to satisfy the last equation.

7. Many applications in engineering require the solution of differential equations. In this application a special problem will be discussed which is known as the linear boundary value problem. We are looking for a twice continously differentiable function y: $[a, b] \rightarrow R$ which satisfies the second order linear ordinary differential equation

$$y''(x) + p(x)y'(x) + q(x)y(x) = r(x) \tag{3.35}$$

with boundary conditions $y(a) = A$ and $y(b) = B$. Here, p, q, r are given functions.

In applying the difference method, we divide interval $[a, b]$ into n equal parts with nodes $x_k = a + hk$ with $h = (b-a)/n$ and at each node x_k ($1 \le k \le n-1$) we use centered difference approximations for the derivatives as

$$y'(x_k) \approx \frac{y(x_{k+1}) - y(x_{k-1})}{2h} \tag{3.36}$$

and

$$y''(x_k) \approx \frac{y(x_{k+1}) - 2y(x_k) + y(x_{k-1})}{h^2} \tag{3.37}$$

Substituting these approximations into the differential equation (3.35) we obtain a system of linear equations for the unknown function values $y(x_1), y(x_2), \ldots, y(x_{n-1})$:

$$\frac{y(x_{k+1}) - 2y(x_k) + y(x_{k-1})}{h^2} + p_k \frac{y(x_{k+1}) - y(x_{k-1})}{2h} + q_k y(x_k) = r_k \quad (3.38)$$

for $k = 1, 2, \ldots, n-1$ where $p_k = p(x_k)$, $q_k = q(x_k)$, $r_k = r(x_k)$, $y(x_0) = y(a) = A$, and $y(x_n) = y(b) = B$.

Example 3.26. Consider problem

$$y'' - \left(1 + x^2\right)y = 1, \qquad y(0) = 1, y(1) = 3.$$

By selecting $n = 10$ we have $h = 0.1$ and system (3.38) has 9 equations for 9 unknowns. The results — obtained by a computer — are shown in Table 3.11.

Table 3.11. Results of Example 3.24.

x	0	0.1	0.2	0.3	0.4	0.5	0.6	0.7	0.8	0.9	1
y(x)	1.0000	1.0746	1.1701	1.2877	1.4294	1.5976	1.7958	2.0285	2.3013	2.6219	3.0000

♦

3.9 Exercises

1. Check if vector $\mathbf{x}$ can be expressed as a linear combination of $\mathbf{x}_1$, $\mathbf{x}_2$ and $\mathbf{x}_3$ if

a) $\mathbf{x}_1 = \begin{pmatrix} 1 \\ 1 \\ 1 \end{pmatrix}$, $\mathbf{x}_2 = \begin{pmatrix} 1 \\ 2 \\ 2 \end{pmatrix}$, $\mathbf{x}_3 = \begin{pmatrix} 2 \\ 3 \\ 3 \end{pmatrix}$, $\mathbf{x} = \begin{pmatrix} 1 \\ 4 \\ 4 \end{pmatrix}$;

b) $\mathbf{x}_1 = \begin{pmatrix} 1 \\ 1 \\ 1 \end{pmatrix}$, $\mathbf{x}_2 = \begin{pmatrix} 1 \\ 2 \\ 2 \end{pmatrix}$, $\mathbf{x}_3 = \begin{pmatrix} 2 \\ 3 \\ 3 \end{pmatrix}$, $\mathbf{x} = \begin{pmatrix} 1 \\ 4 \\ 5 \end{pmatrix}$.

2. Find the solutions, if they exist, of the following systems of equations:

a) $\begin{aligned} 2x_1 - x_2 + x_3 &= 3 \\ x_1 + x_2 + x_3 &= 2 \\ 3x_1 + x_2 - x_3 &= 2 \end{aligned}$

b) $\begin{aligned} x_1 + x_2 + x_3 + x_4 &= 2 \\ x_1 - x_2 + x_3 + 2x_4 &= 0 \\ 2x_1 + x_2 + x_3 - x_4 &= 3 \end{aligned}$

c) $\begin{aligned} x_1 + x_2 &= 2 \\ 2x_1 - x_2 &= 1 \\ 2x_1 + x_2 &= 3 \end{aligned}$

d) $\begin{aligned} x_1 + x_2 + x_3 + x_4 &= 2 \\ x_1 - x_2 + x_3 + 2x_4 &= 0. \end{aligned}$

3. Rewrite the systems of the previous Exercise in matrix form $\mathbf{Ax} = \mathbf{b}$.

4. Determine which of the following systems of linear equations are solvable:

a) $\begin{aligned} x_1 + 2x_2 + x_3 + x_4 &= 3 \\ -x_1 + x_2 - x_3 - x_4 &= 0 \end{aligned}$

b) $\begin{aligned} x_1 + x_2 + x_4 &= 2 \\ x_1 + x_3 + x_4 &= 2 \\ x_2 + x_3 + x_4 &= 1 \\ x_1 + x_2 + x_3 + x_4 &= 2 \end{aligned}$

c)
$$x_1 + 2x_2 + x_3 = 4$$
$$x_1 + x_2 + 2x_3 = 4$$
$$2x_1 + x_2 + x_3 = 4$$
$$x_1 + x_2 + x_3 = 2$$

d)
$$x_1 + x_2 + x_3 + x_4 = 1$$
$$x_1 - x_2 + x_3 - x_4 = 1$$
$$x_1 + 2x_2 + x_3 + 2x_4 = 2.$$

5. Check the solvability of the equations of Exercise 2 by using Theorem 3.3.

6. Check the solvability of the equations of Exercise 4 by using Theorem 3.3.

7. Find a basis of the solution space of the following homogeneous systems:

a)
$$x_1 + x_2 + x_3 = 0$$
$$2x_1 + x_2 + 2x_3 = 0$$

b)
$$x_1 + x_2 - x_3 - x_4 = 0$$
$$2x_1 + x_2 + x_3 - 2x_4 = 0$$
$$3x_1 + x_2 - x_3 - 3x_4 = 0$$

c)
$$x_1 + 2x_2 + x_3 + x_4 = 0$$
$$-x_1 + x_2 - x_3 - x_4 = 0.$$

8. What is the dimension of the solution space of the homogeneous system $\mathbf{Ax} = \mathbf{0}$, where $\mathbf{A}$ is a $2 \times n$ matrix?

9. Find the ranks of the coefficient matrices of the systems of Exercises 2 and 4.

10. Let **A** be an $m \times n$ real matrix, and assume that **U** has m columns and **V** has n rows. Are the following statements true?
    a) If the rows of **U** are linearly independent, then

$$\text{rank}(\mathbf{A}) = \text{rank}(\mathbf{UA});$$

    b) If the columns of **V** are linearly independent, then

$$\text{rank}(\mathbf{A}) = \text{rank}(\mathbf{AV}).$$

11. Solve the following systems of equations by using the elimination method

a) $\begin{aligned} x_1 + x_2 + x_3 &= 3 \\ x_1 - x_2 - x_3 &= 3 \\ 2x_1 + x_2 + 7x_3 &= 6 \end{aligned}$

b) $\begin{aligned} x_1 + 2x_2 + 3x_3 &= 4 \\ 2x_1 + 6x_2 + 10x_3 &= 14 \\ -x_1 - 3x_2 - 6x_3 &= 9 \end{aligned}$

c) $\begin{aligned} x_1 + 2x_2 + 3x_3 &= 4 \\ 2x_1 + 4x_2 + 9x_3 &= 4 \\ 4x_1 + 3x_2 + 2x_3 &= 1. \end{aligned}$

12. Solve the systems of the previous Exercise by the formula $\mathbf{x} = \mathbf{A}^{-1}\mathbf{b}$, where **A** is the coefficient matrix and **b** is the right hand side vector.

13. Illustrate Theorem 3.2 with the systems of Exercise 11.

14. Illustrate Theorem 3.13 with the coefficient matrices of Exercise 11.

15. Solve the following matrix equations:

a) $\begin{pmatrix} 1 & 1 & 1 \\ 1 & -1 & -1 \\ 2 & 1 & 7 \end{pmatrix} \mathbf{X} = \begin{pmatrix} 3 & 0 \\ 3 & 0 \\ 6 & 0 \end{pmatrix}$

b) $\begin{pmatrix} 1 & 2 & 3 \\ 2 & 6 & 10 \\ -1 & -3 & -6 \end{pmatrix} \mathbf{X} = \begin{pmatrix} 4 & 8 \\ 14 & 28 \\ 9 & 18 \end{pmatrix}$

c) $\begin{pmatrix} 1 & 2 & 3 \\ 2 & 4 & 9 \\ 4 & 3 & 2 \end{pmatrix} \mathbf{X} = \begin{pmatrix} 4 & 12 \\ 4 & 12 \\ 1 & 3 \end{pmatrix}.$

16. Solve the previous problems by using the solution formula $\mathbf{X} = \mathbf{A}^{-1}\mathbf{B}$, where $\mathbf{A}$ is the coefficient matrix and $\mathbf{B}$ is the right-hand side matrix.

17. Invert the coefficient matrices of Exercise 11.

18. Illustrate Theorem 3.11 with the matrix equations of Exercise 15.

19. Repeat Example 3.20 with input coefficient matrix

$$\mathbf{A} = \begin{pmatrix} 0.05 & 0.10 & 0.05 \\ 0.10 & 0.20 & 0.10 \\ 0.05 & 0.05 & 0.20 \end{pmatrix}$$

and output vector

$$\mathbf{y} = \begin{pmatrix} 100 \\ 200 \\ 300 \end{pmatrix}.$$

20. Find the interpolation polynomial based on data values

$$x_1 = -2, \; x_2 = -1, \; x_3 = 0, \; x_4 = 1$$

and

$$f_1 = 5, f_2 = 2, f_3 = 1, f_4 = 2.$$

21. By solving system (3.22) it is known from numerical analysis that the unique interpolation polynomial based on data set (x_i, f_i), $i = 1, 2, ..., n$, can be obtained as

$$p(x) = f_1 l_1(x) + f_2 l_2(x) + ... + f_n l_n(x),$$

where

$$l_i(x) = \frac{(x - x_1)...(x - x_{i-1})(x - x_{i+1})...(x - x_n)}{(x_i - x_1)...(x_i - x_{i-1})(x_i - x_{i+1})...(x_i - x_n)}$$

(see, for example, Szidarovszky and Yakowitz, 1978). Solve the numerical example of the previous Exercise using this formula.

22. (Continuation of Exercise 21). Explain why the existence of a unique interpolation polynomial is guaranteed when the coefficient matrix of system (3.22) is nonsingular which is the case if the values $x_1, x_2, ..., x_n$ are distinct.

23. Find the cubic spline base on the data values:

$$x_1 = -2, \ x_2 = 1, \ x_3 = 0, \ x_4 = 1, \ x_5 = 2,$$
$$f_1 = 3, \ f_2 = 1, \ f_3 = 2, \ f_4 = 1, \ f_5 = 2.$$

24. Solve the boundary value problem

$$y''(x) + y'(x) + y(x) = 1, \ y(0) = y(1) = 0,$$

Select $h = 0.1$ as stepsize.

25. The least squares problem discussed in Section 3.8 can be generalized as follows. Let $\mathbf{x}_1, \mathbf{x}_2, ..., \mathbf{x}_k$ be given n-element real vectors, and let $\mathbf{x} \in R^n$ be given. Find the linear combination $c_1 \mathbf{x}_1 + c_2 \mathbf{x}_2 + ... + c_k \mathbf{x}_k$ that has minimal distance from $\mathbf{x}$. This problem is usually solved as

follows. Define matrix $\mathbf{X} = (\mathbf{x}_1, \mathbf{x}_2, \ldots, \mathbf{x}_k)$, then the vector of coefficients $\mathbf{c} = (c_1, c_2, \ldots, c_k)^T$ is obtained as the solution of equation (3.24) as

$$\left(\mathbf{X}^T \mathbf{X}\right) \mathbf{c} = \mathbf{X}^T \mathbf{x}.$$

Repeat Example 2.33 and Exercise 2.28 by using this method.

Chapter 4

Determinants

4.1 Introduction

The most simple "system" of linear equations has only one equation and one unknown:

$$a_{11}x = b_1,$$

where a_{11} and b_1 are real or complex scalars. If $a_{11} \neq 0$, then a unique solution exists:

$$x = \frac{b_1}{a_{11}}. \qquad (4.1)$$

If $a_{11} = 0$ and $b_{11} \neq 0$, then there is no solution; and if $a_{11} = b_{11} = 0$, then x can take any arbitrary real or complex value.

Consider next the case of two equations with two unknowns:

$$a_{11}x_1 + a_{12}x_2 = b_1,$$
$$a_{21}x_1 + a_{22}x_2 = b_2,$$

which can be summarized as

$$\mathbf{Ax} = \mathbf{b}$$

with the 2×2 matrix $\mathbf{A} = (a_{ij})$ and the 2-element vector $\mathbf{b} = (b_i)$. The application of the elimination method gives the solution:

$$x_1 = \frac{b_1 a_{22} - a_{12} b_2}{a_{11} a_{22} - a_{12} a_{21}} \quad \text{and} \quad x_2 = \frac{a_{11} b_2 - b_1 a_{21}}{a_{11} a_{22} - a_{12} a_{21}} \tag{4.2}$$

if the denominator $a_{11} a_{22} - a_{12} a_{21}$ is nonzero. Otherwise, there is no solution if at least one of the numerators is nonzero, and there are infinitely many solutions if both numerators are equal to zero. Notice that the denominator depends on only the coefficient matrix. The numerator of x_1 can be obtained from the denominator if the first column of the coefficient matrix is replaced by the right-hand side vector. Similarly, the numerator of x_2 can be obtained from the denominator by replacing the second column of the coefficient matrix by the right-hand side vector. Since the numerator and denominator have the same structure, it will be useful in analysing systems of linear equations to examine these quantities and their main properties in detail. Notice first that the denominators are scalars, and are defined for square (1×1, 2×2) matrices. If $\mathbf{A}$ denotes a square matrix, then we will denote this quantity by $\det(\mathbf{A})$ and call it the *determinant* of matrix $\mathbf{A}$. From equations (4.1) and (4.2) we see that for 1×1 and 2×2 matrices

$$\det(a_{11}) = a_{11}$$

and

$$\det \begin{pmatrix} a_{11} & a_{12} \\ a_{21} & a_{22} \end{pmatrix} = a_{11} a_{22} - a_{12} a_{21}.$$

Consider next the case of three linear equations with three unknowns. Similar expressions are obtained for the solutions with a common denominator, which has to be therefore the definition of the determinant of 3×3 matrices:

$$\det \begin{pmatrix} a_{11} & a_{12} & a_{13} \\ a_{21} & a_{22} & a_{23} \\ a_{31} & a_{32} & a_{33} \end{pmatrix} = a_{11}\,a_{22}\,a_{33} + a_{13}\,a_{21}\,a_{32} + a_{12}\,a_{23}\,a_{31} - $$

$$a_{13}\,a_{22}\,a_{31} - a_{11}\,a_{23}\,a_{32} - a_{12}\,a_{21}\,a_{33}.$$

In this chapter the concept of determinants of square matrices will be first introduced for the general $n \times n$ case, and the main properties of determinants will be examined including their applications in solving systems of linear equations.

The general definition of the determinants of $n \times n$ matrices will be based on the common properties and characteristics of the above special cases.

Notice first that each term has n factors. In the 1×1 case only one element, a_{11}, gives the determinant; in the 2×2 case there are two terms, each term is a product of two matrix elements; and in the 3×3 case, there are six terms and each of them is the product of 3 matrix elements. Two questions have to be answered here: what rule decides which matrix elements are multiplied in each term, and how to determine the sign of each term. Consider first an arbitrary term of the above special determinant expressions. In the 1×1 case, the first index of the only term is 1, the second index is also 1. In the 2×2 case two matrix elements are multiplied in each term. The first indices of the two elements are in order 1 and 2 in both terms. The second indices in the first (positive) term are 1 and 2, and those is the second (negative) term are 2 and 1 as shown is the following diagram:

second indices

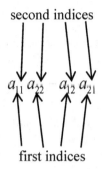

$$a_{11}\,a_{22}\quad a_{12}\,a_{21}$$

first indices

A similar structure is observed in the 3×3 case. In each term, the first indices are always 1, 2, and 3 in this natural order. In the positive terms, the second indices are

$$1,2,3;\ 3,1,2;\quad \text{and}\quad 2,3,1,$$

and in the negative terms they are

$$3,2,1;\ 1,3,2;\quad \text{and}\quad 2,1,3.$$

From the above description we realize that in each term, the first indices are the integers 1, or 1,2 or 1,2,3 always in increasing order, however the second indices form all permutations of these integers. In the 2×2 case, the two possible permutations $<1,2>$ and $<2,1>$ form the sequences of the second indices. In the 3×3 case there are $3! = 6$ possible permutations of the numbers 1,2,3, and the above six sequences of the second indices coincide with these permutations. We can summarize the above discussion by noticing that in the determinants of 2×2 and 3×3 matrices there are 2! and 3! terms, respectively, and each term has the form

$$a_{1i_1}a_{2i_2}\quad \text{or}\quad a_{1i_1}a_{2i_2}a_{3i_3},$$

where $<i_1, i_2>$ or $<i_1, i_2, i_3>$ is a permutation of the integers 1,2 or 1,2,3, respectively. In addition, each permutation of the integers 1,2 or 1,2,3 can be found exactly ones as a sequence of the second indices. In order to see how the sign of each term is selected, we have to examine the set of permutations associated to positive terms as well as that associated to the negative terms. In the 2×2 case, the permutation $<1,2>$

gives the positive term and <2,1> corresponds to the negative term. Notice that the first permutation is the natural (increasing) order of the integers 1,2. In the second permutation we have to interchange the elements 2 and 1 in order to obtain the natural order. In any permutation, the interchange of any two elements is called *inversion*. So, in the positive term no inversion is needed to obtain the natural order, and in the negative term one inversion is required. Consider next the positive terms in the 3×3 case. The first such permutation is <1,2,3> with no inversion required, since the elements are already in the natural order. The second such permutation is <3,1,2>. We need one inversion to move 1 to the first position. After interchanging 1 and 3, the resulting permutation is <1,3,2>, and an additional inversion (interchanging 3 and 2) is needed to obtain the natural order. In the case of the third such permutation two inversions are needed again:

<2,3,1> (interchange 2 and 1) $\rightarrow$ <1,3,2> (interchange 3 and 2) $\rightarrow$ <1,2,3>.

Consider next the negative terms. Permutation <3,2,1> needs only one inversion (interchanging 3 and 1) to obtain the natural order. The same number of inversions are needed in the case of the two other such permutations:

<1,3,2> (interchange 3 and 2) $\rightarrow$ <1,2,3>

and

<2,1,3> (interchange 2 and 1) $\rightarrow$ <1,2,3>.

Notice that in the case of positive terms 0 or 2 inversions are needed, and for negative terms only one is required. This notion can be generally formulated as follows.

Definition 4.1. Let $i_1, i_2, ..., i_n$ be a permutation of the integers $1, 2, ..., n$. Let π denote this permutation and let $I(\pi)$ be the minimum number of inversions needed to obtain the natural order from the given permutation. The *sign* of permutation π is denoted by sign (π) and is defined as

$$\text{sign}(\pi) = \begin{cases} 1 & \text{if } I(\pi) \text{ is even} \\ -1 & \text{if } I(\pi) \text{ is odd} \end{cases}$$

We mention that if $I(\pi)$ is even, then any (not necessarily minimal) rearrangement of permutation π into the increasing order always requires even number of inversions, and if $I(\pi)$ is odd, then all rearrangements require an odd number of inversions. Therefore, in determining the sign of a permutation we do not need to find the minimum number of inversions required to rearrange the numbers into natural order. It is sufficient to find one of such rearrangements and check if the number of inversions is even or odd.

Definition 4.2. The *determinant* of an $n \times n$ real or complex matrix $\mathbf{A} = (a_{ij})$ is defined as

$$\det(\mathbf{A}) = \sum_{\pi = <i_1, \ldots, i_n> \in P_n} \text{sign}(\pi) a_{1i_1} a_{2i_2} \ldots a_{ni_n}, \tag{4.3}$$

where P_n is the set all $n!$ permutations of the integers $1, 2, \ldots, n$. This definition is illustrated by three examples.

Example 4.1. Assume that $\mathbf{A}$ is a lower triangular matrix:

$$\mathbf{A} = \begin{pmatrix} a_{11} & 0 & 0 & \cdots & \cdots & 0 \\ a_{21} & a_{22} & 0 & \cdots & \cdots & 0 \\ \cdots & \cdots & \cdots & \cdots & \cdots & \cdots \\ a_{n-1,1} & a_{n-1,2} & a_{n-1,3} & & a_{n-1,n-1} & 0 \\ a_{n1} & a_{n2} & a_{n3} & \cdots & a_{n,n-1} & a_{nn} \end{pmatrix}.$$

We will show that $\det(\mathbf{A})$ is the product of the diagonal elements. Consider an arbitrary term of the sum (4.3):

$$\text{sign}(\pi) \; a_{1i_1} a_{2i_2} \ldots a_{ni_n}.$$

Notice first that $a_{12} = a_{13} = \ldots = a_{1n} = 0$, therefore a_{1i_1} is zero for $i_1 \neq 1$. Therefore, all such terms can be omitted. We may therefore assume that

$i_1 = 1$ Consider next the factor a_{2i_2}. Since $i_2 \neq i_1 = 1$ and with $i_2 = 3, 4,$ ..., n this matrix element is zero, all terms for $i_2 \neq 2$ are equal to zero. We may therefore assume that $i_2 = 2$.

Continue this argument for the third factor, then for the fourth factor, and so on, until the last factor to see that the only nonzero term is

$$a_{11} a_{22} ... a_{nn},$$

since sign $(\pi) = 1$ for the natural order $<1, 2, ..., n>$. Hence this is the value of det($\mathbf{A}$).

One may similarly prove that the determinant of an upper triangular matrix is also the product of the diagonal elements. In this case we have to consider the factors in the reverse order $a_{ni_n}, a_{n-1,i_{n-1}}, ..., a_{1i_1}$.

(This case will be also a simple consequence of Theorem 4.1 to be given in the next section, since the transpose of an upper triangular matrix is lower triangular).

◆

Example 4.2. Assume that $\mathbf{A}$ is a diagonal matrix. Then

$$\det(\mathbf{A}) = a_{11} a_{22} ... a_{nn},$$

since $\mathbf{A}$ is a special lower triangular matrix with zero elements above the diagonal. For example, det($\mathbf{I}_n$) = 1, where $\mathbf{I}_n$ denotes the $n \times n$ identity matrix.

◆

Example 4.3. Assume that the kth row of matrix $\mathbf{A}$ contains only zero elements. Then in each term of (4.3) the factor a_{ki_k} is zero, therefore the entire sum is zero. Hence, det($\mathbf{A}$)=0. One may similarly prove that det($\mathbf{A}$) = 0, if the kth column of $\mathbf{A}$ consists of only zero elements. Then the factor a_{1i_l} (with $i_l = k$) is zero in each term. (We may also apply Theorem 4.1 of the next Section.)

◆

The main properties of determinants will be next discussed and their main applications will be outlined.

4.2 Properties of Determinants

The first result to be discussed in this section can be stated as follows. If we interchange the columns of a matrix by its rows then the value of the determinant remains the same.

Theorem 4.1. $\det(\mathbf{A}) = \det(\mathbf{A}^T)$.

Proof. Let $\pi = <i_1, i_2, \ldots, i_n>$ be a permutation of the integers $1, 2, \ldots, n$. The inverse of permutation π is defined as the reordering of the numbers $i_1, i_2, \ldots, i_n$ into the natural order $1, 2, \ldots, n$, and it is denoted by π^{-1}. For example, the inverse of $\pi = <3,2,1>$ is itself. The inverse of $\pi = <3,2,1>$ is $<3,1,2>$, since in π, 1 is on the 3rd place, 2 is in the 1st place, and 3 is on the 2nd place. In general, the inverse of permutation $<i_1, i_2, \ldots, i_n>$ is $<j_1, j_2, \ldots, j_n>$, where $i_{j_1} = 1$, $i_{j_2} = 2$, $\ldots$, $i_{j_n} = n$. It is easy to see that if π runs through all permutations P_n, then the same is true for π^{-1}, furthermore $\text{sign}(\pi) = \text{sign}(\pi^{-1})$. The first statement follows from the facts that the inverses of different permutations are different, and there are exactly $n!$ different permutations of the integers $1, 2, \ldots, n$. The second statement is the consequence of the following simple observation. Consider a sequence of pair-wise inversions that transforms permutation π into natural order. If we apply the same sequence in the opposite order to the inverse permutation π^{-1}, then the resulting order of the integers becomes increasing again. Therefore

$$\det\left(\mathbf{A}^T\right) = \sum_{\pi = <i_1, i_2, \ldots, i_n> \in P_n} \text{sign}(\pi) a_{i_1 1} a_{i_2 2} \ldots a_{i_n n},$$

since the (k, i_k) element of $\mathbf{A}^T$ is $a_{i_k k}$. Let $<j_1, \ldots, j_n>$ denote the inverse of π, then by rearranging the summation with respect to π^{-1} we have

$$\det\left(\mathbf{A}^T\right) = \sum_{\pi^{-1} = <j_1, \ldots, j_n> \in P_n} \text{sign}(\pi^{-1}) a_{1 j_1} a_{2 j_2} \ldots a_{n j_n} = \det(\mathbf{A}).$$

♣

In many cases, as it will be illustrated in the next theorem, it is convenient to manipulate with the rows or the columns of the matrix the determinant of which is under consideration.

Let $\mathbf{A}$ be an $n \times n$ real or complex matrix. Let $\mathbf{r}_1^T, ..., \mathbf{r}_n^T$ be the rows of $\mathbf{A}$, and let $\mathbf{a}_1, \mathbf{a}_2, ..., \mathbf{a}_n$ denote the columns of $\mathbf{A}$. Then instead of $\det(\mathbf{A})$ we

may write $\det \begin{pmatrix} \mathbf{r}_1^T \\ \vdots \\ \mathbf{r}_n^T \end{pmatrix}$ or $\det(\mathbf{a}_1, ..., \mathbf{a}_n)$.

Theorem 4.2. The following properties of $\det(\mathbf{A})$ are valid:
(a) If two rows (or columns) of matrix $\mathbf{A}$ are interchanged, then the determinant of the new matrix is $-\det(\mathbf{A})$;
(b) If two rows (or columns) of matrix $\mathbf{A}$ are identical, then $\det(\mathbf{A}) = 0$;
(c) The value of $\det(\mathbf{A})$ is multiplied by c if a row (or column) of $\mathbf{A}$ is multiplied by c;
(d)

$$
\det \begin{pmatrix} \mathbf{r}_1^T \\ \cdots \\ \mathbf{r}_{k-1}^T \\ \mathbf{r}_k^T + \mathbf{r}_k'^T \\ \mathbf{r}_{k+1}^T \\ \cdots \\ \mathbf{r}_n^T \end{pmatrix} = \det \left(\begin{pmatrix} \mathbf{r}_1^T \\ \cdots \\ \mathbf{r}_{k-1}^T \\ \mathbf{r}_k^T \\ \mathbf{r}_{k+1}^T \\ \cdots \\ \mathbf{r}_n^T \end{pmatrix} + \begin{pmatrix} \mathbf{r}_1^T \\ \cdots \\ \mathbf{r}_{k-1}^T \\ \mathbf{r}_k'^T \\ \mathbf{r}_{k+1}^T \\ \cdots \\ \mathbf{r}_n^T \end{pmatrix} \right),
$$

and

$$
\det(\mathbf{a}_1, ..., \mathbf{a}_{k-1}, \mathbf{a}_k + \mathbf{a}_k', \mathbf{a}_{k+1}, ..., \mathbf{a}_n)
$$
$$
= \det(\mathbf{a}_1, ..., \mathbf{a}_{k-1}, \mathbf{a}_k, \mathbf{a}_{k+1}, ..., \mathbf{a}_n) + \det(\mathbf{a}_1, ..., \mathbf{a}_{k-1}, \mathbf{a}_k', \mathbf{a}_{k+1}, ..., \mathbf{a}_n);
$$

(e)The value of det($\mathbf{A}$) does not change, if a constant multiple of a row (or column) is added to another row (or column);

(f) det($\mathbf{A}$) $\neq$ 0 if and only if the rows (or columns) of $\mathbf{A}$ are linearly independent.

Proof. From Theorem 4.1 we know that interchanging the rows of a matrix by its columns the value of the determinant does not change. Therefore, it is sufficient to prove the assertions for rows or for columns only. We will next present the proofs for rows.

(a) If two rows if $\mathbf{A}$ are interchanged, then each term of sum (4.3) becomes

$$\text{sign}\left(\pi\right) a_{1i_1}...a_{li_l}...a_{ki_k}...a_{ni_n},$$

where $l > k$, $\pi = <i_1, i_2, ..., i_n>$, and permutation $<1, ..., k, ..., l, ..., n>$ is obtained from the increasing order $<1, 2, ..., n>$ by interchanging k and l. Since one pair-wise inversion changes the sign of the permutation, this term can be rewritten as

$$-\text{sign}\left(\pi'\right) a_{1i_1}...a_{ki_l}...a_{li_k}...a_{ni_n},$$

where $\pi' = <i_1,...,i_l,...,i_k,...,i_n>$ (that is, π' is obtained by interchanging i_l and i_k in π). Adding up all such terms the value of $-$ det($\mathbf{A}$) is obtained.

(b) By interchanging the identical rows, the matrix, as well as the value of its determinant remains the same. However, from property (a) we know that the value of the determinant changes sign. Therefore

$$\det(\mathbf{A}) = -\det(\mathbf{A}),$$

that is,

$$\det(\mathbf{A}) = 0.$$

(c) If row k of matrix $\mathbf{A}$ is multiplied by c, then each term of the sum (4.3) is multiplied by c, since the factor a_{ki_k} is replaced by $c \cdot a_{ki_k}$. Therefore, the entire sum is multiplied by c.

(d) Consider an arbitrary term of the sum (4.3):

$$\text{sign}(\pi)a_{1i_1}\ldots\left(a_{ki_k}+a'_{ki_k}\right)\ldots a_{ni_n}$$
$$=\text{sign}(\pi)a_{1i_1}\ldots a_{ki_k}\ldots a_{ni_n}+\text{sign}(\pi)a_{1i_1}\ldots a'_{ki_k}\ldots a_{ni_n}.$$

Adding up the first terms for all $\pi \in P_n$ gives

$$\det\begin{pmatrix}\mathbf{r}_1^T\\ \ldots\\ \mathbf{r}_k^T\\ \ldots\\ \mathbf{r}_n^T\end{pmatrix},$$

and the sum of the second terms for all $\pi \in P_n$ equals

$$\det\begin{pmatrix}\mathbf{r}_1^T\\ \ldots\\ \mathbf{r'}_k^T\\ \ldots\\ \mathbf{r}_n^T\end{pmatrix}.$$

(e) Using properties (d) and (b) we see that

$$\det\begin{pmatrix}\mathbf{r}_1^T\\ \ldots\\ \mathbf{r}_k^T+c\mathbf{r}_l^T\\ \ldots\\ \mathbf{r}_n^T\end{pmatrix}=\det\begin{pmatrix}\mathbf{r}_1^T\\ \ldots\\ \mathbf{r}_k^T\\ \ldots\\ \mathbf{r}_n^T\end{pmatrix}+c\cdot\det\begin{pmatrix}\mathbf{r}_1^T\\ \ldots\\ \mathbf{r}_l^T\\ \ldots\\ \mathbf{r}_n^T\end{pmatrix}.$$

The first term is the determinant of the original matrix. In the second term, $\mathbf{r}_l^T$ is in the kth row however the same vector $\mathbf{r}_l^T$ is also present in the lth row, therefore two rows are identical. Then property (b) implies that the second term is zero.

(f) Assume first that the rows of $\mathbf{A}$ are linearly dependent. Then with suitable coefficients

$$c_1 \mathbf{r}_1^T + c_2 \mathbf{r}_2^T + ... + c_n \mathbf{r}_n^T = \mathbf{0}^T,$$

where at least one coefficient, say c_1, is nonzero. Then

$$\mathbf{r}_1^T + d_2 \mathbf{r}_2^T + ... + d_n \mathbf{r}_n^T = \mathbf{0}^T,$$

where $d_k = c_k / c_1$, $(k = 2, ..., n)$. Then the repeated application of property (e) implies that

$$\det(\mathbf{A}) = \det \begin{pmatrix} \mathbf{r}_1^T + d_2 \mathbf{r}_2^T \\ \mathbf{r}_2^T \\ ... \\ \mathbf{r}_n^T \end{pmatrix} = \det \begin{pmatrix} \mathbf{r}_1^T + d_2 \mathbf{r}_2^T + d_3 \mathbf{r}_3^T \\ \mathbf{r}_2^T \\ ... \\ \mathbf{r}_n^T \end{pmatrix} = ...$$

$$= \det \begin{pmatrix} \mathbf{r}_1^T + d_2 \mathbf{r}_2^T + ... + d_n \mathbf{r}_n^T \\ \mathbf{r}_2^T \\ ... \\ \mathbf{r}_n^T \end{pmatrix} = \det \begin{pmatrix} \mathbf{0}^T \\ \mathbf{r}_2^T \\ ... \\ \mathbf{r}_n^T \end{pmatrix} = 0,$$

since in each term of the sum (4.3) applied to the last determinant, the factor a_{1i_1} is always zero.

Assume next that the rows of matrix $\mathbf{A}$ are linearly independent. Then rank$(\mathbf{A}) = n$, and the system of homogeneous linear equations $\mathbf{Ax} = \mathbf{0}$ has the unique solution $\mathbf{x} = \mathbf{0}$. Apply the elimination procedure of Section 3.7 to solve this system. At each elimination step one of the following operations is performed:

(i) Interchanging two rows and/or columns only may change the sign of the determinant.

(ii) Dividing any row by the pivot coefficient results in dividing the value of the determinant by the nonzero pivot coefficient.

(iii) Adding a constant-multiple of a row to another row does not change the value of the determinant.

Therefore det(**A**) is the nonzero-multiple of the determinant of the last coefficient matrix, which is the identity matrix. Then Example 4.1 implies that the determinant is nonzero.

♣

The last part of the proof of the theorem suggests the use of the elimination method to find the values of determinants. When the elimination process terminates then we have the following possibilities:

Case 1. All variables are eliminated, then the final matrix is the identity matrix. Then the value of the determinant is the product of the pivot elements time $(-1)^K$, where K is the number of times we had to interchange two rows or two columns.

Case 2. There is at least one zero row in the last matrix in which case the determinant is zero.

For example, the determinant of the coefficient matrix of Example 3.12. is $1 \cdot 3 \cdot 3 = 9$, since we divided by 1, 3 and 3 to get the pivot rows and there was no need to interchange rows or colums.

Consider next a scalar-valued function f which assigns to each n-tuple $(\mathbf{a}_1, ..., \mathbf{a}_n)$ of n-element vectors a scalar, and the following conditions are satisfied:

(A) $f(\mathbf{e}_1,...,\mathbf{e}_n) = 1$ where $\mathbf{e}_1, ..., \mathbf{e}_n$ are the natural basis vectors;

(B) $f(\mathbf{a}_1,...,\mathbf{a}_n)$ changes sign if two of the vectors $\mathbf{a}_i$ and $\mathbf{a}_j$ are interchanged;

(C) $f\left(\mathbf{a}_1,...,\mathbf{a}_{k-1},\mathbf{a}_k + \mathbf{a}'_k ,\mathbf{a}_{k+1},...,\mathbf{a}_n\right)$

$= f\left(\mathbf{a}_1,...,\mathbf{a}_{k-1},\mathbf{a}_k,\mathbf{a}_{k+1},...,\mathbf{a}_n\right) + f\left(\mathbf{a}_1,...,\mathbf{a}_{k-1},\mathbf{a}'_k,\mathbf{a}_{k+1},...,\mathbf{a}_n\right);$

(D) $f\left(\mathbf{a}_1,...,\mathbf{a}_{k-1},c\mathbf{a}_k,\mathbf{a}_{k+1},...,\mathbf{a}_n\right) = cf\left(\mathbf{a}_1,...,\mathbf{a}_{k-1},\mathbf{a}_k,\mathbf{a}_{k+1},...,\mathbf{a}_n\right).$

Theorem 4.3. If conditions (A), (B), (C) and (D) are satisfied, then for all n-element vectors $\mathbf{a}_1, ..., \mathbf{a}_n$,

$$f\left(\mathbf{a}_1, ..., \mathbf{a}_n\right) = \det\left(\mathbf{a}_1, ..., \mathbf{a}_n\right).$$

Proof. Consider the function

$$\Delta\left(\mathbf{a}_1, ..., \mathbf{a}_n\right) = f\left(\mathbf{a}_1, ..., \mathbf{a}_n\right) - \det\left(\mathbf{a}_1, ..., \mathbf{a}_n\right).$$

Then $\Delta(\mathbf{e}_1, ..., \mathbf{e}_n) = 0$ and function Δ satisfies properties (B), (C) and (D). Repeating the proof of part (b) of Theorem 4.2 gives that $\Delta(\mathbf{a}_1, ..., \mathbf{a}_n) = 0$ if $\mathbf{a}_i = \mathbf{a}_j$ for some $i \neq j$. Let a_{ij} denote the ith element of vector $\mathbf{a}_j$ $(j = 1, 2, ..., n; i = 1, 2, ..., n)$, then for all j,

$$\mathbf{a}_j = a_{1j}\mathbf{e}_1 + ... + a_{nj}\mathbf{e}_n,$$

therefore

$$\Delta\left(\mathbf{a}_1, ..., \mathbf{a}_n\right) = \Delta\left(\sum_{l_1} a_{l_1 1}\mathbf{e}_{l_1}, \mathbf{a}_2, ..., \mathbf{a}_n\right)$$

$$= \sum_{l_1} a_{l_1 1}\Delta\left(\mathbf{e}_{l_1}, \mathbf{a}_2, ..., \mathbf{a}_n\right) = \sum_{l_1} a_{l_1 1}\Delta\left(\mathbf{e}_{l_1}, \sum_{l_2} a_{l_2 2}\mathbf{e}_{l_2}, \mathbf{a}_3, ..., \mathbf{a}_n\right) \quad (4.4)$$

$$= \sum_{l_1} a_{l_1 1}\sum_{l_2} a_{l_2 2}\Delta\left(\mathbf{e}_{l_1}, \mathbf{e}_{l_2}, \mathbf{a}_3, ..., \mathbf{a}_n\right) = ...$$

$$= \sum_{l_1}\sum_{l_2}...\sum_{l_n} a_{l_1 1}a_{l_2 2}... a_{l_n n}\Delta\left(\mathbf{e}_{l_1}, ..., \mathbf{e}_{l_n}\right).$$

We will next prove that each term is zero.

Notice first that $\Delta(\mathbf{e}_{l_1}, ..., \mathbf{e}_{l_n}) = 0$ if two numbers l_i and l_j are equal.

If the numbers $l_1, l_2, ..., l_n$ are different, then they form a permutation of the integers $1, 2, ..., n$, therefore

$$\Delta\left(\mathbf{e}_{l_{1}},...,\mathbf{e}_{l_{n}}\right) = \begin{cases} \Delta\left(\mathbf{e}_{1},...,\mathbf{e}_{n}\right) \text{ if an even number of inversions are needed} \\ \text{to bring permutation } <l_{1},...,l_{n}> \text{ to natural order} \\ -\Delta\left(\mathbf{e}_{1},...,\mathbf{e}_{n}\right) \text{ if an odd number of inversions are needed} \\ \text{to bring permutation } <l_{1},...,l_{n}> \text{ to natural order} \end{cases}$$

Since $\Delta\left(\mathbf{e}_{1},...,\mathbf{e}_{n}\right) = 0$, all terms in (4.4) are equal to zero. Therefore that $\Delta\left(\mathbf{a}_{1},...,\mathbf{a}_{n}\right) = 0$, is

$$f\left(\mathbf{a}_{1}, ..., \mathbf{a}_{n}\right) = \det\left(\mathbf{a}_{1}, ..., \mathbf{a}_{n}\right).$$

Thus, the proof is completed.

♣

Corollary. Assume that function f satisfies only conditions (B), (C), and (D). Then for all n-vectors $\mathbf{a}_1, ..., \mathbf{a}_n$

$$f\left(\mathbf{a}_{1}, ..., \mathbf{a}_{n}\right) = \det\left(\mathbf{a}_{1}, ..., \mathbf{a}_{n}\right) \cdot f\left(\mathbf{e}_{1}, ..., \mathbf{e}_{n}\right) \qquad (4.5)$$

Proof. If $f\left(\mathbf{e}_{1},...,\mathbf{e}_{n}\right) = 1$, then the assertion follows from the theorem. Otherwise, consider the function

$$g\left(\mathbf{a}_{1}, ..., \mathbf{a}_{n}\right) = \frac{\Delta\left(\mathbf{a}_{1},...,\mathbf{a}_{n}\right)}{f\left(\mathbf{e}_{1},...,\mathbf{e}_{n}\right) - 1},$$

which satisfies all conditions (A), (B), (C), and (D). Therefore

$$g\left(\mathbf{a}_{1}, ..., \mathbf{a}_{n}\right) = \det\left(\mathbf{a}_{1}, ..., \mathbf{a}_{n}\right),$$

and from the last equation we see that

$$\det\left(\mathbf{a}_{1}, ..., \mathbf{a}_{n}\right) = \frac{f\left(\mathbf{a}_{1}, ..., \mathbf{a}_{n}\right) - \det\left(\mathbf{a}_{1}, ..., \mathbf{a}_{n}\right)}{f\left(\mathbf{e}_{1}, ..., \mathbf{e}_{n}\right) - 1},$$

which is equivalent to the assertion.

♣

One of the most frequently used properties of determinants is known as the multiplication theorem, which is formulated next.

Theorem 4.4. Let $\mathbf{B}$ and $\mathbf{A}$ be $n \times n$ real or complex matrices. Then

$$\det\left(\mathbf{BA}\right) = \det\left(\mathbf{B}\right) \cdot \det\left(\mathbf{A}\right).$$

Proof. Define the function

$$f\left(\mathbf{a}_1, \ldots, \mathbf{a}_n\right) = \det\left(\mathbf{Ba}_1, \ldots, \mathbf{Ba}_n\right)$$

which obviously satisfies properties (B), (C), and (D). Therefore, the Corollary of Theorem 4.3 implies that

$$\det\left(\mathbf{Ba}_1, \ldots, \mathbf{Ba}_n\right) = \det\left(\mathbf{a}_1, \ldots, \mathbf{a}_n\right) \cdot \det\left(\mathbf{Be}_1, \ldots, \mathbf{Be}_n\right).$$

Let now $\mathbf{a}_1, \ldots, \mathbf{a}_n$ be the columns of $\mathbf{A}$. Then

$$\left(\mathbf{a}_1, \ldots, \mathbf{a}_n\right) = \mathbf{A},$$
$$\left(\mathbf{Ba}_1, \ldots, \mathbf{Ba}_n\right) = \mathbf{B} \cdot \left(\mathbf{a}_1, \ldots, \mathbf{a}_n\right) = \mathbf{B} \cdot \mathbf{A},$$
$$\left(\mathbf{Be}_1, \ldots, \mathbf{Be}_n\right) = \mathbf{B} \cdot \left(\mathbf{e}_1, \ldots, \mathbf{e}_n\right) = \mathbf{B} \cdot \mathbf{I} = \mathbf{B}$$

and hence, the proof is complete.

♣

Corollary. If $\mathbf{A}^{-1}$ exists, then

$$\det\left(\mathbf{A}^{-1}\right) = \frac{1}{\det\left(\mathbf{A}\right)},$$

since

$$1 = \det\left(\mathbf{I}_n\right) = \det\left(\mathbf{A}^{-1} \cdot \mathbf{A}\right) = \det\left(\mathbf{A}^{-1}\right)\det\left(\mathbf{A}\right).$$

Remark. Recall that **AB** and **BA** are usually different matrices. The assertion of the theorem implies that their determinants are the same even if they are different.

4.3 Cofactors and Expansion by Cofactors

The value of det(**A**) for 2×2 or 3×3 matrices can be obtained easily from the definition. In the 2×2 case,

$$\det \begin{pmatrix} a_{11} & a_{12} \\ a_{21} & a_{22} \end{pmatrix} = a_{11}a_{22} - a_{12}a_{21},$$

where the two terms can be obtained by the method illustrated in Figure 4.1.

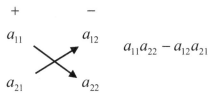

Figure 4.1. Diagram for evaluating 2×2 determinants

The six terms of the determinants of a 3×3 matrix can be obtained by a similar, but more complicated diagram, which is shown in Figure 4.2. The reader should be cautioned, since there is no similar way to find the values of determinants of higher order. In this section a general method will be introduced to find the values of $n \times n$ determinants in general. Our method will be based on the following concept.

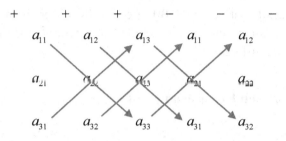

Figure 4.2. Diagram for evaluating 3 × 3 determinants

Definition 4.3. Let $\mathbf{A} = (a_{ij})$ be an $n \times n$ matrix. The (i, j) *cofactor A_{ij}* of $\mathbf{A}$ is defined as

$$A_{ij} = (-1)^{i+j} D_{ij},$$

where D_{ij} is the determinant of the $(n-1) \times (n-1)$ matrix that is obtained by deleting the ith row and j^{th} column of matrix $\mathbf{A}$.
The main result of this section can be formulated as follows.

Theorem 4.5. Let $\mathbf{A} = (a_{ij})$ be an $n \times n$ matrix. Then for $i = 1, 2, ..., n$,

$$\det(\mathbf{A}) = \sum_{j=1}^{n} a_{ij} A_{ij}, \tag{4.6}$$

where A_{ij} is the (i, j) cofactor of $\mathbf{A}$.

Proof. Fix the value of i and define function

$$f(\mathbf{a}_1, ..., \mathbf{a}_n) = \sum_{j=1}^{n} a_{ij} A_{ij}$$

where a_{ij} is the ith element of vector $\mathbf{a}_j$ and A_{ij} is the (i, j) cofactor of matrix $(\mathbf{a}_1, ..., \mathbf{a}_n)$. It is easy to prove that f satisfies all conditions of Theorem 4.3:
(A) If $\mathbf{A} = \mathbf{I}_n$, then from (4.6) we see that

$$\det(\mathbf{A}) = a_{ii}A_{ii} = 1,$$

because all other terms are equal to zero (since $a_{ij} = 0$ for $i \neq j$) and $a_{ii} = 1$ and $A_{ii} = \det(\mathbf{I}_{n-1}) = 1$.

(B) Assume that columns k and l ($l < k$) are interchanged. Let a'_{ij} and A'_{ij} denote the elements and cofactors of the new matrix. Then it is easy to see that

$$a'_{ij} = \begin{cases} a_{il} & \text{if } j = k \\ a_{il} & \text{if } j = k \\ a_{ij} & \text{otherwise,} \end{cases}$$

and

$$A'_{ij} = \begin{cases} -A_{il} & \text{if } j = k \\ -A_{ik} & \text{if } j = l \\ -A_{ij} & \text{otherwise.} \end{cases}$$

Therefore, each term changes sign.

(C) Assume that column $\mathbf{a}_k$ is replaced by $\mathbf{a}_k + \mathbf{a}'_k$. If $j = k$, then for all i, the A_{ij} cofactor remains the same but the element a_{ij} has to be replaced by the sum $a_{ij} + a'_{ij}$. If $j \neq k$, then the element a_{ij} remains the same and the cofactor is replaced by $A_{ij} + A'_{ij}$, where A'_{ij}, is obtained from A_{ij}, by changing the elements a_{lk} to $\mathbf{a}'_{lk}$ for $l \neq i$. In both cases each term breaks up as the sum of the corresponding terms with k^{th} column $\mathbf{a}_k$ and $\mathbf{a}'_k$, respectively.

(D) Assume next that column $\mathbf{a}_k$ is multiplied by a scalar c. Then each term in the definition of f is multiplied by c: for $j = k$, a_{ik} is multiplied by c; and for $j \neq k$, the cofactor A_{ij} is multiplied by c, since one of its columns is multiplied by this scalar.

These properties and Theorem 4.3 imply that f coincides with the determinant function.

♣

Corollary 1. Combining Theorems 4.1 and 4.5 we conclude that for all j,

$$\det(\mathbf{A}) = \sum_{i=1}^{n} a_{ij} A_{ij}. \tag{4.7}$$

Formulas (4.6) and (4.7) are called the *row* and *column expansions* of the determinant, respectively.

Corollary 2. For $i \neq k$,

$$\sum_{j=1}^{n} a_{ij} A_{kj} = 0 \tag{4.8}$$

and for $j \neq l$,

$$\sum_{i=1}^{n} a_{ij} A_{il} = 0. \tag{4.9}$$

The left-hand side of equation (4.8) is the determinant of the matrix which is obtained from $\mathbf{A}$ by replacing its kth row by its ith row. The ith and k^{th} rows of the resulting matrix therefore become identical, and the determinant of this matrix is necessarily zero. The left-hand side of equation (4.9) is the determinant of the matrix which is obtained from $\mathbf{A}$ by replacing its lth column by its jth column, and since this matrix has two identical columns, its determinant is zero.

The expansion formulas are illustrated next.

Example 4.4. We will expand the determinant of the 4×4 matrix

$$\mathbf{A} = \begin{pmatrix} 2 & 1 & 1 & 1 \\ 1 & 2 & 1 & 1 \\ 1 & 1 & 2 & 1 \\ 1 & 1 & 1 & 2 \end{pmatrix}$$

by the repeated application of the row expansion formula. Selecting always the first row,

$$\det\left(\mathbf{A}\right) = 2\cdot\left(-1\right)^{1+1}\det\begin{pmatrix} 2 & 1 & 1 \\ 1 & 2 & 1 \\ 1 & 1 & 2 \end{pmatrix} + 1\cdot\left(-1\right)^{1+2}\det\begin{pmatrix} 1 & 1 & 1 \\ 1 & 2 & 1 \\ 1 & 1 & 2 \end{pmatrix}$$

$$+ 1\cdot\left(-1\right)^{1+3}\det\begin{pmatrix} 1 & 2 & 1 \\ 1 & 1 & 1 \\ 1 & 1 & 2 \end{pmatrix} + 1\cdot\left(-1\right)^{1+4}\det\begin{pmatrix} 1 & 2 & 1 \\ 1 & 1 & 2 \\ 1 & 1 & 1 \end{pmatrix}.$$

The 3×3 determinants are obtained in the same way:

$$\det\begin{pmatrix} 2 & 1 & 1 \\ 1 & 2 & 1 \\ 1 & 1 & 2 \end{pmatrix}$$

$$= 2\cdot\left(-1\right)^{1+1}\det\begin{pmatrix} 2 & 1 \\ 1 & 2 \end{pmatrix} + 1\cdot\left(-1\right)^{1+2}\det\begin{pmatrix} 1 & 1 \\ 1 & 2 \end{pmatrix} + 1\cdot\left(-1\right)^{1+3}\det\begin{pmatrix} 1 & 2 \\ 1 & 1 \end{pmatrix}$$

$$= 2\cdot\left(4-1\right) - 1\cdot\left(2-1\right) + 1\cdot\left(1-2\right) = 6-1-1 = 4,$$

$$\det\begin{pmatrix} 1 & 1 & 1 \\ 1 & 2 & 1 \\ 1 & 1 & 2 \end{pmatrix}$$

$$= 1\cdot\left(-1\right)^{1+1}\det\begin{pmatrix} 2 & 1 \\ 1 & 2 \end{pmatrix} + 1\cdot\left(-1\right)^{1+2}\det\begin{pmatrix} 1 & 1 \\ 1 & 2 \end{pmatrix} + 1\cdot\left(-1\right)^{1+3}\det\begin{pmatrix} 1 & 2 \\ 1 & 1 \end{pmatrix}$$

$$= 1\cdot\left(4-1\right) - 1\cdot\left(2-1\right) + 1\cdot\left(1-2\right) = 3-1-1 = 1,$$

$$\det\begin{pmatrix} 1 & 2 & 1 \\ 1 & 1 & 1 \\ 1 & 1 & 2 \end{pmatrix}$$

$$= 1 \cdot (-1)^{1+1} \det\begin{pmatrix} 1 & 1 \\ 1 & 2 \end{pmatrix} + 2 \cdot (-1)^{1+2} \det\begin{pmatrix} 1 & 1 \\ 1 & 2 \end{pmatrix} + 1 \cdot (-1)^{1+3} \det\begin{pmatrix} 1 & 1 \\ 1 & 1 \end{pmatrix}$$

$$= 1 \cdot (2-1) - 2 \cdot (2-1) + 1 \cdot (1-1) = 1 - 2 + 0 = -1$$

(which result can be immediately obtained from the previous determinant, since the first two rows are interchanged),

$$\det\begin{pmatrix} 1 & 2 & 1 \\ 1 & 1 & 2 \\ 1 & 1 & 1 \end{pmatrix}$$

$$= 1 \cdot (-1)^{1+1} \det\begin{pmatrix} 1 & 2 \\ 1 & 1 \end{pmatrix} + 2 \cdot (-1)^{1+2} \det\begin{pmatrix} 1 & 2 \\ 1 & 1 \end{pmatrix} + 1 \cdot (-1)^{1+3} \det\begin{pmatrix} 1 & 1 \\ 1 & 1 \end{pmatrix}$$

$$= 1 \cdot (1-2) - 2 \cdot (1-2) + 1 \cdot (1-1) = -1 + 2 + 0 = 1$$

(which result is obvious from the value of the previous determinant, since the last two rows are interchanged). Therefore

$$\det(\mathbf{A}) = 2 \cdot 4 - 1 \cdot 1 + 1 \cdot (-1) - 1 \cdot 1 = 5.$$

♦

4.4 Determinants and Systems of Linear Equations

We start this section with a theorem, which is the restatement and combination of Theorems 3.3 and 3.13.

Theorem 4.6. Let $\mathbf{A}$ be an $n \times n$ matrix. The following conditions are equivalent to each other:
(a) The system $\mathbf{Ax} = \mathbf{b}$ has a unique solution for all n-element vectors $\mathbf{b}$;
(b) $\mathbf{A}^{-1}$ exists;
(c) The columns of $\mathbf{A}$ are linearly independent;
(d) The rows of $\mathbf{A}$ are linearly independent;
(e) rank($\mathbf{A}$) = n;
(f) det($\mathbf{A}$) $\neq 0$.

Notice that condition (f) gives a sufficient and necessary condition for the existence of a unique solution of system $\mathbf{Ax} = \mathbf{b}$. By using the concept of determinants, we can even solve this system.
Assume that det($\mathbf{A}$) $\neq 0$. For $k = 1, 2, \ldots, n$, introduce the notation

$$D_k = \det\left(\mathbf{a}_1, \ldots, \mathbf{a}_{k-1}, \mathbf{b}, \mathbf{a}_{k+1}, \ldots, \mathbf{a}_n\right).$$

Here D_k is the determinant of the matrix which is obtained from $\mathbf{A}$ by replacing its kth column by the right-hand side vector. The columns of $\mathbf{A}$ are denoted again by $\mathbf{a}_1, \ldots, \mathbf{a}_n$.

Theorem 4.7. If det($\mathbf{A}$) $\neq 0$, then for $k = 1, 2, \ldots, n$, the kth element of the solution of system $\mathbf{Ax} = \mathbf{b}$ is obtained as

$$x_k = \frac{D_k}{\det(\mathbf{A})}. \tag{4.10}$$

Proof. If det($\mathbf{A}$) $\neq 0$ then there is a unique solution, $\mathbf{x} = (x_i)$. Using the properties of determinant, for $k = 1, 2, \ldots, n$, we have

$$D_k = \det\left(\mathbf{a}_1, \ldots, \mathbf{a}_{k-1}, \sum_{i=1}^{n} x_i \mathbf{a}_i, \mathbf{a}_{k+1}, \ldots, \mathbf{a}_n\right)$$

$$= \sum_{i=1}^{n} x_i \det\left(\mathbf{a}_1, \ldots, \mathbf{a}_{k-1}, \mathbf{a}_i, \mathbf{a}_{k+1}, \ldots, \mathbf{a}_n\right).$$

If $i \neq k$, then the ith and k^{th} columns of matrix ($\mathbf{a}_1, \ldots, \mathbf{a}_{k-1}, \mathbf{a}_i, \mathbf{a}_{k+1}, \ldots, \mathbf{a}_n$) are identical, so its determinant is zero. Therefore

$$D_k = x_k \det\left(\mathbf{a}_1, ..., \mathbf{a}_{k-1}, \mathbf{a}_k, \mathbf{a}_{k+1}, ..., \mathbf{a}_n\right) = x_k \cdot \det\left(\mathbf{A}\right),$$

which implies the assertion.

♣

Remark. Relations (4.10) are called the *Cramer's rule*.

Example 4.5. We will apply Cramer's rule to solve the system

$$x_1 + x_2 + x_3 = 3$$
$$2x_1 + 3x_2 - x_3 = 4$$
$$3x_1 + 5x_2 + x_3 = 9,$$

which was the subject of our earlier Example 3.11. Expanding the 3×3 determinants with respect to their first rows we have

$$\det\left(\mathbf{A}\right) = \det\begin{pmatrix} 1 & 1 & 1 \\ 2 & 3 & -1 \\ 3 & 5 & 1 \end{pmatrix} = 1 \cdot (3+5) - 1 \cdot (2+3) + 1 \cdot (10-9) = 4,$$

$$D_1 = \det\begin{pmatrix} 3 & 1 & 1 \\ 4 & 3 & -1 \\ 9 & 5 & 1 \end{pmatrix} = 3 \cdot (3+5) - 1 \cdot (4+9) + 1 \cdot (20-27) = 4,$$

$$D_2 = \det\begin{pmatrix} 1 & 3 & 1 \\ 2 & 4 & -1 \\ 3 & 9 & 1 \end{pmatrix} = 1 \cdot (4+9) - 3 \cdot (2+3) + 1 \cdot (18-12) = 4,$$

and

$$D_3 = \det\begin{pmatrix} 1 & 1 & 3 \\ 2 & 3 & 4 \\ 3 & 5 & 9 \end{pmatrix} = 1 \cdot (27-20) - 1 \cdot (18-12) + 3 \cdot (10-9) = 4.$$

Therefore

$$x_1 = \frac{D_1}{\det(A)} = 1, \; x_2 = \frac{D_2}{\det(A)} = 1, \quad \text{and} \quad x_3 = \frac{D_3}{\det(A)} = 1.$$

♦

In Theorem 4.6 we observed a strong relation between the ranks and determinants of $n \times n$ matrices by noticing that rank$(A) = n$ if and only if $\det(A) \neq 0$. We can largely extend this result for general $m \times n$ matrices by using a new concept known as minor.

Definition 4.4. Let A be an $m \times n$ real or complex matrix. An *r-rowed minor (determinant)* of A is the determinant of an $r \times r$ matrix obtained from A by deleting $m - r$ rows and $n - r$ columns.

An $m \times n$ matrix therefore has $\binom{m}{r} \cdot \binom{n}{r}$ r-rowed minors, since the remaining rows can be selected in $\binom{m}{r}$ different ways, and the remaining columns can be selected in $\binom{n}{r}$ different ways.

The construction of minors is illustrated first, then we will prove an important theorem relating matrix ranks to basic properties of minors.

Example 4.6. Matrix

$$\begin{pmatrix} 1 & 2 & 3 \\ 2 & 1 & 2 \end{pmatrix}$$

has three 2-rowed minors

$$\det\begin{pmatrix} 1 & 2 \\ 2 & 1 \end{pmatrix}, \; \det\begin{pmatrix} 1 & 3 \\ 2 & 2 \end{pmatrix}, \quad \text{and} \quad \det\begin{pmatrix} 2 & 3 \\ 1 & 2 \end{pmatrix}.$$

Similarly, matrix

$$\begin{pmatrix} 1 & 2 & 3 \\ 2 & 1 & 2 \\ 1 & 2 & 2 \end{pmatrix}$$

has nine 2-rowed minors:

$$\det\begin{pmatrix} 1 & 2 \\ 2 & 2 \end{pmatrix}, \ \det\begin{pmatrix} 2 & 2 \\ 1 & 2 \end{pmatrix}, \ \det\begin{pmatrix} 2 & 1 \\ 1 & 2 \end{pmatrix},$$

$$\det\begin{pmatrix} 2 & 3 \\ 2 & 2 \end{pmatrix}, \ \det\begin{pmatrix} 1 & 3 \\ 1 & 2 \end{pmatrix}, \ \det\begin{pmatrix} 1 & 2 \\ 1 & 2 \end{pmatrix},$$

$$\det\begin{pmatrix} 2 & 3 \\ 1 & 2 \end{pmatrix}, \ \det\begin{pmatrix} 1 & 3 \\ 2 & 2 \end{pmatrix}, \ \det\begin{pmatrix} 1 & 2 \\ 2 & 1 \end{pmatrix}.$$

Notice that in these cases the numbers of the 2×2 minors are given as

$$\begin{pmatrix} 2 \\ 2 \end{pmatrix} \cdot \begin{pmatrix} 3 \\ 2 \end{pmatrix} = 1 \cdot 3 = 3$$

and

$$\begin{pmatrix} 3 \\ 2 \end{pmatrix} \cdot \begin{pmatrix} 3 \\ 2 \end{pmatrix} = 3 \cdot 3 = 9$$

in the 2×3 and 3×3 matrices, respectively.

$\blacklozenge$

Theorem 4.8. Let $\mathbf{A}$ be an $m \times n$ nonzero matrix. The rank of $\mathbf{A}$ is r if and only if there is at least one nonzero r-rowed minor, and all higher-rowed minors are zero.

Proof. Assume that rank$(\mathbf{A}) = r$, and the size of the largest nonzero minor is $s \times s$. We will prove that $r = s$. Since rank$(\mathbf{A}) = r$, there are r linearly independent rows of $\mathbf{A}$. Let $\mathbf{A}'$ denote the submatrix of $\mathbf{A}$ after deleting all other rows. The fact that rank$(\mathbf{A}') = r$ implies that there are r linearly independent columns of $\mathbf{A}'$. Let $\mathbf{A}''$ denote the submatrix of $\mathbf{A}'$

after deleting all other columns. Then $\mathbf{A}''$ is $r \times r$ and its columns are linearly independent implying that $\det(\mathbf{A}'') \neq 0$. Since s denotes the size of the largest nonzero minor, $s \geq r$. Consider next the nonzero $s \times s$ minor. Its rows are linearly independent which implies that the corresponding entire rows of $\mathbf{A}$ are also linearly independent. Hence $r \geq s$, which completes the proof.

♣

Example 4.7. We will now determine the rank of matrix

$$\mathbf{A} = \begin{pmatrix} 1 & 1 & 1 \\ 2 & 2 & 2 \\ 1 & 2 & 3 \end{pmatrix}$$

by applying the above theorem. The only 3-rowed minor is the determinant of $\mathbf{A}$:

$$\det(\mathbf{A}) = 1 \cdot (6-4) - 1 \cdot (6-2) + 1 \cdot (4-2) = 2 - 4 + 2 = 0,$$

where we used expansion with respect to the first row. Therefore, $\text{rank}(\mathbf{A}) < 3$. However, there is at least one 2-rowed nonzero minor

$$\det \begin{pmatrix} 1 & 1 \\ 1 & 2 \end{pmatrix} = 2 - 1 = 1 \neq 0,$$

which is obtained by deleting the last column and the second row of $\mathbf{A}$. Hence, $\text{rank}(\mathbf{A}) = 2$.

♦

4.5 Further Examples and Applications

The most important applications of determinants in solving systems of linear equations and in finding matrix ranks were already discussed in this chapter. Another application will be introduced later in Section 6.2, where a parametric determinant will be expanded in order to find the characteristics equation of a square matrix. In Section 7.4, several special matrix classes will be introduced based on the signs of the principal minors (which are the determinants of certain parts of the matrices in question). In this section some additional applications will be introduced.

1. We give first a *closed form representation* to find the *inverse* of a nonsingular $n \times n$ matrix $\mathbf{A}$:

$$\mathbf{A}^{-1} = \frac{1}{\det(\mathbf{A})} \begin{pmatrix} A_{11} & A_{21} & \cdots & A_{n1} \\ A_{12} & A_{22} & \cdots & A_{n2} \\ \cdots & \cdots & \cdots & \cdots \\ A_{1n} & A_{2n} & \cdots & A_{nn} \end{pmatrix} \tag{4.11}$$

Notice that the matrix of the right-hand side can be obtained from $\mathbf{A}$ by replacing each element by its cofactor and transposing the resulting matrix.

This formula can be proved in the following way. Consider the following matrix product:

$$\begin{pmatrix} a_{11} & a_{12} & \cdots & a_{1n} \\ a_{21} & a_{22} & \cdots & a_{2n} \\ \cdots & \cdots & \cdots & \cdots \\ a_{n1} & a_{n2} & \cdots & a_{nn} \end{pmatrix} \begin{pmatrix} A_{11} & A_{21} & \cdots & A_{n1} \\ A_{12} & A_{22} & \cdots & A_{n2} \\ \cdots & \cdots & \cdots & \cdots \\ A_{1n} & A_{2n} & \cdots & A_{nn} \end{pmatrix}.$$

The (i, j) element of the product can be obtained as

$$\sum_{k=1}^{n} a_{ik} A_{jk} = \begin{cases} 0 & \text{if } i \neq j, \\ \det(\mathbf{A}) & \text{if } i = j, \end{cases}$$

which follows from relations (4.6) and (4.8). Therefore, the above matrix product is $\det(\mathbf{A}) \cdot \mathbf{I}_n$, which implies equality (4.11).

Example 4.8. Consider first a 2×2 nonsingular matrix

$$\mathbf{A} = \begin{pmatrix} a_{11} & a_{12} \\ a_{21} & a_{22} \end{pmatrix}.$$

Then Definition 4.3 implies that

$$A_{11} = (-1)^{1+1} a_{22} = a_{22}, \; A_{12} = (-1)^{1+2} a_{21} = -a_{21},$$
$$A_{21} = (-1)^{2+1} a_{12} = -a_{12}, \text{ and } A_{22} = (-1)^{2+2} a_{11} = a_{11}.$$

Therefore equality (4.11) shows that

$$\mathbf{A}^{-1} = \frac{1}{a_{11}a_{22} - a_{12}a_{21}} \begin{pmatrix} a_{22} & -a_{12} \\ -a_{21} & a_{11} \end{pmatrix}.$$

◆

Example 4.9. We will next find the inverse of the 3×3 matrix

$$\mathbf{A} = \begin{pmatrix} 2 & 1 & 3 \\ 4 & 4 & 7 \\ 2 & 5 & 9 \end{pmatrix}$$

which was the subject of our earlier Example 3.9. Simple calculation shows that

$$A_{11} = \det \begin{pmatrix} 4 & 7 \\ 5 & 9 \end{pmatrix} = 36 - 35 = 1, \; A_{12} = -\det \begin{pmatrix} 4 & 7 \\ 2 & 9 \end{pmatrix} = -22,$$

$$A_{13} = \det \begin{pmatrix} 4 & 4 \\ 2 & 5 \end{pmatrix} = 12, \qquad A_{21} = -\det \begin{pmatrix} 1 & 3 \\ 5 & 9 \end{pmatrix} = 6,$$

$$A_{22} = \det \begin{pmatrix} 2 & 3 \\ 2 & 9 \end{pmatrix} = 12, \qquad A_{23} = -\det \begin{pmatrix} 2 & 1 \\ 2 & 5 \end{pmatrix} = -8,$$

$$A_{31} = \det \begin{pmatrix} 1 & 3 \\ 4 & 7 \end{pmatrix} = -5, \qquad A_{32} = -\det \begin{pmatrix} 2 & 3 \\ 4 & 7 \end{pmatrix} = -2,$$

and

$$A_{33} = \det \begin{pmatrix} 2 & 1 \\ 4 & 4 \end{pmatrix} = 4.$$

Expanding $\det(\mathbf{A})$ with respect to the first row we have

$$\det(\mathbf{A}) = a_{11}A_{11} + a_{12}A_{12} + a_{13}A_{13} = 2(1) + 1(-22) + 3(12) = 16,$$

and therefore,

$$\mathbf{A}^{-1} = \frac{1}{16}\begin{pmatrix} 1 & 6 & -5 \\ -22 & 12 & -2 \\ 12 & -8 & 4 \end{pmatrix} = \begin{pmatrix} \dfrac{1}{16} & \dfrac{3}{8} & -\dfrac{5}{16} \\ -\dfrac{11}{8} & \dfrac{3}{4} & -\dfrac{1}{8} \\ \dfrac{3}{4} & -\dfrac{1}{2} & \dfrac{1}{4} \end{pmatrix}.$$

This result coincides with that obtained in Example 3.9 as it should.

♦

2. Consider a single-input linear discrete dynamic system with state transition equation

$$\mathbf{x}(t+1) = \mathbf{A}\mathbf{x}(t) + \mathbf{b}u(t), \quad \mathbf{x}(0) = \mathbf{x}_0 \qquad (4.12)$$

where t denotes the time periods 0, 1, 2, ..., furthemore $\mathbf{x}(t)$ and $u(t)$ are the n-element real state vector and single-dimensional real input at time period t, $\mathbf{A}$ is a given $n \times n$ constant real matrix and $\mathbf{b}$ is a given real n-vector. An important problem of systems theory is the controllability of this system. We say that system (4.12) is completely controllable if for arbitrary time period $T \geq n$ and vector $\mathbf{x}_T \in R^n$ there exists an input sequence $u(0)$, $u(1)$, ..., $u(T-1)$ such that at time period T, the state of the system coincides with the given vector $\mathbf{x}_T$.

It is known from the theory of linear systems (see, for example Szidarovszky and Bahill, 1992) that the above system is completely controllable if and only if the Kalman matrix

$$\mathbf{K} = \left(\mathbf{b}, \ \mathbf{A}\mathbf{b}, \ \mathbf{A}^2\mathbf{b}, \ ..., \ \mathbf{A}^{n-1}\mathbf{b}\right)$$

is nonsingular. We know from Theorem 4.6 that $\mathbf{K}$ is nonsingular if and only if $\det(\mathbf{K}) \neq 0$ This observation gives a simple controllability check, which is illustrated in the following example.

Example 4.10. We will now show that system

$$\mathbf{x}(t+1) = \begin{pmatrix} 1 & 2 & 1 \\ 1 & 1 & 1 \\ 0 & 1 & 1 \end{pmatrix} \mathbf{x}(t) + \begin{pmatrix} 1 \\ 0 \\ 0 \end{pmatrix} u(t)$$

is completely controllable. Simple calculation shows that

$$\mathbf{Ab} = \begin{pmatrix} 1 & 2 & 1 \\ 1 & 1 & 1 \\ 0 & 1 & 1 \end{pmatrix} \begin{pmatrix} 1 \\ 0 \\ 0 \end{pmatrix} = \begin{pmatrix} 1 \\ 1 \\ 0 \end{pmatrix},$$

$$\mathbf{A}^2\mathbf{b} = \mathbf{A}(\mathbf{Ab}) = \begin{pmatrix} 1 & 2 & 1 \\ 1 & 1 & 1 \\ 0 & 1 & 1 \end{pmatrix} \begin{pmatrix} 1 \\ 1 \\ 0 \end{pmatrix} = \begin{pmatrix} 3 \\ 2 \\ 1 \end{pmatrix}.$$

Therefore

$$\mathbf{K} = \begin{pmatrix} 1 & 1 & 3 \\ 0 & 1 & 2 \\ 0 & 0 & 1 \end{pmatrix},$$

which is an upper triangular matrix. From Example 4.1 we know that $\det(\mathbf{K})$ equals the product of the diagonal elements of $\mathbf{K}$, that is, $\det(\mathbf{K}) = 1 \neq 0$. Consequently, the system is completely controllable.

♦

3. The controllability of single-input linear continuous systems

$$\dot{\mathbf{x}}(t) = \mathbf{Ax}(t) + \mathbf{b}u(t), \, \mathbf{x}(0) = \mathbf{x}_0 \tag{4.13}$$

also depends on the Kalman matrix

$$\mathbf{K} = \left(\mathbf{b},\ \mathbf{Ab},\ \mathbf{A^2b},\ ...,\ \mathbf{A}^{n-1}\mathbf{b}\right).$$

It is also known from systems theory that the system is completely controllable if and only if $\mathbf{K}$ is nonsingular. Hence $\det(\mathbf{K}) \neq 0$ is a sufficient and necessary condition for complete controllability.

4. In the cases of multiple input discrete and continous linear systems similar controllability conditions are applied. Consider system

$$\mathbf{x}(t+1) = \mathbf{Ax}(t) + \mathbf{Bu}(t)$$

or

$$\dot{\mathbf{x}}(t) = \mathbf{Ax}(t) + \mathbf{Bu}(t)$$

where $\mathbf{B}$ is an $n \times m$ real matrix and $\mathbf{u}(t)$ is an m-element input vector. In this case the Kalman-matrix

$$\mathbf{K} = \left(\mathbf{B},\ \mathbf{AB},\ \mathbf{A^2B},\ ...,\ \mathbf{A}^{n-1}\mathbf{B}\right)$$

is $n \times (nm)$ and either system is completely controllable if and only if rank$(\mathbf{K}) = n$, that is, the Kalman-matrix has full rank.

This condition can be checked by either verifying that the rows of $\mathbf{K}$ are linearly independent or systematically checking the $\begin{pmatrix} nm \\ n \end{pmatrix}$ $n \times m$ minors of $\mathbf{K}$ until a non-zero value is obtained.

Example 4.11. Consider the 2-dimensional continous system

$$\dot{\mathbf{x}}(t) = \begin{pmatrix} 1 & 2 \\ 1 & 1 \end{pmatrix} x(t) + \begin{pmatrix} -1 & 1 \\ 2 & 2 \end{pmatrix} \mathbf{u}(t),$$

where $n = m = 2$,

$$A = \begin{pmatrix} 1 & 2 \\ 1 & 1 \end{pmatrix}, \quad B = \begin{pmatrix} -1 & 1 \\ 2 & 2 \end{pmatrix},$$

so

$$AB = \begin{pmatrix} 1 & 2 \\ 1 & 1 \end{pmatrix} \cdot \begin{pmatrix} -1 & 1 \\ 2 & 2 \end{pmatrix} = \begin{pmatrix} 3 & 5 \\ 1 & 3 \end{pmatrix}$$

and the Kalman-matrix becomes

$$K = \left(\begin{array}{cc|cc} -1 & 1 & 3 & 5 \\ 2 & 2 & 1 & 3 \end{array} \right).$$

Since the very first minor

$$\begin{pmatrix} -1 & 1 \\ 2 & 2 \end{pmatrix}$$

is nonsingular with determinant $-2-2 = -4 \neq 0$, the rank of K equals $n = 2$, and consequently the system is completely controllable.

$\blacklozenge$

5. In Chapter 6 we will see that the eigenvalues λ of an $n \times n$ matrix A are defined by the equation

$$Ax = \lambda x \tag{4.14}$$

where $x \in R^n$ and $x \neq 0$. Since this equation can be rewritten as

$$(A - \lambda I)x = 0$$

and this homogeneous equation has nonzero solution if and only if the determinant of the coefficient matrix is zero, we conclude that all eigenvalues must satisfy the determinant equation

$$\det(A - \lambda I) = 0. \tag{4.15}$$

Expanding this deerminant an nth-degree polynomial is obtained, which can be solved by computer methods. Polynomial (4.15) is called the characteristic polynomial of matrix **A**. This idea is illustrated next.

Example 4.12. Consider matrix

$$\mathbf{A} = \begin{pmatrix} 1 & 2 & 1 \\ 2 & 2 & 3 \\ 3 & 4 & 4 \end{pmatrix}.$$

Since **I** is the identity matrix with unit diagonal elements, equation (4.15) can be written now as

$$\det \begin{pmatrix} 1-\lambda & 2 & 1 \\ 2 & 2-\lambda & 3 \\ 3 & 4 & 4-\lambda \end{pmatrix} = 0.$$

By expanding the determinant by the cofactors of the first row we have

$$\det \begin{pmatrix} 1-\lambda & 2 & 1 \\ 2 & 2-\lambda & 3 \\ 3 & 4 & 4-\lambda \end{pmatrix} = (1-\lambda)\det \begin{pmatrix} 2-\lambda & 3 \\ 4 & 4-\lambda \end{pmatrix} - 2\det \begin{pmatrix} 2 & 3 \\ 3 & 4-\lambda \end{pmatrix}$$

$$+ 1 \cdot \det \begin{pmatrix} 2 & 2-\lambda \\ 3 & 4 \end{pmatrix} = (1-\lambda)\left[(2-\lambda)(4-\lambda)-12\right] - 2\left[2(4-\lambda)-9\right]$$

$$+ \left[8-3(2-\lambda)\right] = -\lambda^3 + 7\lambda^2 + 5\lambda = 0.$$

The roots of this cubic polynomial are

$$\lambda_1 = 0, \ \lambda_2 = \frac{7+\sqrt{69}}{2}, \quad \text{and} \quad \lambda_3 = \frac{7-\sqrt{69}}{2}.$$

◆

4.6 Exercises

1. Evaluate the following determinants:

a) $\det \begin{pmatrix} a+b & a-b \\ a-b & a+b \end{pmatrix}$

b) $\det \begin{pmatrix} 1 & 5 & 2 \\ 4 & 7 & 1 \\ 1 & 1 & 1 \end{pmatrix}$

c) $\det \begin{pmatrix} 1 & 1 & 1 & 0 \\ 1 & 1 & 0 & 1 \\ 1 & 0 & 1 & 1 \\ 0 & 1 & 1 & 1 \end{pmatrix}$

d) $\det \begin{pmatrix} x & 1 & 1 \\ 1 & x & 1 \\ 1 & 1 & x \end{pmatrix}$

using the definition of determinants.

2. Solve the previous Exercise by using cofactor expansions.

3. Show that with distinct real numbers $x_1, x_2, ..., x_n$,

$$\det \begin{pmatrix} 1 & x_1 & x_1^2 & \cdots & x_1^{n-1} \\ 1 & x_2 & x_2^2 & \cdots & x_2^{n-1} \\ \cdots & \cdots & \cdots & \cdots & \cdots \\ 1 & x_n & x_n^2 & \cdots & x_n^{n-1} \end{pmatrix} \neq 0.$$

4. Solve equation

$$\det \begin{pmatrix} x & 2 & 1 \\ x & 1 & 2 \\ x & 1 & 3 \end{pmatrix} = 1.$$

5. Evaluate

$$\det \begin{pmatrix} x & 1 & 2 \\ 0 & y & 3 \\ 0 & 0 & z \end{pmatrix}.$$

6. Find

$$\det \begin{pmatrix} a & b & 0 & 0 \\ b & a & b & 0 \\ 0 & b & a & b \\ 0 & 0 & b & a \end{pmatrix}.$$

7. Is $\det(\mathbf{A} + \mathbf{B}) = \det(\mathbf{A}) + \det(\mathbf{B})$?

8. Is $\det(\mathbf{A} - \mathbf{B}) = \det(\mathbf{A}) - \det(\mathbf{B})$?

9. Is $\det(a\mathbf{A}) = a \det(\mathbf{A})$?

10. Prove that if two columns of a square matrix $\mathbf{A}$ are proportional, then $\det(\mathbf{A}) = 0$.

11. Prove that the converse of the statement of the previous Exercise is false.

12. Use determinants to solve the following systems of linear equations:

a) $x + y = 2$

$$x - y = 3$$

b) $2x - y = 2$

$$x + y = 1$$

c) $x + y + z = 2$

$$-y + 2z = -1$$

$$2x + 2y = 2$$

d) $x + 2y + z = 1$

$$x - 2y + z = 1$$

$$x + y - z = -1$$

13. Find the inverses of the coefficient matrices of the previous Exercise by solving the matrix equation $\mathbf{AX} = \mathbf{I}$ column-wise, and solving each system by the Cramer's rule.

14. Let $\mathbf{A}$ be a block-diagonal matrix:

$$\mathbf{A} = \begin{pmatrix} \mathbf{A}_1 & & & \mathbf{O} \\ & \mathbf{A}_2 & & \\ & & \ddots & \\ \mathbf{O} & & & \mathbf{A}_k \end{pmatrix},$$

where all blocks $\mathbf{A}_i$ ($i = 1, 2, \ldots, k$) are square matrices. Prove that $\det(\mathbf{A}) = \det(\mathbf{A}_1) \cdot \det(\mathbf{A}_2) \cdot \ldots \cdot \det(\mathbf{A}_k)$.

15. Assume that for an $n \times n$ real matrix, $\mathbf{A} \cdot \mathbf{A}^T = \mathbf{I}_n$ (such matrices are called orthogonal). Prove that $\det(\mathbf{A}) = \pm 1$.

16. Assume that for an $n \times n$ real matrix, $\mathbf{A}^k = \mathbf{0}$ (such matrices are called nilpotent). Prove that $\det(\mathbf{A}) = 0$.

17. Check the determinant multiplication theorem for 2×2 real matrices.

18. Apply the determinant multiplication theorem to prove that $\det(\alpha A) = \alpha^n \det(A)$, if A is an $n \times n$ real matrix and α is a real scalar.

19. Illustrate Theorem 4.1 and parts (a), (c), (e), and (f) of Theorem 4.2 with matrices

$$A = \begin{pmatrix} 1 & 2 & 1 \\ 0 & 1 & 1 \\ 3 & 1 & 2 \end{pmatrix}, \quad B = \begin{pmatrix} 1 & -1 & -1 \\ 2 & 1 & 2 \\ 0 & 1 & 3 \end{pmatrix}.$$

20. Illustrate relations (4.8) and (4.9) for matrix A of the previous Exercise.

21. Check the conditions of Theorem 4.6 for the linear systems of Exercise 12.

22. Find the ranks of the following matrices by using the definition of ranks and also by using Theorem 4.8:

a) $\begin{pmatrix} 1 & 2 & 2 & 3 \\ 2 & 1 & 1 & 0 \end{pmatrix}$

b) $\begin{pmatrix} 1 & 2 & 1 \\ 1 & 3 & 2 \\ 0 & 1 & 4 \end{pmatrix}$

c) $\begin{pmatrix} 1 & 2 & 3 & 4 \\ 0 & 1 & 2 & 3 \\ 4 & 3 & 2 & 1 \end{pmatrix}.$

23. Repeat Example 4.9 with matrices

a) $\begin{pmatrix} 1 & 1 & 1 \\ 1 & -1 & -1 \\ 2 & 1 & 7 \end{pmatrix}$

b) $\begin{pmatrix} 1 & 2 & 3 \\ 2 & 6 & 10 \\ -1 & -3 & -6 \end{pmatrix}$

c) $\begin{pmatrix} 1 & 2 & 3 \\ 2 & 4 & 9 \\ 4 & 3 & 2 \end{pmatrix}$.

Compare these results with those of Exercise 3.17.

24. If **A** is an $n \times n$ matrix, and A_{ij} denotes its (i, j) cofactor, then show that relation (4.11) implies that

$$\det \begin{pmatrix} A_{11} & A_{21} & \cdots & A_{n1} \\ A_{12} & A_{22} & \cdots & A_{n2} \\ \cdots & \cdots & \cdots & \cdots \\ A_{1n} & A_{2n} & \cdots & A_{nn} \end{pmatrix} = \left[\det(\mathbf{A}) \right]^{n-1}.$$

25. Illustrate the relation of the previous Exercise for matrices of Exercise 23.

26. Examine the complete controllability of system

$$\mathbf{x}(t+1) = \begin{pmatrix} 1 & -1 & 1 \\ 1 & 0 & 1 \\ 0 & 1 & 1 \end{pmatrix} \mathbf{x}(t) + \begin{pmatrix} 1 \\ 1 \\ 0 \end{pmatrix} u(t).$$

27. Repeat the previous Exercise with system

$$\mathbf{x}(t+1) = \begin{pmatrix} 1 & -1 & 1 \\ 1 & 0 & 1 \\ 0 & 1 & 1 \end{pmatrix} \mathbf{x}(t) + \begin{pmatrix} 1 & 0 \\ 1 & 1 \\ 0 & 1 \end{pmatrix} u(t).$$

28. Find the characteristic polynomials of the matrices of Exercise 23.

Chapter 5

Linear Mappings and Matrices

5.1 Introduction

The interrelation of different mathematical systems of similar types is usually studied by examining some mappings from one system to another. Such mappings are especially important to be examined that preserves some important characteristics of the systems. In earlier chapters the linear structure of vector spaces played the most significant role, therefore mappings preserving the linear structure will be the most important ones for future studies. This chapter is devoted to the examination of such mappings.

5.2 Vector Coordinates

Assume that V is a finitely generated vector space of arbitrary elements and let $B = \{\mathbf{u}_1, \mathbf{u}_2, \ldots, \mathbf{u}_n\}$ be a basis in V. From Theorem 2.7 we know that an arbitrary vector $\mathbf{x} \in V$ can be uniquely represented as the linear combination of the basis vectors. That is, there are unique scalars $x_1, \ldots, x_n$ such that

$$\mathbf{x} = x_1\mathbf{u}_1 + x_2\mathbf{u}_2 + \ldots + x_n\mathbf{u}_n. \tag{5.1}$$

The coefficients $x_1, \ldots, x_n$ are called the *coordinates* of $\mathbf{x}$ in basis B. We can therefore identify vector $\mathbf{x}$ with the n-element column vector

$$\mathbf{x}_B = \begin{pmatrix} x_1 \\ x_2 \\ \cdots \\ x_n \end{pmatrix},$$

which is called the *coordinate-vector* of $\mathbf{x}$ in basis B. Notice that in the case when a different basis is selected, the coefficients x_i as well as the coordinate-vector $\mathbf{x}_B$ may become different. This observation is illustrated in the following example.

Example 5.1. Select first the natural basis $\mathbf{B} = \{\mathbf{e}_1, \mathbf{e}_2, \mathbf{e}_3\}$ in R^3, and let

$$\mathbf{x} = \begin{pmatrix} 1 \\ 1 \\ 1 \end{pmatrix}.$$

Then obviously,

$$\mathbf{x} = 1 \cdot \mathbf{e}_1 + 1 \cdot \mathbf{e}_2 + 1 \cdot \mathbf{e}_3,$$

therefore,

$$\mathbf{x}_B = \begin{pmatrix} 1 \\ 1 \\ 1 \end{pmatrix}.$$

Assume next that the basis $B' = \{\mathbf{u}_1, \mathbf{u}_2, \mathbf{u}_3\}$ is selected, with

$$\mathbf{u}_1 = \begin{pmatrix} 1 \\ 1 \\ 1 \end{pmatrix}, \ \mathbf{u}_2 = \begin{pmatrix} 1 \\ 1 \\ 0 \end{pmatrix}, \ \mathbf{u}_3 = \begin{pmatrix} 1 \\ 0 \\ 0 \end{pmatrix}.$$

Then $\mathbf{x} = \mathbf{u}_1$, that is, $\mathbf{x} = 1 \cdot \mathbf{u}_1 + 0 \cdot \mathbf{u}_2 + 0 \cdot \mathbf{u}_3$ consequently,

$$\mathbf{x}_{B'} = \begin{pmatrix} 1 \\ 0 \\ 0 \end{pmatrix}.$$

♦

The uniqueness of the scalars $\mathbf{x}_1$, ..., $\mathbf{x}_n$ in representation (5.1) implies that mapping $\mathbf{x} \mapsto \mathbf{x}_B$ is one-to-one and it maps V into the space of n-element real (or complex) vectors. We will next show that the mapping preserves the linear structures of these vector spaces by showing that for arbitrary $\mathbf{x}, \mathbf{y} \in V$,

$$(\mathbf{x} + \mathbf{y})_B = \mathbf{x}_B + \mathbf{y}_B, \tag{5.2}$$

and for arbitrary $\mathbf{x} \in V$ and scalar a,

$$(a \cdot \mathbf{x})_B = a \cdot \mathbf{x}_B. \tag{5.3}$$

The first relation can be proved as follows. Assume that $\mathbf{x}_B = (x_i)$ and $\mathbf{y}_B = (y_i)$, then

$$\mathbf{x} = x_1 \mathbf{u}_1 + ... + x_n \mathbf{u}_n \quad \text{and} \quad \mathbf{y} = y_1 \mathbf{u}_1 + ... + y_n \mathbf{u}_n,$$

and so,

$$\mathbf{x} + \mathbf{y} = (x_1 + y_1) \mathbf{u}_1 + ... + (x_n + y_n) \mathbf{u}_n.$$

Hence, the coordinates of $\mathbf{x} + \mathbf{y}$ in basis B are $x_1 + y_1$, ..., $x_n + y_n$. The second relation can be proved in a similar way. If $\mathbf{x} = (x_i)$, then

$$a\mathbf{x} = a(x_1 \mathbf{u}_1 + ... + x_n \mathbf{u}_n) = (ax_1) \mathbf{u}_1 + ... + (ax_n) \mathbf{u}_n.$$

That is, the coordinates of $\mathbf{x}$ in basis B are ax_1, ..., ax_n, which proves the assertion. Equations (5.2) and (5.3) show that there is a special relation between V and the vector space of n-dimensional real (or complex) vectors. In the next section this concept will be generalized and examined in detail.

5.3 Linear Mappings

Let V and V' be two (not necessarily different) vector spaces, and assume that either both are real or both are complex, furthermore mapping A maps V into V'. That is, $A(\mathbf{x})$ is defined for all $\mathbf{x} \in V$, and the images $A(\mathbf{x})$ are in V' for all $\mathbf{x} \in V$. This fact is denoted as $A:V \mapsto V'$. It is not required that each element of V' can be obtained as an image $A(\mathbf{x})$ with some $\mathbf{x} \in V$. The set of all images is denoted by $R(A)$ and called the *range* of mapping A:

$$R(A) = \left\{ \mathbf{x}' \in V' \mid \text{ there exists } \mathbf{x} \in V \text{ such that } \mathbf{x}' = A(\mathbf{x}) \right\}$$

In this section a special class of mappings between vector spaces will be discussed.

Definition 5.1. Mapping $A:V \mapsto V'$ is called a *linear mapping*, if for all $\mathbf{x}, \mathbf{y} \in V$ and scalars a,

(i) $A(\mathbf{x} + \mathbf{y}) = A(\mathbf{x}) + A(\mathbf{y})$;

(ii) $A(a\mathbf{x}) = a \cdot A(\mathbf{x})$.

Notice that in property (ii), elements of V as well as elements of V' are multiplied by scalar a. This is the reason why we assumed that either both vector spaces are real (that is, only real numbers are considered as scalars), or both are complex (when complex numbers can be also considered to be scalars).

Notice that the mapping $\mathbf{x} \mapsto \mathbf{x}_B$ satisfies conditions (i) and (ii) which are equivalent to requirements (5.2) and (5.3). Thus, it is a linear mapping. As it was mentioned earlier, the range of a linear mapping does not need to be the entire space V'. For example, if one defines $A(\mathbf{x}) = \mathbf{0}'$ for all $\mathbf{x} \in V$, where $\mathbf{0}'$ is the zero element of V', then this is a special linear mapping, the range of which consists of only one element. In all cases, the range of a linear mapping has to be a subspace of V', as it is stated in the following theorem.

Theorem 5.1. Let $A:V \mapsto V'$ be a linear mapping. Then $R(A)$ is a subspace of V'.

Proof. By applying Theorem 2.2 we have to show that if $\mathbf{x}'$ and $\mathbf{y}'$ are in $R(A)$, then with arbitrary scalars a and b, $a\mathbf{x}'+b\mathbf{y}' \in R(A)$. Since $\mathbf{x}'$ and $\mathbf{y}'$ are in $R(A)$, there exist elements $\mathbf{x}$ and $\mathbf{y}$ in V such that $\mathbf{x}' = A(\mathbf{x})$ and $\mathbf{y}' = A(\mathbf{y})$. Since V is a vector space, $a\mathbf{x}+b\mathbf{y} \in V$, and the linearity of mapping A implies that

$$A(a\mathbf{x}+b\mathbf{y}) = A(a\mathbf{x}) + A(b\mathbf{y}) = aA(\mathbf{x}) + bA(\mathbf{y}) = a\mathbf{x}' + b\mathbf{y}' \in R(A).$$

♣

In verifying that a mapping is linear we have to check if conditions (i) and (ii) are satisfied. The following result gives only one condition as the sufficient and necessary condition for a mapping to be linear. So, in practical cases only one condition has to be checked.

Theorem 5.2. *A* mapping $A:V \mapsto V'$ is linear if and only if for all $\mathbf{x}, \mathbf{y} \in V$ and scalars a and b,

$$A(a\mathbf{x}+b\mathbf{y}) = a \cdot A(\mathbf{x}) + b \cdot A(\mathbf{y}). \tag{5.4}$$

Proof. Assume first that conditions (i) and (ii) hold. Then

$$A(a\mathbf{x}+b\mathbf{y}) = A(a\mathbf{x}) + A(b\mathbf{y}) = a \cdot A(\mathbf{x}) + b \cdot A(\mathbf{y}),$$

so (5.4) is satisfied. Suppose next that equation (5.4) holds for all $\mathbf{x}$, $\mathbf{y}$, a and b. First we show that $A(\mathbf{0}) = \mathbf{0}'$, where $\mathbf{0}$ and $\mathbf{0}'$ are the zero elements in V and V', respectively.
Select $\mathbf{x} = \mathbf{y} = \mathbf{0}$ and $a = b = 1$. Then

$$A(\mathbf{0}) = A(\mathbf{0} + \mathbf{0}) = A(\mathbf{0}) + A(\mathbf{0})$$

and adding $-A(\mathbf{0})$ to both sides we have

$$\mathbf{0}' = A(\mathbf{0}).$$

Select next $a = b = 1$. Then

$$A(\mathbf{x}+\mathbf{y}) = A(1\cdot\mathbf{x}+1\cdot\mathbf{y}) = 1\cdot A(\mathbf{x})+1\cdot A(\mathbf{y}) = A(\mathbf{x})+A(\mathbf{y})$$

that is, condition (i) is satisfied. Select next $\mathbf{y} = \mathbf{0}$, then

$$A(a\mathbf{x}) = A(a\mathbf{x}+b\mathbf{0}) = aA(\mathbf{x})+bA(\mathbf{0}) = aA(\mathbf{x})+b\mathbf{0}'$$
$$= aA(\mathbf{x})+\mathbf{0}' = aA(\mathbf{x}).$$

Hence condition (ii) also holds showing that A is a linear mapping.

♣

In the special case when $R(A) = V'$, we say that A is a mapping of V onto V'. A special but important class of such linear mappings is defined next.

Definition 5.2. Let A be a one-to-one linear mapping of V onto V'. Then A is called an *isomorphism*, and vector spaces V and V' are called *isomorphic*.

We will first show that finitely generated vector spaces are isomorphic if and only if they have the same dimension.

Theorem 5.3. Let V and V' be vector spaces, and assume that V is finitely generated. Then they are isomorphic if and only if V' is also finitely generated and $\dim(V) = \dim(V')$.

Proof. We first show that an isomorphism preserves linear independence and linear dependence by proving that the images of linearly independent elements are also linearly independent, and the images of linearly dependent elements are also linearly dependent. It is sufficient to show that elements $\mathbf{x}_1, \ldots, \mathbf{x}_k \in V$ are linearly dependent if and only if $A(\mathbf{x}_1), \ldots, A(\mathbf{x}_k)$ are linearly dependent in V'. Assume first that $\mathbf{x}_1, \ldots, \mathbf{x}_k$ are linearly dependent, then with some scalars,

$$a_1\mathbf{x}_1 + \ldots + a_k\mathbf{x}_k = \mathbf{0},$$

where at least one coefficient is nonzero. Then

$$a_1 A(\mathbf{x}_1) + \ldots + a_k A(\mathbf{x}_k) = A(a_1 \mathbf{x}_1) + \ldots + A(a_k \mathbf{x}_k)$$
$$= A(a_1 \mathbf{x}_1 + \ldots + a_k \mathbf{x}_k) = A(\mathbf{0}) = \mathbf{0}',$$

therefore $A(\mathbf{x}_1)$, ..., $A(\mathbf{x}_k)$ are also linearly dependent. Assume next that vectors $A(\mathbf{x}_1)$, ..., $A(\mathbf{x}_k)$ are linearly dependent, then with some scalars

$$c_1 A(\mathbf{x}_1) + \ldots + c_k A(\mathbf{x}_k) = \mathbf{0}'$$

and at least one coefficient is nonzero. This equation implies that

$$A(c_1 \mathbf{x}_1 + \ldots + c_k \mathbf{x}_k) = A(c_1 \mathbf{x}_1) + \ldots + A(c_k \mathbf{x}_k) = c_1 A(\mathbf{x}_1) + \ldots + c_k A(\mathbf{x}_k) = \mathbf{0}'.$$

Since A is one-to-one and $A(\mathbf{0}) = \mathbf{0}'$, necessarily

$$c_1 \mathbf{x}_1 + \ldots + c_k \mathbf{x}_k = \mathbf{0}$$

showing that elements $\mathbf{x}_1$, ..., $\mathbf{x}_k$ are linearly dependent.

The above observation immidiately implies that if V and V' are isomorphic and V is finitely generated, then the same is true for V', and $\dim(V') = \dim(V)$. We will next prove that if V and V' are finitely generated vector spaces and $\dim(V') = \dim(V)$, then they are isomorphic. Let $\{\mathbf{x}_1, \ldots, \mathbf{x}_n\}$ be a basis of V and let $\{\mathbf{z}_1, \ldots, \mathbf{z}_n\}$ be a basis in V'. If $\mathbf{x}$ is an arbitrary element of V, then $\mathbf{x}$ can be written as

$$\mathbf{x} = a_1 \mathbf{x}_1 + \ldots + a_n \mathbf{x}_n$$

with some scalars $a_i (i = 1, 2, \ldots, n)$ Define a mapping $A : V \mapsto V'$ as

$$A(\mathbf{x}) = a_1 \mathbf{z}_1 + \ldots + a_n \mathbf{z}_n.$$

In order to show that mapping A is an isomorphism between V and V' we have to show that A is a one-to-one linear mapping from V onto V'. It is clear that mapping A is defined for all $\mathbf{x} \in V$, since an arbitrary element of V can be expressed as a linear combination of the basis elements. To show that A is one-to-one assume that for some $\mathbf{x}$ and $\mathbf{y}$, $A(\mathbf{x}) = A(\mathbf{y})$. With some scalars $a_1, \ldots, a_n$ and $b_1, \ldots, b_n$,

$$\mathbf{x} = a_1\mathbf{x}_1 + ... + a_n\mathbf{x}_n \quad \text{and} \quad \mathbf{y} = b_1\mathbf{x}_1 + ... + b_n\mathbf{x}_n.$$

Then

$$A(\mathbf{x}) = a_1\mathbf{z}_1 + ... + a_n\mathbf{z}_n \quad \text{and} \quad A(\mathbf{y}) = b_1\mathbf{z}_1 + ... + b_n\mathbf{z}_n,$$

and since $A(\mathbf{x}) = A(\mathbf{y})$,

$$a_1\mathbf{z}_1 + ... + a_n\mathbf{z}_n = b_1\mathbf{z}_1 + ... + b_n\mathbf{z}_n.$$

That is,

$$(a_1 - b_1)\mathbf{z}_1 + ... + (a_n - b_n)\mathbf{z}_n = \mathbf{0}'.$$

The linear independence of the basis elements implies that $a_1 = b_1$, ..., $a_n = b_n$, consequently $\mathbf{x} = \mathbf{y}$, and therefore A is one-to-one. The linearity of mapping A can be proved by using Theorem 5.2. Let a and b be two scalars. Then

$$ a\mathbf{x} + b\mathbf{y} = a(a_1\mathbf{x}_1 + ... + a_n\mathbf{x}_n) + b(b_1\mathbf{x}_1 + ... + b_n\mathbf{x}_n) $$
$$ = (aa_1 + bb_1)\mathbf{x}_1 + ... + (aa_n + bb_n)\mathbf{x}_n, $$

therefore, the definition of mapping A implies that

$$ A(a\mathbf{x} + b\mathbf{y}) = (aa_1 + bb_1)\mathbf{z}_1 + ... + (aa_n + bb_n)\mathbf{z}_n $$
$$ = a(a_1\mathbf{z}_1 + ... + a_n\mathbf{z}_n) + b(b_1\mathbf{z}_1 + ... + b_n\mathbf{z}_n) = aA(\mathbf{x}) + bA(\mathbf{y}). $$

We will finally show that $R(A) = V'$, that is, A is a mapping from V onto V'. Let $\mathbf{z} \in V'$ be arbitrary, then with some scalars c_i ($i = 1, 2, ..., n$),

$$\mathbf{z} = c_1\mathbf{z}_1 + ... + c_n\mathbf{z}_n.$$

Consider the element $\mathbf{x} = c_1\mathbf{x}_1 + ... + c_n\mathbf{x}_n$. Then the definition of A implies that $A(\mathbf{x}) = c_1\mathbf{z}_1 + ... + c_n\mathbf{z}_n = \mathbf{z}$, therefore $\mathbf{z} \in R(A)$, which completes the proof.

♣

Some special linear mappings are examined in the next few examples.

Example 5.2. Select $V = R^n$ and $V' = R^m$, and let $\mathbf{A}$ be an $m \times n$ real matrix. For all $\mathbf{x} \in V$, define mapping A as $A(\mathbf{x}) = \mathbf{A} \cdot \mathbf{x}$. It is clear that $A(\mathbf{x})$ is defined for all $\mathbf{x} \in V$, since the product $\mathbf{A} \cdot \mathbf{x}$ exists for all $\mathbf{x} \in R^n$. Notice that $R(A)$ is not necessarily the entire space R^m. For example, if $\mathbf{A}$ is the zero matrix, then $\mathbf{A} \cdot \mathbf{x} = \mathbf{0}$ for all $\mathbf{x} \in V$, therefore in this case, $R(A) = \{\mathbf{0}\}$. Mapping A is linear, since for all $\mathbf{x}, \mathbf{y} \in V$ and scalars a and b,

$$A(a\mathbf{x} + b\mathbf{y}) = \mathbf{A} \cdot (a\mathbf{x} + b\mathbf{y}) = \mathbf{A} \cdot (a\mathbf{x}) + \mathbf{A} \cdot (b\mathbf{y}) = a\mathbf{A}\mathbf{x} + b\mathbf{A}\mathbf{y}$$
$$= aA(\mathbf{x}) + bA(\mathbf{y}).$$

Next we show that $R(A) = V'$ if and only if $\operatorname{rank}(\mathbf{A}) = m$. This is the consequence of the simple fact that if $\mathbf{x} = (x_i)$, then

$$A(\mathbf{x}) = \mathbf{A} \cdot \mathbf{x} = x_1 \mathbf{a}_1 + \ldots + x_n \mathbf{a}_n$$

where $\mathbf{a}_1, \mathbf{a}_2, \ldots, \mathbf{a}_n$ are the columns of $\mathbf{A}$. Therefore $R(A)$ coincides with the set of all linear combinations of the columns of $\mathbf{A}$, which is the subspace generated by the columns of $\mathbf{A}$. Since the dimension of this subspace equals $\operatorname{rank}(\mathbf{A})$,

$$\dim(R(A)) = \operatorname{rank}(\mathbf{A}).$$

Hence, $R(A) = V'$ if and only if $\dim(R(A)) = m$, which is equivalent to the condition that $\operatorname{rank}(\mathbf{A}) = m$.

Assume next that $\mathbf{A}$ is $n \times n$ and nonsingular. Then $\operatorname{rank}(\mathbf{A}) = n$, therefore $R(A) = R^n = V'$. That is, A is a mapping of R^n onto itself. It can also be shown that the mapping is one-to-one. In contrary to this assertion assume that with some $\mathbf{x}, \mathbf{y} \in R^n$, $A(\mathbf{x}) = A(\mathbf{y})$. Then $\mathbf{A} \cdot \mathbf{x} = \mathbf{A} \cdot \mathbf{y}$, which is equivalent to equality

$$\mathbf{A} \cdot (\mathbf{x} - \mathbf{y}) = \mathbf{0}.$$

Multiply both sides from the left by $\mathbf{A}^{-1}$ to have

$$\mathbf{x} - \mathbf{y} = \mathbf{0},$$

since $\mathbf{A}^{-1}\mathbf{A} = \mathbf{I}$ and $\mathbf{A}^{-1}\mathbf{0} = \mathbf{0}$. Thus, $\mathbf{x} = \mathbf{y}$.

Let $a \neq 0$ be a scalar, and select the special $n \times n$ nonsingular matrix $\mathbf{A} = a \cdot \mathbf{I}$, where $\mathbf{I}$ is the $n \times n$ identity matrix. In this special case,

$$A(\mathbf{x}) = \mathbf{A} \cdot \mathbf{x} = a \cdot \mathbf{Ix} = a \cdot \mathbf{x},$$

that is, the mapping multiplies each vector by scalar a.

◆

Example 5.3. Select again $V = R^n$ and let V' be the set of real numbers. Let the real numbers $a_1, \ldots, a_n$ be given. If $\mathbf{x} \in V$ is an arbitrary vector, then define

$$A(\mathbf{x}) = a_1 x_1 + \ldots + a_n x_n$$

where $x_1, \ldots, x_n$ are the elements of $\mathbf{x}$. This mapping is a special case of the one examined in the previous example with $m = 1$. Notice that $A(\mathbf{x})$ is a real number. In the literature such linear mappings are called *linear functionals*.

◆

Example 5.4. Let V be the set of all single variable real polynomials of degree not exceeding n. In Example 2.3 we verified that V is a vector space. If $p \in V$, then define mapping A as taking the derivative of p:

$$A(p) = \frac{d}{dt} p.$$

Notice first that $A(p)$ is a polynomial of degree not exceeding $n - 1$. Therefore, define V' as the set of all real polynomials of degree at most $n - 1$. Then A is a linear mapping of V onto V'. The linearity of the mapping follows from the elementary properties of differentiation:

$$A(ap + bq) = \frac{d}{dt}(ap + bq) = a\frac{d}{dt}p + b\frac{d}{dt}q$$

for all $p, q \in V$ and scalars a and b. Let

$$r(t) = c_0 + c_1 t + \ldots + c_{n-1} t^{n-1} \in V'$$

be arbitrary. Define

$$p(t) = C + c_0 t + \frac{c_1}{2} t^2 + \frac{c_2}{3} t^3 + \ldots + \frac{c_{n-1}}{n} t^n$$

with arbitrary constant term C. Simple differentiation shows that

$$A(p) = \frac{d}{dt} p = r.$$

Since C is not unique, the mapping is not one-to-one.

♦

Example 5.5. Let V be the set $C[a, b]$ of all continuous real functions defined on a finite, closed interval $[a, b]$. Define now mapping A as integrating the functions on interval $[a, b]$. That is, for all $f \in C[a,b]$,

$$A(f) = \int_a^b f(t)\,dt.$$

Since $A(f)$ is a real number, select $V' = R$. Mapping A is a linear mapping of $C[a, b]$ onto R. The linearity follows immidiately from the linearity of the integral, since for all $f, g \in C[a,b]$ and real constants c and d,

$$A(cf + dg) = \int_a^b \left(cf(t) + dg(t)\right)dt = c\int_a^b f(t)\,dt + d\int_a^b g(t)\,dt = cA(f) + dA(g).$$

The fact that the range of this mapping is the entire real line follows from the observation, that if for any arbitrary real number c we select

$$p(t) = \frac{c}{b-a} \quad \text{for all } t \in [a, b], \text{ then}$$

$$A(p) = \int_a^b \frac{c}{b-a}\,dt = c.$$

♦

We will next give a simple sufficient and necessary condition that a linear mapping is one-to-one. The condition will be based on the following concept.

Definition 5.3. Let $A: V \mapsto V'$ be a linear mapping. The *nullspace* of A is defined as

$$N(A) = \left\{ \mathbf{x} \in V \,\middle|\, A(\mathbf{x}) = \mathbf{0'} \right\}.$$

First, we show that $N(A)$ is always a subspace in V.

Theorem 5.4. Let $A: V \mapsto V'$ be a linear mapping. Then $N(A)$ is a subspace of V.

Proof. We will show that $N(A)$ satisfies the condition of Theorem 2.2. Let $\mathbf{x}, \mathbf{y} \in N(A)$ and a, b be scalars. Then

$$A(a\mathbf{x} + b\mathbf{y}) = aA(\mathbf{x}) + bA(\mathbf{y}) = a \cdot \mathbf{0'} + b \cdot \mathbf{0'} = \mathbf{0'},$$

therefore $a\mathbf{x} + b\mathbf{y} \in N(A)$. Hence $N(A)$ is a subspace.

♣

We are now ready to present the simple condition that guarantees that a linear mapping is one-to-one.

Theorem 5.5. *A* linear mapping A is one-to-one if and only if $N(A) = \{\mathbf{0}\}$.

Proof. Assume first that $N(A)$ consists of only the zero element. Assume furthermore that with some $\mathbf{x}$ and $\mathbf{y} \in V$, $A(\mathbf{x}) = A(\mathbf{y})$. Then

$$\mathbf{0'} = A(\mathbf{x}) - A(\mathbf{y}) = 1 \cdot A(\mathbf{x}) + (-1) \cdot A(\mathbf{y}) = A(1 \cdot \mathbf{x} + (-1) \cdot \mathbf{y}) = A(\mathbf{x} - \mathbf{y}),$$

therefore $\mathbf{x} - \mathbf{y} \in N(A)$. Consequently $\mathbf{x} - \mathbf{y} = \mathbf{0}$, that is, $\mathbf{x} = \mathbf{y}$ showing that mapping A is one-to-one.

Assume next that A is one-to-one. Let $\mathbf{z} \in N(A)$ be arbitrary, then $A(\mathbf{z}) = \mathbf{0}'$. Since $A(\mathbf{0}) = \mathbf{0}'$ (from the proof of Theorem 5.2) and A is one-to-one, $\mathbf{z} = \mathbf{0}$. Hence $N(A) = \{\mathbf{0}\}$.

♣

Assume next that V is a finitely generated vector space. If a linear mapping $A : V \mapsto V'$ is one-to-one, then V and $R(A)$ are *isomorphic*. If $\dim(V) = n$, then $\dim(R(A)) = n$. Since $N(A)$ consists of only the zero element, $\dim(N(A)) = 0$. We may therefore notice that in this case, $\dim(V) = \dim(R(A)) + \dim(N(A))$. As we will next show, this relation always holds regardless of the fact that the mapping is one-to-one or not.

Theorem 5.6. Let $A : V \mapsto V'$ be a linear mapping, and assume that V is finitely generated. Then

$$\dim(V) = \dim(R(A)) + \dim(N(A)).$$

Proof. Let $\{\mathbf{x}_1, \ldots, \mathbf{x}_k\}$ be a basis in $N(A)$ and assume that $\{A(\mathbf{y}_1), \ldots, A(\mathbf{y}_l)\}$ is a basis in $R(A)$. In order to prove the assertion, it is sufficient to show that system $\{\mathbf{x}_1, \ldots, \mathbf{x}_k, \mathbf{y}_1, \ldots, \mathbf{y}_l\}$ is a basis of V. To do so, we have to prove two things. First we have to verify that the system consists of linearly independent elements, and second, we have to show that the system generates V.

To prove linear independence, assume that

$$a_1 \mathbf{x}_1 + \ldots + a_k \mathbf{x}_k + b_1 \mathbf{y}_1 + \ldots + b_l \mathbf{y}_l = \mathbf{0} \tag{5.5}$$

with some scalars a_i and b_j ($i = 1, 2, \ldots, k; j = 1, 2, \ldots, l$). Apply mapping A to both sides of the equality. The linearity of A implies that

$$a_1 A(\mathbf{x}_1) + \ldots + a_k A(\mathbf{x}_k) + b_1 A(\mathbf{y}_1) + \ldots + b_l A(\mathbf{y}_l) = \mathbf{0}',$$

where $\mathbf{0}'$ is the zero element of V'. Since $\mathbf{x}_i$ ($i = 1, 2, \ldots, k$) is in the null-space of A, $A(\mathbf{x}_i) = \mathbf{0}'$ for all i.

Therefore,

$$b_1 A(\mathbf{y}_1) + \ldots + b_l A(\mathbf{y}_l) = \mathbf{0}'.$$

Notice that $\{A(\mathbf{y}_1), \ldots, A(\mathbf{y}_l)\}$ is a basis of $R(A)$, and the linear combination of the basis elements is zero only if all coefficients are equal to zero: $b_1 = \ldots = b_l = 0$. Then from (5.5) we have

$$a_1\mathbf{x}_1 + \ldots + a_k\mathbf{x}_k = \mathbf{0},$$

and by using the fact that $\{\mathbf{x}_1, \ldots, \mathbf{x}_k\}$ is a basis in $N(A)$, we conclude that $a_1 = \ldots = a_k = 0$. (Since a linear combination of basis elements is zero only if all coefficients equal zero.) Hence, all coefficients in (5.5) are equal to zero implying that vectors $\mathbf{x}_1, \ldots, \mathbf{x}_k, \mathbf{y}_1, \ldots, \mathbf{y}_l$ are linearly independent.

We have to show next that this system generates V. Let $\mathbf{v} \in V$ be an arbitrary element. Since $A(\mathbf{v}) \in R(A)$ and $\{A(\mathbf{y}_1), \ldots, A(\mathbf{y}_l)\}$ is a basis of $R(A)$,

$$A(\mathbf{v}) = c_1 A(\mathbf{y}_1) + \ldots + c_l A(\mathbf{y}_l)$$

with some scalars $c_1, \ldots, c_l$. The linearity of A implies that

$$A(\mathbf{v} - (c_1\mathbf{y}_1 + \ldots + c_l\mathbf{y}_l)) = \mathbf{0}',$$

from which we see that

$$\mathbf{v} - (c_1\mathbf{y}_1 + \ldots + c_l\mathbf{y}_l) \in N(A).$$

Since $\{\mathbf{x}_1, \ldots, \mathbf{x}_k\}$ is a basis of $N(A)$, this vector can be expressed as a linear combination of the basis elements:

$$\mathbf{v} - (c_1\mathbf{y}_1 + \ldots + c_l\mathbf{y}_l) = d_1\mathbf{x}_1 + \ldots + d_k\mathbf{x}_k$$

with some scalars $d_1, \ldots, d_k$. That is,

$$\mathbf{v} = d_1\mathbf{x}_1 + \ldots + d_k\mathbf{x}_k + c_1\mathbf{y}_1 + \ldots + c_l\mathbf{y}_l,$$

which shows that $\mathbf{v}$ is a linear combination of elements $\mathbf{x}_1, \ldots, \mathbf{x}_k, \mathbf{y}_1, \ldots, \mathbf{y}_l$.

Therefore system $\{x_1, \ldots, x_k, y_1, \ldots, y_l\}$ generates V. This fact and the linear independence of these vectors imply that the system is a basis in V.

♣

Corollary. Assume that V is finitely generated. A linear mapping $A : V \mapsto V'$ is one-to-one if and only if V and $R(A)$ are isomorphic, which holds if and only if $\dim(V) = \dim(R(A))$. From the assertion of the theorem we know that this relation holds if and only if $\dim(N(A)) = 0$ which is equvalent to the condition that $N(A)$ consists of only the zero vector. This observation shows that Theorem 5.5 is implied by Theorem 5.6.

Let $A : V \mapsto V'$ be a linear mapping. If A is one-to-one, then for all $z \in R(A)$, there is a unique $x \in V$ such that $A(x) = z$. If A is not one-to-one, then such an element x is not necessarily unique. For all $z \in R(A)$ denote

$$C(z) = \left\{ x \in V \,\middle|\, A(x) = z \right\}.$$

Set $C(z)$ containts all elements $x \in V$ which have the same image z in V' with respect to mapping A. We can easily show that if x_0 is an arbitrary element of $C(z)$, then

$$C(z) = \left\{ x_0 + y \,\middle|\, y \in N(A) \right\}, \tag{5.6}$$

that is, $C(z)$ is a linear manifold with directing space $N(A)$ (which follows from Definition 3.5). To show that relation (5.6) holds assume first that x and x_0 are in $C(z)$. Then $A(x) = A(x_0) = z$, which implies that

$$0' = A(x) - A(x_0) = A(x - x_0),$$

and so, $x - x_0 \in N(A)$.

Assume next that $y \in N(A)$ is arbitrary, then $A(x_0 + y) = A(x_0) + A(y) = z + 0' = z$, that is, $x_0 + y \in C(z)$.

From the definition of set $C(z)$ it is also obvious that for $z \neq z'$,

$$C(z) \cap C(z') = \varnothing,$$

where $\varnothing$ denotes the empty set, since for any common element $\mathbf{x}$ of $C(\mathbf{z})$ and $C(\mathbf{z}')$, the image $A(\mathbf{x})$ cannot be $\mathbf{z}$ and $\mathbf{z}'$ at the same time. Furthermore,

$$\bigcup_{\mathbf{z} \in R(A)} C(\mathbf{z}) = V,$$

since mapping A is defined for all elements of V. The last two observations can be summarized by saying that V is the union of the disjoint sets $C(\mathbf{z})$ for all $\mathbf{z} \in R(A)$. For simple reference, $C(\mathbf{z})$ will be called the *equivalence class associated to* $\mathbf{z}$.

A linear structure will be next defined on the set of the equivalence classes generated by a linear mapping A. The sum of classes $C(\mathbf{z})$ and $C(\mathbf{z}')$ is defined as the class $C(\mathbf{z} + \mathbf{z}')$ associated to $\mathbf{z} + \mathbf{z}'$, and for any scalar a, the a-multiple of $C(\mathbf{z})$ is defined as the class $C(a\mathbf{z})$ associated to $a\mathbf{z}$. We will prove that these two operations satisfy all conditions of Definition 2.1 implying that the equivalence classes form a vector space. Conditions (i) and (ii) follow from the commutativity and associativity of the addition defined in V. If the null element among the equivalence classes is defined as $C(\mathbf{0}')$, and $-C(\mathbf{z})$ is defined as $C(-\mathbf{z})$, then properties (iii) and (iv) are obviously satisfied. Relations (v), (vi), and (vii) follow from the fact that vector space V' satisfies the same conditions, and (viii) is implied by the simple fact that for all $\mathbf{z} \in R(A)$,

$$1 \cdot C(\mathbf{z}) = C(1 \cdot \mathbf{z}) = C(\mathbf{z}).$$

This vector space of the equivalence classes is called the *factor space of* A and denoted by V/A. Notice that C can be also considered as a mapping which maps $R(A)$ onto V/A. From the above observations it follows that C is linear and one-to-one, therefore is an isomophism, so vector spaces $R(A)$ and V/A are isomorphic.

5.4 The Vector Space of Linear Mappings

Let V and V' be (not necessarily different) vector spaces. Assume that either both are real or both are complex. Let $L(V, V')$ denote the set of all linear mappings from V into V'. If $A, B \in L(V, V')$, then we can define the sum of mappings A and B as the mapping assigning the sum $A(\mathbf{x}) + B(\mathbf{x})$ to all $\mathbf{x} \in V$. That is, $A + B$ is defined as

$$(A+B)(\mathbf{x}) = A(\mathbf{x}) + B(\mathbf{x}) \tag{5.7}$$

for all $\mathbf{x} \in V$. We can similarly define the scalar-multiple of a linear mapping. The c-multiple of A is the mapping that assigns $cA(\mathbf{x})$ to every $\mathbf{x} \in V$, that is,

$$(cA)(\mathbf{x}) = cA(\mathbf{x}). \tag{5.8}$$

Theorem 5.7. Mappings $A + B$ and aA are linear.

Proof. We will apply Theorem 5.2. Let a and b be scalars and assume that $\mathbf{x}$ and $\mathbf{y}$ are arbitrary elements of V. Then the linearity of A and B implies that

$$\begin{aligned}
(A+B)(a\mathbf{x}+b\mathbf{y}) &= A(a\mathbf{x}+b\mathbf{y}) + B(a\mathbf{x}+b\mathbf{y}) \\
&= aA(\mathbf{x}) + bA(\mathbf{y}) + aB(\mathbf{x}) + bB(\mathbf{y}) \\
&= a(A(\mathbf{x}) + B(\mathbf{x})) + b(A(\mathbf{y}) + B(\mathbf{y})) = a(A+B)(\mathbf{x}) + b(A+B)(\mathbf{y}),
\end{aligned}$$

and

$$\begin{aligned}
(cA)(a\mathbf{x}+b\mathbf{y}) &= c \cdot A(a\mathbf{x}+b\mathbf{y}) = c(aA(\mathbf{x}) + bA(\mathbf{y})) \\
&= a(cA(\mathbf{x})) + b(cA(\mathbf{y})) = a(cA)(\mathbf{x}) + b(cA)(\mathbf{y})
\end{aligned}$$

showing that both $A + B$ and cA are linear mappings.

♣

Next we will show that $L(V, V')$ is a vector space if addition and multiplication by scalars are defined as above. We will verify that all conditions of Definition 2.1 are satisfied:

(i) Since V' is a vector space, for all $\mathbf{x} \in V$,

$$(A+B)(\mathbf{x}) = A(\mathbf{x}) + B(\mathbf{x}) = B(\mathbf{x}) + A(\mathbf{x}) = (B+A)(\mathbf{x});$$

(ii) Similarly, for all $\mathbf{x} \in V$,

$$((A+B)+C)(\mathbf{x}) = (A+B)(\mathbf{x}) + C(\mathbf{x}) = (A(\mathbf{x}) + B(\mathbf{x})) + C(\mathbf{x})$$
$$= A(\mathbf{x}) + (B(\mathbf{x}) + C(\mathbf{x})) = A(\mathbf{x}) + (B+C)\mathbf{x} = (A+(B+C))(\mathbf{x});$$

(iii) The null element O in L(V, V') is defined as $O(\mathbf{x}) = 0'$ for all $\mathbf{x} \in V$. Then

$$(A+O)(\mathbf{x}) = A(\mathbf{x}) + O(\mathbf{x}) = A(\mathbf{x}) + 0' = A(\mathbf{x});$$

(iv) Define $-A$ as $(-1) \cdot A$ then

$$(A+(-A))(\mathbf{x}) = A(\mathbf{x}) + (-A)(\mathbf{x}) = A(\mathbf{x}) + (-1)A(\mathbf{x})$$
$$= 1 \cdot A(\mathbf{x}) + (-1) \cdot A(\mathbf{x}) = (1+(-1))A(\mathbf{x}) = 0 \cdot A(\mathbf{x}) = 0' = O(\mathbf{x})$$

for all $\mathbf{x} \in V$;

Conditions (v), (vi), and (vii) follow from the observations that for all $\mathbf{x} \in V$,

$$(a(A+B))(\mathbf{x}) = a(A+B)(\mathbf{x}) = a(A(\mathbf{x}) + B(\mathbf{x}))$$
$$= aA(\mathbf{x}) + aB(\mathbf{x}) = (aA)(\mathbf{x}) + (aB)(\mathbf{x}) = (aA+aB)(\mathbf{x}),$$

and

$$((a+b)A)(\mathbf{x}) = (a+b)A(\mathbf{x}) = aA(\mathbf{x}) + bA(\mathbf{x})$$
$$= (aA)(\mathbf{x}) + (bA)(\mathbf{x}) = (aA+bA)(\mathbf{x}),$$

and

$$((ab)A)(\mathbf{x}) = (ab)A(\mathbf{x}) = a(bA(\mathbf{x})) = a(bA)(\mathbf{x}) = (a(bA))(\mathbf{x}).$$

And finally, notice that property (viii) is obvious, since for all $\mathbf{x} \in V$,

$$(1 \cdot A)(\mathbf{x}) = 1 \cdot A(\mathbf{x}) = A(\mathbf{x}).$$

Consider next the special case when V' is a set of real numbers if V is a real vector space, or V' is the set of complex numbers if V is a complex vector space. Then the vector space $L(V, V')$ is called the *dual space of* V, and is denoted by V^*.

5.5 Multiplication of Linear Mappings, and Inverses

Let V, V' and V'' be vector spaces. It is assumed that either all are real or all are complex. Let $B \in L(V, V')$ and $A \in L(V', V'')$ be two linear mappings. The product of A and B is denoted by $A \cdot B$ and is defined as

$$(A \cdot B)(\mathbf{x}) = A(B(\mathbf{x}))$$

for all $\mathbf{x} \in V$. That is, mapping B maps $\mathbf{x}$ into $B(\mathbf{x})$, and then this image is mapped by A into $A(B(\mathbf{x}))$.

This definition is the same as the composition of functions A and B known from calculus, where the composition was denoted by $A \circ B$. In this book we will use the notation $A \cdot B$ to emphasize the algebraic structure of the set of linear mappings. Mapping $A \cdot B$ is defined on the entire space V and its range is a subset of V''. Furthermore, it is a linear mapping, since for all elements $\mathbf{x}, \mathbf{y} \in V$ and scalars a, b,

$$(A \cdot B)(a\mathbf{x} + b\mathbf{y}) = A(B(a\mathbf{x} + b\mathbf{y})) = A(a \cdot B(\mathbf{x}) + b \cdot B(\mathbf{y}))$$
$$= a \cdot A(B(\mathbf{x})) + b \cdot A(B(\mathbf{y})) = a \cdot (A \cdot B)(\mathbf{x}) + b \cdot (A \cdot B)(\mathbf{y}).$$

Then Theorem 5.2 implies that $A \cdot B \in L(V, V'')$.
The multiplication of linear mappings is not commutative in general, since if spaces V and V'' are different, then product $B \cdot A$ cannot be defined.
Notice that A maps V' into V'', and B maps V into V', therefore mapping B, in general, cannot be applied to the element $A(\mathbf{z}) \in V''$. If $V = V''$, then both $A \cdot B$ and $B \cdot A$ are defined, but they are not necessarily equal. Such an example is given next.

Example 5.6. Let $V = V' = V'' = R^2$, and defined mappings A and B as

$$A(\mathbf{x}) = \begin{pmatrix} 1 & 1 \\ -1 & -1 \end{pmatrix} \cdot \mathbf{x} \quad \text{and} \quad B(\mathbf{x}) = \begin{pmatrix} 1 & 1 \\ 1 & 1 \end{pmatrix} \cdot \mathbf{x}$$

In Example 5.2 we have seen that multiplication by a matrix is a linear mapping. If x_1 and x_2 denote the elements of $\mathbf{x}$, then

$$(A \cdot B)\mathbf{x} = A(B(\mathbf{x})) = A\left(\begin{pmatrix} 1 & 1 \\ 1 & 1 \end{pmatrix}\begin{pmatrix} x_1 \\ x_2 \end{pmatrix}\right)$$

$$= A\begin{pmatrix} x_1 + x_2 \\ x_1 + x_2 \end{pmatrix} = \begin{pmatrix} 1 & 1 \\ -1 & -1 \end{pmatrix} \cdot \begin{pmatrix} x_1 + x_2 \\ x_1 + x_2 \end{pmatrix} = \begin{pmatrix} 2x_1 + 2x_2 \\ -2x_1 - 2x_2 \end{pmatrix},$$

and

$$(B \cdot A)\mathbf{x} = B(A(\mathbf{x})) = B\left(\begin{pmatrix} 1 & 1 \\ -1 & -1 \end{pmatrix} \cdot \begin{pmatrix} x_1 \\ x_2 \end{pmatrix}\right)$$

$$= B\begin{pmatrix} x_1 + x_2 \\ -x_1 - x_2 \end{pmatrix} = \begin{pmatrix} 1 & 1 \\ 1 & 1 \end{pmatrix} \cdot \begin{pmatrix} x_1 + x_2 \\ -x_1 - x_2 \end{pmatrix} = \begin{pmatrix} 0 \\ 0 \end{pmatrix}.$$

Notice that $(A \cdot B)(\mathbf{x}) = (B \cdot A)(\mathbf{x})$ if and only if $x_1 + x_2 = 0$.

$\blacklozenge$

We will next show that the multiplication of mappings is associative and distributive.

Let $C \in L(V, V')$, $B \in L(V', V'')$, $A \in L(V'', V''')$, where V''' is a vector space of the same type as V, V', and V''. Then for all $\mathbf{x} \in V$,

$$((A \cdot B) \cdot C)(\mathbf{x}) = (A \cdot B)(C(\mathbf{x})) = A(B(C(\mathbf{x})))$$

$$= A((B \cdot C)(\mathbf{x})) = (A \cdot (B \cdot C))(\mathbf{x}),$$

and so,

$$(A \cdot B) \cdot C = A \cdot (B \cdot C).$$

Assume next that $A, B \in L(V', V'')$ and $C \in L(V, V')$. Then

$$((A+B) \cdot C)(\mathbf{x}) = (A+B)(C(\mathbf{x})) = A(C(\mathbf{x})) + B(C(\mathbf{x}))$$
$$= (A \cdot C)(\mathbf{x}) + (B \cdot C)\mathbf{x} = (A \cdot C + B \cdot C)(\mathbf{x}),$$

that is,

$$(A + B) \cdot C = A \cdot C + B \cdot C.$$

If $A \in L(V', V'')$ and $B, C \in L(V, V')$, then similarly,

$$(A \cdot (B+C))(\mathbf{x}) = A((B+C)(\mathbf{x})) = A(B(\mathbf{x}) + C(\mathbf{x}))$$
$$= A(B(\mathbf{x})) + A(C(\mathbf{x})) = (A \cdot B)(\mathbf{x}) + (A \cdot C)(\mathbf{x}) = (A \cdot B + A \cdot C)(\mathbf{x}),$$

which shows that

$$A \cdot (B + C) = A \cdot B + A \cdot C.$$

Consider next the special case when $V = V' = V''$. Instead of the general notation $L(V, V)$ we may now use the simplified notation $L(V)$ for the set of all linear mappings from V into itself. Such mappings are called *linear transformations* on V. If both A and B are linear transformations in $L(V)$, then $A \cdot B$ is also in $L(V)$. This simple observation gives the possibility to define polynomials of linear transformations. If $A \in L(V)$, then the square A^2 of A is defined as $A^2 = A \cdot A$, and in general, A^n is defined by the recursive relation $A^k = A \cdot A^{k-1}$. We also define $A^0 = I$ where I is the identity transformation for which $I(\mathbf{x}) = \mathbf{x}$ for all $\mathbf{x} \in V$.

Let

$$p(t) = a_0 + a_1 t + a_2 t^2 + \dots + a_n t^n \tag{5.9}$$

be next a single variable polynomial. If V is a real vector space, then we assume that the polynomial coefficients are all real, and if V is complex, then all coefficients are assumed to be real or complex. Then $p(A)$ is defined as

$$p(A) = a_0 I + a_1 A + a_2 A^2 + \dots + a_n A^n.$$

Powers and polynomials of linear transformations are illustrated in the following two examples.

Example 5.7. Assume that $V - R^n$ and $\mathbf{A}$ is an $n \times n$ real matrix. Define mapping A as $A(\mathbf{x}) = \mathbf{A} \cdot \mathbf{x}$ for all n-element real vectors $\mathbf{x}$. Then

$$A^2(\mathbf{x}) = A(A(\mathbf{x})) = A(\mathbf{A} \cdot \mathbf{x}) = \mathbf{A} \cdot (\mathbf{A} \cdot \mathbf{x}) = \mathbf{A}^2 \cdot \mathbf{x},$$

where $\mathbf{A}^2$ is the square of matrix $\mathbf{A}$. By using finite induction, we can prove that in general,

$$A^n(\mathbf{x}) = \mathbf{A}^n \cdot \mathbf{x}. \tag{5.10}$$

For $n = 1$ this relation is true. Assume that it holds for a positive integer n. Then it is true for $n + 1$, since

$$A^{n+1}(\mathbf{x}) = A(A^n(\mathbf{x})) = A(\mathbf{A}^n \cdot \mathbf{x}) = \mathbf{A} \cdot (\mathbf{A}^n \cdot \mathbf{x}) = \mathbf{A}^{n+1} \cdot \mathbf{x}.$$

Assume that a real polynomial is given by equation (5.9). Then

$$\begin{aligned}
p(A)(\mathbf{x}) &= \left(a_0 I + a_1 A + a_2 A^2 + \ldots + a_n A^n\right)(\mathbf{x}) \\
&= a_0 I(\mathbf{x}) + a_1 A(\mathbf{x}) + a_2 A^2(\mathbf{x}) + \ldots + a_n A^n(\mathbf{x}) \\
&= a_0 \mathbf{I} \cdot \mathbf{x} + a_1 \mathbf{A} \cdot \mathbf{x} + a_2 \mathbf{A}^2 \cdot \mathbf{x} + \ldots + a_n \mathbf{A}^n \cdot \mathbf{x} \\
&= \left(a_0 \mathbf{I} + a_1 \mathbf{A} + a_2 \mathbf{A}^2 + \ldots + a_n \mathbf{A}^n\right) \cdot \mathbf{x}.
\end{aligned}$$

$\blacklozenge$

Example 5.8. Select V as the set of all continuously differentiable real functions on an interval $[a, b]$. For all $f \in V$, define $A(f)$ as the derivative of f:

$$A(f) = \frac{d}{dt} f.$$

The square of mapping A cannot be defined on the entire V, since the selection of V implies that although $\dfrac{d}{dt}f$ is continuous but is not necessarily differentiable. That is, mapping A cannot be necessarily applied again to the derivative function $\dfrac{d}{dt}f$.

Select V next as the set of all single variable real polynomials, then $A(f) = \dfrac{d}{dt}f$ is also a real polynomial for all $f \in V$.

Therefore, transformation A can be applied again, and

$$A^2(f) = \frac{d^2}{dt^2}f,$$

and in general

$$A^n(f) = \frac{d^n}{dt^n}f.$$

◆

Before defining inverses of linear transformations we recall that the identity transformation I on an arbitrary vector space V was defined as

$$I(\mathbf{x}) = \mathbf{x}$$

for all $\mathbf{x} \in V$. If $A \in L(V)$ is any linear transformation, then from this definition it is clear that

$$A \cdot I = A \quad \text{and} \quad I \cdot A = A.$$

Let $A \in L(V)$ be a linear transformation on a vector space V. The inverse of A is defined as follows.

Definition 5.4. Linear transformation $A \in L(V)$ is called *invertible* if $R(A) = V$ and the transformation is one-to-one. The inverse A^{-1} of A is defined as for all $\mathbf{x} \in V$,

$$A^{-1}(\mathbf{x}) = \{\mathbf{y} \in V \mid \mathbf{x} = A(\mathbf{y})\}. \tag{5.11}$$

Notice first that $A^{-1}(\mathbf{x})$ has exactly one element, since A is one-to-one. Therefore A^{-1} is a mapping of V into itself. Using Theorem 5.2 we can easily prove that A^{-1} is also a linear transformation. Let a, b be two scalars and $\mathbf{x}, \mathbf{y} \in V$. Then

$$A^{-1}(a\mathbf{x} + b\mathbf{y}) = aA^{-1}(\mathbf{x}) + bA^{-1}(\mathbf{y}),$$

since

$$A\left(aA^{-1}(\mathbf{x}) + bA^{-1}(\mathbf{y})\right) = a \cdot A\left(A^{-1}(\mathbf{x})\right) + b \cdot A\left(A^{-1}(\mathbf{y})\right) = a\mathbf{x} + b\mathbf{y}.$$

Here we used the fact that for all $\mathbf{x} \in V$,

$$A\left(A^{-1}(\mathbf{x})\right) = \mathbf{x} \tag{5.12}$$

which follows from relation (5.11) by selecting $\mathbf{y} = A^{-1}(\mathbf{x})$.
Inverse transformations have the following properties.

Theorem 5.8. Properties of inverse transformations.
Let $A \in L(V)$ be invertible. Then
(i) $(A^{-1})^{-1} = A$;
(ii) $A \cdot A^{-1} = A^{-1} \cdot A = I$;
If $A, B, C \in L(V)$, then
(iii) $A \cdot B = C \cdot A = I$ implies that A is invertible and $B = C = A^{-1}$;
If $A, B \in L(V)$ are invertible, then
(iv) $A \cdot B$ is invertible and $(A \cdot B)^{-1} = B^{-1} \cdot A^{-1}$.

Proof.   (i) Transformation A^{-1} is invertible, since it satisfies the conditions of Definition 5.4.
First, $R(A^{-1}) = V$, since $R(A) = V$. Next, we show that A^{-1} is one-to-one. Assume that $A^{-1}(\mathbf{x}) = A^{-1}(\mathbf{y})$. Let $\mathbf{z}$ denote this common element, then from (5.11) we conclude that

$$\mathbf{x} = A(\mathbf{z}) \quad \text{and} \quad \mathbf{y} = A(\mathbf{z}),$$

that is, $\mathbf{x} = \mathbf{y}$. Therefore $(A^{-1})^{-1}$ exists. Let $\mathbf{u} \in V$ be arbitrary, and denote

$$\mathbf{v} = \left(A^{-1}\right)^{-1}(\mathbf{u}).$$

Denote A^{-1} by B, then

$$\mathbf{v} = B^{-1}(\mathbf{u}).$$

Applying B on both sides of this equality and using relation (5.12) we have

$$B(\mathbf{v}) = B\left(B^{-1}(\mathbf{u})\right) = \mathbf{u},$$

that is

$$A^{-1}(\mathbf{v}) = \mathbf{u}.$$

Apply next A on both sides and use relation (5.12) again to see that

$$\mathbf{v} = A\left(A^{-1}(\mathbf{v})\right) = A(\mathbf{u}),$$

that is,

$$\left(A^{-1}\right)^{-1}(\mathbf{u}) = A(\mathbf{u}),$$

which is equivalent to the assertion.

(ii) The first part, $A \cdot A^{-1} = I$, is equivalent to equality (5.12), and the second part is a consequence of the first part by replacing A by A^{-1} and using property (i).

(iii) First, we prove that A is invertible by showing that A is one-to-one and $R(A) = V$. Assume first that $A(\mathbf{x}) = A(\mathbf{y})$. Then

$$\mathbf{x} = I(\mathbf{x}) = (C \cdot A)\mathbf{x} = C\left(A(\mathbf{x})\right) = C\left(A(\mathbf{y})\right) = (C \cdot A)(\mathbf{y}) = I(\mathbf{y}) = \mathbf{y},$$

that is, A is one-to-one. Assume now that $\mathbf{x} \in V$ is an arbitrary element. Then $B(\mathbf{x}) \in V$, and

$$A(B(\mathbf{x})) = (A \cdot B)(\mathbf{x}) = I(\mathbf{x}) = \mathbf{x},$$

therefore $\mathbf{x} \in R(A)$, and so $R(A) = V$.

We prove next that $B = C$. For all $\mathbf{x} \in V$,

$$C(\mathbf{x}) = C(I(\mathbf{x})) = C((A \cdot B)(\mathbf{x})) = (C \cdot (A \cdot B))(\mathbf{x}) = ((C \cdot A) \cdot B)(\mathbf{x})$$
$$= (I \cdot B)(\mathbf{x}) = I(B(\mathbf{x})) = B(\mathbf{x}).$$

And finally, we verify that $B = C = A^{-1}$. For all $\mathbf{x} \in V$, similarly to the previous calculation we have

$$C(\mathbf{x}) = C(I(\mathbf{x})) = C((A \cdot A^{-1})(\mathbf{x})) = (C \cdot (A \cdot A^{-1}))(\mathbf{x})$$
$$= ((C \cdot A) \cdot A^{-1})(\mathbf{x})$$
$$= (I \cdot A^{-1})(\mathbf{x}) = I(A^{-1}(\mathbf{x})) = A^{-1}(\mathbf{x}).$$

(iv) Simple calculation shows that

$$(B^{-1} \cdot A^{-1}) \cdot (A \cdot B) = B^{-1}(A^{-1} \cdot (A \cdot B))$$
$$= B^{-1}((A^{-1} \cdot A) \cdot B) = B^{-1} \cdot (I \cdot B) = B^{-1} \cdot B = I,$$

similarly,

$$(A \cdot B) \cdot (B^{-1} \cdot A^{-1}) = A \cdot (B \cdot (B^{-1} \cdot A^{-1}))$$
$$= A \cdot ((B \cdot B^{-1}) \cdot A^{-1}) = A \cdot (I \cdot A^{-1}) = A \cdot A^{-1} = I.$$

Thus, the proof is completed.

♣

In the case of finitely generated subspaces property (iii) can be significantly simplified.

Theorem 5.9. Let V be a finitely generated vector space and $A \in L(V)$. Then the followings are equivalent to each other:

(a) Transformation A is invertible;

(b) There is a $B \in L(V)$ such that $A \cdot B = I$;

(c) There is a $C \in L(V)$ such that $C \cdot A = I$.

Proof. Notice first that (a) implies (b) and (c), since in the case of an invertible A, the selections $B = A^{-1}$ and $C = A^{-1}$ are suitable. Next, we show that (b) implies (a) by verifying that condition (b) implies that $R(A) = V$ and A is one-to-one. Let $\mathbf{v} \in V$ be arbitrary. Then $B(\mathbf{v}) \in V$, and

$$A\big(B(\mathbf{v})\big) = \big(A \cdot B\big)(\mathbf{v}) = I(\mathbf{v}) = \mathbf{v}$$

which shows that $\mathbf{v} \in R(A)$. That is, $R(A) = V$. To prove that A is one-to-one assume that $A(\mathbf{x}) = A(\mathbf{y})$ with some $\mathbf{x}, \mathbf{y} \in V, \mathbf{x} \neq \mathbf{y}$. Then $A(\mathbf{x} - \mathbf{y}) = \mathbf{0}$, therefore $N(A)$ has at least one nonzero element, and so $\dim(N(A)) \geq 1$. Then Theorem 5.6 implies that $\dim(R(A)) < n$, where n is the dimension of V. Since $(A \cdot B)(\mathbf{x}) = A(B(\mathbf{x})) \in R(A)$, $\dim(R(A \cdot B)) \leq \dim(R(A))$, therefore $\dim(R(A \cdot B)) < n$.

Since $R(I) = V$ and $\dim(R(I)) = n$, $A \cdot B \neq I$ and this contradiction implies that A^{-1} exists, and relations $\mathbf{x} = A(B(\mathbf{x}))$ (for all $\mathbf{x} \in V$) and (5.11) imply that $B(\mathbf{x}) = A^{-1}(\mathbf{x})$ for all $\mathbf{x} \in V$, that is, $B = A^{-1}$.

The fact that condition (c) implies (a) can be proved in a similar way, the details are left as an exercise.

♣

Remark. The assertion of the theorem does not hold if V is infinite dimensional, as it is illustrated in the following example.

Example 5.9. Let V be the set of all single variable real polynomials. Let $p(t) = a_0 + a_1 t + a_2 t^2 + \ldots + a_n t^n \in V$. Define

$$A(p)(t) = a_1 + 2a_2 t + \ldots + na_n t^{n-1} \ \big(= p'(t)\big)$$

and

$$B(p)(t) = a_0 t + \frac{a_1 t^2}{2} + \frac{a_2 t^3}{3} + \ldots + \frac{a_n t^{n+1}}{n+1} \quad \left(= \int_0^t p(\tau) d\tau \right).$$

It is easy to see that

$$(A \cdot B)(p)(t) = A(B(p))(t)$$

$$= \frac{d}{dt} \left(a_0 t + \frac{a_1 t^2}{2} + \frac{a_2 t^3}{3} + \ldots + \frac{a_n t^{n+1}}{n+1} \right) = p(t),$$

that is, $A \cdot B = I$, however A^{-1} does not exist, since A is not one-to-one. This fact follows from the observation that for an arbitrary real number a,

$$\frac{d}{dt}(p(t) + a) = \frac{d}{dt} p(t).$$

♦

5.6   Matrix Representations of Linear Mappings

In this section we assume that V and V' are finitely generated vector spaces, and either both are real or both are complex. Suppose that $B = \{\mathbf{u}_1, \ldots, \mathbf{u}_n\}$ and $B' = \{\mathbf{v}_1, \ldots, \mathbf{v}_m\}$ are bases in V and V', respectively. Let $A \in L(V, V')$ be a linear mapping of V into V'. If $\mathbf{x} \in V$ is an arbitrary element, then

$$\mathbf{x} = x_1 \mathbf{u}_1 + \ldots + x_n \mathbf{u}_n,$$

where the scalars $x_1, \ldots, x_n$ are the coordinates of $\mathbf{x}$ in basis B. Then the linearity of mapping A implies that

$$A(\mathbf{x}) = x_1 A(\mathbf{u}_1) + \ldots + x_n A(\mathbf{u}_n). \tag{5.13}$$

Notice that for all $i = 1, 2, \ldots, n$, $A(\mathbf{u}_i) \in V'$, therefore this vector can be expressed as the linear combination of the basis elements of V':

$$A\left(\mathbf{u}_i\right) = a_{1i}\mathbf{v}_1 + \ldots + a_{mi}\mathbf{v}_m \tag{5.14}$$

with some scalars $a_{1i}, \ldots, a_{mi}$. Substituting this equation into (5.13) we have

$$
\begin{aligned}
A(\mathbf{x}) &= x_1\left(a_{11}\mathbf{v}_1 + \ldots + a_{m1}\mathbf{v}_m\right) + \ldots + x_n\left(a_{1n}\mathbf{v}_1 + \ldots + a_{mn}\mathbf{v}_m\right) \\
&= \left(a_{11}x_1 + \ldots + a_{1n}x_n\right)\mathbf{v}_1 + \ldots + \left(a_{m1}x_1 + \ldots + a_{mn}x_n\right)\mathbf{v}_m.
\end{aligned}
$$

This equation shows that the coordinates of the image element $A(\mathbf{x})$ in basis B' are

$$a_{11}x_1 + \ldots + a_{1n}x_n, \ldots, a_{m1}x_1 + \ldots + a_{mn}x_n,$$

that is, its coordinate vector is the following:

$$
A(\mathbf{x})_{B'} = \begin{pmatrix} a_{11}x_1 + \ldots + a_{1n}x_n \\ \ldots \\ a_{m1}x_1 + \ldots + a_{mn}x_n \end{pmatrix}.
$$

Introduce the $m \times n$ matrix

$$
\mathbf{A}_{B, B'} = \begin{pmatrix} a_{11} & \cdots & a_{1n} \\ & \cdots & \\ a_{m1} & \cdots & a_{mn} \end{pmatrix}
$$

and coordinate vector

$$
\mathbf{x}_B = \begin{pmatrix} x_1 \\ \ldots \\ x_n \end{pmatrix}
$$

to see that

$$A(\mathbf{x})_{B'} = \mathbf{A}_{B, B'} \cdot \mathbf{x}_B. \tag{5.15}$$

This equality and constant matrix $\mathbf{A}_{B,B'}$ are called the *matrix-representation* of mapping A. In Example 5.2 we have seen that multiplying by an $m \times n$ real matrix is a linear mapping of R^n into R^m. This matrix representation shows that all linear mappings between finitely generated vector spaces can be represented as multiplication by constant matrices. Since the columns of $\mathbf{A}_{B,B'}$ are the coordinate vectors of the images $A(\mathbf{u}_1), ..., A(\mathbf{u}_n)$ of the basis elements of B, it is easy to see that $\dim(R(A)) = \mathrm{rank}(\mathbf{A}_{B,B'})$. Equalities (5.13) and (5.14) show that matrix $\mathbf{A}_{B,B'}$ can be constructed in the following way:

Step 1. Find the images $A(\mathbf{u}_1), ..., A(\mathbf{u}_n)$ of the basis elements;

Step 2. Compute the coordinate vectors of these images in basis B';

Step 3. Construct the $m \times n$ matrix the columns of which are the coordinate vectors of $A(\mathbf{u}_1), ..., A(\mathbf{u}_n)$, respectively.

This procedure is illustrated in the following examples.

Example 5.10. Let V and V' be two finitely generated vector spaces and assume that $A(\mathbf{x}) = \mathbf{0}'$ for all $\mathbf{x} \in V$, where $\mathbf{0}'$ denotes the zero element of V'. Then for all basis elements $\mathbf{u}_i$, $A(\mathbf{u}_i) = \mathbf{0}'$. Notice that all coordinates of the zero vector $\mathbf{0}'$ are equal to zero, since

$$\mathbf{0}' = 0 \cdot \mathbf{v}_1 + 0 \cdot \mathbf{v}_2 + ... + 0 \cdot \mathbf{v}_m.$$

Therefore, all columns of matrix $\mathbf{A}_{B,B'}$ are equal to the zero vector, consequently $\mathbf{A}_{B,B'}$ is the $m \times n$ zero matrix.

♦

Example 5.11. Let V be a finitely generated vector space with basis $B = \{\mathbf{u}_1, ..., \mathbf{u}_n\}$ and select $V' = V$ and $B' = B$. Define A as the identity mapping I, then for all basis elements

$$A(\mathbf{u}_i) = \mathbf{u}_i$$

with coordinate vector e_i, since

$$\mathbf{u}_i = 0 \cdot \mathbf{u}_1 + ... + 0 \cdot \mathbf{u}_{i-1} + 1 \cdot \mathbf{u}_i + 0 \cdot \mathbf{u}_{i+1} + ... + 0 \cdot \mathbf{u}_n.$$

Therefore, the columns of $\mathbf{A}_{B,B'}$ are $\mathbf{e}_1$, $\mathbf{e}_2$, ..., $\mathbf{e}_n$, respectively. That is, $\mathbf{A}_{B,B'}$ is the $n \times n$ identity matrix.

♦

Example 5.12. Let V be the set of single variable real polynomials of degree at most n, and select $V' = V$. Assume that mapping A is defined as $A(p) = \dfrac{d}{dt}p$ for all $p \in V$. From Example 2.20 we conclude that vector space V is $n + 1$ dimensional with a basis $B = \{1, t, t^2, ..., t^n\}$, since each polynomial of degree at most n can be uniquely represented as a linear combination of these elements. Select $B' = B$, which means that $\mathbf{u}_1 = \mathbf{v}_1 = 1$, $\mathbf{u}_2 = \mathbf{v}_2 = t,..., \mathbf{u}_{n+1} = \mathbf{v}_{n+1} = t^n$. The images of the basis elements are as follows:

$$A(\mathbf{u}_1) = 0, \ A(\mathbf{u}_2) = 1, \ A(\mathbf{u}_3) = 2t, \ ..., \ A(\mathbf{u}_{n+1}) = nt^{n-1}.$$

Using the same basis elements, we see that

$$A(\mathbf{u}_1) = 0 \cdot \mathbf{u}_1 + 0 \cdot \mathbf{u}_2 + ... + 0 \cdot \mathbf{u}_n + 0 \cdot \mathbf{u}_{n+1}$$
$$A(\mathbf{u}_2) = 1 \cdot \mathbf{u}_1 + 0 \cdot \mathbf{u}_2 + ... + 0 \cdot \mathbf{u}_n + 0 \cdot \mathbf{u}_{n+1}$$
$$A(\mathbf{u}_3) = 0 \cdot \mathbf{u}_1 + 2 \cdot \mathbf{u}_2 + ... + 0 \cdot \mathbf{u}_n + 0 \cdot \mathbf{u}_{n+1}$$
$$...............$$
$$A(\mathbf{u}_{n+1}) = 0 \cdot \mathbf{u}_1 + 0 \cdot \mathbf{u}_2 + ... + n \cdot \mathbf{u}_n + 0 \cdot \mathbf{u}_{n+1}.$$

That is, the coordinate vectors of these images are obtained as

$$A(\mathbf{u}_1)_{B'} = \begin{pmatrix} 0 \\ 0 \\ \vdots \\ 0 \\ 0 \end{pmatrix}, \ A(\mathbf{u}_2)_{B'} = \begin{pmatrix} 1 \\ 0 \\ \vdots \\ 0 \\ 0 \end{pmatrix}, \ A(\mathbf{u}_3)_{B'} = \begin{pmatrix} 0 \\ 2 \\ \vdots \\ 0 \\ 0 \end{pmatrix}, ..., A(\mathbf{u}_{n+1})_{B'} = \begin{pmatrix} 0 \\ 0 \\ \vdots \\ n \\ 0 \end{pmatrix}.$$

And finally, matrix $\mathbf{A}_{B,B'}$ is constructed as the matrix the columns of which are these coordinate vectors in the same order as above:

$$\mathbf{A}_{B,B'} = \begin{pmatrix} 0 & 1 & 0 & \cdots & 0 \\ 0 & 0 & 2 & \cdots & 0 \\ \vdots & \vdots & \vdots & & \\ 0 & 0 & 0 & \cdots & n \\ 0 & 0 & 0 & \cdots & 0 \end{pmatrix}.$$

The elements of this matrix are characterized as

$$a_{ij} = \begin{cases} i & \text{if } 1 \le i \le n \text{ and } j = i+1 \\ 0 & \text{otherwise.} \end{cases}$$

♦

In the previous two sections we have introduced and examined the sums, scalar multiples, products, and inverses of linear mappings. We will now see how their matrix representations are effected by these operations.

Let V and V' be finitely generated vector spaces. As always, it is assumed that either both are real or both are complex. Let $B = \{\mathbf{u}_1, \ldots, \mathbf{u}_n\}$ and $B' = \{\mathbf{v}_1, \ldots, \mathbf{v}_m\}$ be bases in V and V', respectively. Assume that A and C are two linear mappings of V into V'. Use equations (5.13) and (5.14) for both mappings to see that for all $\mathbf{x} \in V$,

$$A(\mathbf{x}) = x_1 A(\mathbf{u}_1) + \ldots + x_n A(\mathbf{u}_n),$$

and

$$C(\mathbf{x}) = x_1 C(\mathbf{u}_1) + \ldots + x_n C(\mathbf{u}_n).$$

Therefore

$$(A+C)(\mathbf{x}) = A(\mathbf{x}) + C(\mathbf{x}) = x_1 \left(A(\mathbf{u}_1) + C(\mathbf{u}_1) \right) + \ldots + x_n \left(A(\mathbf{u}_n) + C(\mathbf{u}_n) \right).$$

From equation (5.2) we know that for all i, the coordinate vector of $A(\mathbf{u}_i)$ + $C(\mathbf{u}_i)$ is the sum of the coordinate vectors of $A(\mathbf{u}_i)$ and $C(\mathbf{u}_i)$. Therefore, the columns in the matrix representation of mapping $A + C$ are the sums of the corresponding columns of the matrix representations of mappings A and C. That is, the matrix representation of $A + C$ can be obtained as the sum of the matrix representations of A and C:

$$(A+C)_{B,B'} = \mathbf{A}_{B,B'} + \mathbf{C}_{B,B'}. \tag{5.16}$$

If A is a linear mapping of V into V' and a is a scalar, then

$$(aA)(\mathbf{x}) = a \cdot A(\mathbf{x}) = a\big(x_1 A(\mathbf{u}_1)\big) + \ldots + a\big(x_n A(\mathbf{u}_n)\big)$$
$$= x_1\big(a \cdot A(\mathbf{u}_1)\big) + \ldots + x_n\big(a \cdot A(\mathbf{u}_n)\big).$$

From equation (5.3) we know that the coordinate vector of a scalar multiple of an element equals the same scalar multiple of the coordinate vector of that element. Therefore, each column of the matrix representation of aA can be obtained by multiplying each column of the matrix representation of A by a. That is,

$$(aA)_{B,B'} = a \cdot \mathbf{A}_{B,B'}. \tag{5.17}$$

In Section 5.4 we have seen that the set $L(V, V')$ of linear mappings of V into V' is a vector space. Notice that the correspondence $A \mapsto \mathbf{A}_{B,B'}$ is one-to-one, and relations (5.16) and (5.17) imply that it can be considered as a one-to-one linear mapping of $L(V, V')$ onto the set of all $m \times n$ real (or complex) matrices. Hence these vector spaces are isomorphic, and

$$\dim\big(L(V, V')\big) = m \cdot n,$$

since in Example 2.19 we have shown that the space of all $m \times n$ real or complex matrices is $m \cdot n$ dimensional.

Assume next that V, V', and V'' are finitely generated vector spaces, either all are real or all are complex. Let B, B' and B'' be a basis of V, V', and V'', respectively. If $C \in L(V, V')$ and $A \in L(V', V'')$, then for all $\mathbf{x} \in V$,

$$(AC)(\mathbf{x}) = A(C(\mathbf{x})).$$

Let $\mathbf{x}_B$ be the coordinate vector of $\mathbf{x}$ in basis B, and let $\mathbf{A}_{B',B''}$ and $\mathbf{C}_{B,B'}$ be the matrix representations of mappings A and C, respectively. The repeated application of equality (5.15) implies that

$$(AC)(\mathbf{x})_{B''} = A(C(\mathbf{x}))_{B''} = \mathbf{A}_{B',B''} \cdot C(\mathbf{x})_{B'}$$
$$= \mathbf{A}_{B',B''} \cdot \mathbf{C}_{B,B'} \cdot \mathbf{x}_B = \left(\mathbf{A}_{B',B''} \cdot \mathbf{C}_{B,B'}\right) \cdot \mathbf{x}_B .$$

Hence, we have shown that relation

$$(A \cdot C)_{B,B''} = \mathbf{A}_{B',B''} \cdot \mathbf{C}_{B,B'}, \tag{5.18}$$

is true, that is the matrix representation of a product of two mappings equals the product of their matrix representations.

Consider finally the special case when $V' = V$. Let $A \in L(V)$ be an invertible linear transformation, and let C denote the inverse of A. Then $A \cdot C = I$, and combining relation (5.18) with the matrix representation of the identity transformation derived in Example 5.11 we immediately see that

$$\mathbf{I} = (I)_{B,B} = (A \cdot C)_{B,B} = \mathbf{A}_{B,B} \cdot \mathbf{C}_{B,B}.$$

That is, $\mathbf{C}_{B,B}$ is the inverse of $\mathbf{A}_{B,B}$. This observation can be summarized as

$$\left(A^{-1}\right)_{B,B} = \mathbf{A}_{B,B}^{-1}, \tag{5.19}$$

which can be interpreted by saying that the matrix representation of the inverse of a linear transformation is the inverse of the matrix representation of the original transformation.

Comparing the results of this and the previous sections with the elements of matrix algebra outlined in Chapter 1, an analogy becomes clear.

Addition, multiplication by scalars, products, and inverses of linear mappings have the same properties as the corresponding operations defined on real or complex matrices.

As the conclusion of this section, the special class of linear mappings from a finitely generated real (or complex) vector space V into the set of real (or complex) numbers will be briefly examined. As we mentioned earlier in Example 5.3, such linear mappings are called *linear functionals*. The set of all real (or complex) valued linear functionals defined on V is called the *dual space of V*. This notion has been already introduced at the end of Section 5.4. Let n denote the dimension of V. Since V' is one dimensional, the matrix representation of any linear functional is a $1 \times n$ matrix, which is actually a row vector. If A denotes a linear functional, and $B = \{\mathbf{u}_1, ..., \mathbf{u}_n\}$ is a given basis of V, then the matrix representation of mapping A in basis B is given as

$$\mathbf{A}_{B,B} = \left(A(\mathbf{u}_1), ..., A(\mathbf{u}_n) \right).$$

This row vector can be viewed as an element of R^n (or C^n), therefore the dual space of an n-dimensional vector space V is isomorphic to R^n (or C^n).

5.7 Coordinates and Matrix Representation in a New Basis

Let V' denote a finitely generated vector space and let $B = \{\mathbf{u}_1, ..., \mathbf{u}_n\}$ be a basis of V. In Section 5.1 we have seen that there is a unique correspondence between an arbitrary element $\mathbf{x} \in V$ and its coordinate vector $\mathbf{x}_B$. In Example 5.1 we have seen that the coordinate vector usually changes if a new basis is selected in V. We will first examine how to obtain the new coordinate vector efficiently.

Assume that $B' = \{\mathbf{v}_1, ..., \mathbf{v}_n\}$ is a new basis in V. From Corollary 1 of Theorem 2.9 we know that B and B' must have the same number of elements. For $j = 1, 2, ..., n$, let $\mathbf{v}_{jB}$ denote the coordinate vector of $\mathbf{v}_j$ in basis B, and define matrix $\mathbf{T}$ as the $n \times n$ matrix with columns $\mathbf{v}_{1B}, \mathbf{v}_{2B}, ..., \mathbf{v}_{nB}$. If t_{ij} denotes the (i, j) element of matrix $\mathbf{T}$, then for all j,

$$\mathbf{v}_{jB} = \begin{pmatrix} t_{1j} \\ t_{2j} \\ \dots \\ t_{nj} \end{pmatrix},$$

and therefore

$$\mathbf{v}_j = t_{1j}\mathbf{u}_1 + \dots + t_{nj}\mathbf{u}_n. \tag{5.20}$$

Assume that $\mathbf{x} \in V$ is an arbitrary element. Let $\mathbf{x}_B = (x_i)$ and $\mathbf{x}_{B'} = (x'_j)$ denote its coordinate vector in bases B and B', respectively. The definition of the coordinates implies that

$$\mathbf{x} = \sum_{i=1}^{n} x_i \mathbf{u}_i = \sum_{j=1}^{n} x'_j \mathbf{v}_j,$$

and by combining this equation with (5.20) we have

$$\sum_{i=1}^{n} x_i \mathbf{u}_i = \sum_{j=1}^{n} x'_j \mathbf{v}_j = \sum_{j=1}^{n} x'_j \left(\sum_{i=1}^{n} t_{ij} \mathbf{u}_i \right) = \sum_{i=1}^{n} \left(\sum_{j=1}^{n} t_{ij} x'_j \right) \mathbf{u}_i.$$

Since $B = \{\mathbf{u}_1, \dots, \mathbf{u}_n\}$ is a basis in B and each element of V can be uniquely represented as the linear combination of the basis elements, we conclude that for $i = 1, 2, \dots, n$,

$$x_i = \sum_{j=1}^{n} t_{ij} x'_j.$$

Notice that x_i is the ith element of the coordinate vector $\mathbf{x}_B$, and $\sum_{j=1}^{n} t_{ij} x'_j$ is the ith element of the product $\mathbf{T} \cdot \mathbf{x}_{B'}$. Therefore, we can summarize the above derivation as

$$\mathbf{x}_B = \mathbf{T} \cdot \mathbf{x}_{B'}. \tag{5.21}$$

Here **T** is called the *transformation matrix*. We will next prove that **T** is a nonsingular matrix. The columns of **T** are the coordinate vectors of the new basis elements $v_1, \ldots, v_n$ in basis B. The mapping $x \mapsto x_B$ is linear and one-to-one, so it is an isomorphism. From the proof of Theorem 5.3 we can conclude that the linear independence of the basis elements $v_1, \ldots, v_n$ implies the same for the coordinate vectors. Therefore, the columns of **T** are linearly independent, which implies that T^{-1} exists. Multiply both sides of equation (5.21) by T^{-1} to obtain an equivalent form:

$$x_{B'} = T^{-1} x_B. \tag{5.22}$$

Relations (5.21) and (5.22) are usually called the *basis transformation equations*, and their use is illustrated in the next example.

Example 5.13. Let $V = R^3$ and $B = \{e_1, e_2, e_3\}$, furthermore

$$x = \begin{pmatrix} 1 \\ 1 \\ 1 \end{pmatrix}$$

as in Example 5.1. Since

$$x = 1 \cdot e_1 + 1 \cdot e_2 + 1 \cdot e_3,$$

the coordinate vector of **x** in basis B is the following:

$$x_B = \begin{pmatrix} 1 \\ 1 \\ 1 \end{pmatrix}.$$

Select the new basis

$$\mathbf{v}_1 = \begin{pmatrix} 1 \\ 1 \\ 1 \end{pmatrix}, \; \mathbf{v}_2 = \begin{pmatrix} 1 \\ 1 \\ 0 \end{pmatrix}, \; \mathbf{v}_3 = \begin{pmatrix} 1 \\ 0 \\ 0 \end{pmatrix}.$$

Then

$$\mathbf{v}_1 = 1 \cdot \mathbf{e}_1 + 1 \cdot \mathbf{e}_2 + 1 \cdot \mathbf{e}_3,$$
$$\mathbf{v}_2 = 1 \cdot \mathbf{e}_1 + 1 \cdot \mathbf{e}_2 + 0 \cdot \mathbf{e}_3,$$
$$\mathbf{v}_3 = 1 \cdot \mathbf{e}_1 + 0 \cdot \mathbf{e}_2 + 0 \cdot \mathbf{e}_3,$$

and so

$$\mathbf{v}_{1B} = \begin{pmatrix} 1 \\ 1 \\ 1 \end{pmatrix}, \; \mathbf{v}_{2B} = \begin{pmatrix} 1 \\ 1 \\ 0 \end{pmatrix}, \; \mathbf{v}_{3B} = \begin{pmatrix} 1 \\ 0 \\ 0 \end{pmatrix}.$$

Therefore

$$\mathbf{T} = \begin{pmatrix} 1 & 1 & 1 \\ 1 & 1 & 0 \\ 1 & 0 & 0 \end{pmatrix},$$

where we have copied coordinate vectors $\mathbf{v}_{1B}, \mathbf{v}_{2B}, \mathbf{v}_{3B}$ into the columns of $\mathbf{T}$. The simple application of the elimination method shows that

$$\mathbf{T}^{-1} = \begin{pmatrix} 0 & 0 & 1 \\ 0 & 1 & -1 \\ 1 & -1 & 0 \end{pmatrix}.$$

Then equality (5.22) implies that

$$\mathbf{x}_{B'} = \begin{pmatrix} 0 & 0 & 1 \\ 0 & 1 & -1 \\ 1 & -1 & 0 \end{pmatrix} \begin{pmatrix} 1 \\ 1 \\ 1 \end{pmatrix} = \begin{pmatrix} 1 \\ 0 \\ 0 \end{pmatrix},$$

which coincides with the result obtained in Example 5.1.

♦

Consider next a linear transformation $A \in L(V)$ on V. If B is a basis of V, then $\mathbf{A}_{B,B}$ denotes the matrix representation of A. Assume now that a new basis B' is selected. Using equations (5.21) and (5.22) we can easily find the matrix representation of A in the new basis. Let $\mathbf{x} \in V$ be an arbitrary element. Then

$$\left(A(\mathbf{x})\right)_B = \mathbf{A}_{B,B} \cdot \mathbf{x}_B$$

and by using relation (5.21) we have

$$\mathbf{T} \cdot \left(A(\mathbf{x})\right)_{B'} = \mathbf{A}_{B,B} \cdot \mathbf{T} \cdot \mathbf{x}_{B'}.$$

Multiply both sides by $\mathbf{T}^{-1}$ to get

$$\left(A(\mathbf{x})\right)_{B'} = \mathbf{T}^{-1} \mathbf{A}_{B,B} \mathbf{T} \mathbf{x}_B,$$

which implies that the matrix representation of A in basis B' is the following:

$$\mathbf{A}_{B',B'} = \mathbf{T}^{-1} \mathbf{A}_{B,B} \mathbf{T}. \tag{5.23}$$

Definition 5.5. The $n \times n$ matrices $\mathbf{A}$ and $\mathbf{C}$ are called *similar* if there exists on invertible matrix $\mathbf{X}$ such that

$$\mathbf{C} = \mathbf{X}^{-1} \mathbf{A} \mathbf{X}.$$

Relation (5.23) shows that the matrix representations of a linear transformation $A \in L(V)$ are similar matrices. It can also be proved that if matrix $\mathbf{A}$ is the matrix representation of a mapping $A \in L(V)$ in a given basis B and matrix $\mathbf{C}$ is similar to $\mathbf{A}$, then there is another basis B' of V such that $\mathbf{C}$ is the matrix representation of A in B'. That is, similar matrices can be viewed as the matrix representations of the same linear transformations in different bases. Therefore, the ranks of similar matrices are equal. The use of equation (5.23) is illustrated in the following example.

Example 5.14. Let $V = R^3$ with the natural basis $B = \{e_1, e_2, e_3\}$. Assume that the matrix representation of a transformation A is given by the 3×3 matrix

$$\mathbf{A}_{B,B} = \begin{pmatrix} 1 & 1 & 1 \\ 1 & 2 & 1 \\ 1 & 1 & 2 \end{pmatrix}.$$

Select the new basis B' with elements

$$\mathbf{v}_1 = \begin{pmatrix} 1 \\ 1 \\ 1 \end{pmatrix}, \ \mathbf{v}_2 = \begin{pmatrix} 1 \\ 1 \\ 0 \end{pmatrix}, \quad \text{and} \quad \mathbf{v}_3 = \begin{pmatrix} 1 \\ 0 \\ 0 \end{pmatrix}.$$

The transformation matrix and its inverse have been determined in the previous example:

$$\mathbf{T} = \begin{pmatrix} 1 & 1 & 1 \\ 1 & 1 & 0 \\ 1 & 0 & 0 \end{pmatrix} \text{ and } \mathbf{T}^{-1} = \begin{pmatrix} 0 & 0 & 1 \\ 0 & 1 & -1 \\ 1 & -1 & 0 \end{pmatrix}.$$

Then the matrix representation of A in the new basis B' is the following:

$$\mathbf{A}_{B',B'} = \mathbf{T}^{-1} \mathbf{A}_{B,B} \mathbf{T} = \begin{pmatrix} 0 & 0 & 1 \\ 0 & 1 & -1 \\ 1 & -1 & 0 \end{pmatrix} \begin{pmatrix} 1 & 1 & 1 \\ 1 & 2 & 1 \\ 1 & 1 & 2 \end{pmatrix} \begin{pmatrix} 1 & 1 & 1 \\ 1 & 1 & 0 \\ 1 & 0 & 0 \end{pmatrix} = \begin{pmatrix} 4 & 2 & 1 \\ 0 & 1 & 0 \\ -1 & -1 & 0 \end{pmatrix}.$$

$\blacklozenge$

5.8 Applications

In this section some linear transformations defined on the two-dimensional plane will be first examined, and then special integral-transformations will be introduced.

1. Vectors are identified with directed line segments in the plane. Each vector therefore corresponds to a point in the plane given its position vector from the origin $\mathbf{0} = \begin{pmatrix} 0 \\ 0 \end{pmatrix}$ as shown in Figure 5.1. If r is the length of the vector and α is the angle between the vector and the positive half of the horizontal axis then

$$x_1 = r \cdot \cos \alpha \text{ and } x_2 = r \cdot \sin \alpha.$$

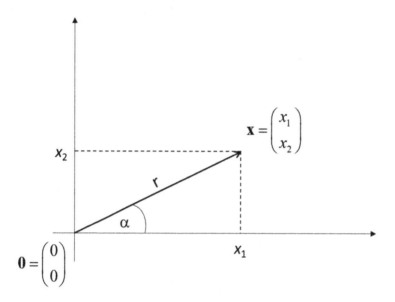

Figure 5.1. Vectors in the plane

Consider first the transformation A which multiplies a vector by a scalar α. Consider the basis $B = \{\mathbf{e}_1, \mathbf{e}_2\}$, where $\mathbf{e}_1 = \begin{pmatrix} 1 \\ 0 \end{pmatrix}$ and $\mathbf{e}_2 = \begin{pmatrix} 0 \\ 1 \end{pmatrix}$. Since

$$A(\mathbf{e}_1) = \begin{pmatrix} \alpha \\ 0 \end{pmatrix} \quad \text{and} \quad A(\mathbf{e}_2) = \begin{pmatrix} 0 \\ \alpha \end{pmatrix},$$

the matrix representation of this transformation is

$$\mathbf{A}_\alpha = \begin{pmatrix} \alpha & 0 \\ 0 & \alpha \end{pmatrix} = \alpha \mathbf{I}.$$

Consider next the transformation B_α which rotates each vector with a given angle α around the origin. As Figure 5.2 shows,

$$B_\alpha(\mathbf{e}_1) = \begin{pmatrix} \cos\alpha \\ \sin\alpha \end{pmatrix} \quad \text{and} \quad B_\alpha(\mathbf{e}_2) = \begin{pmatrix} \cos\left(\alpha + \dfrac{\pi}{2}\right) \\ \sin\left(\alpha + \dfrac{\pi}{2}\right) \end{pmatrix} = \begin{pmatrix} -\sin\alpha \\ \cos\alpha \end{pmatrix}.$$

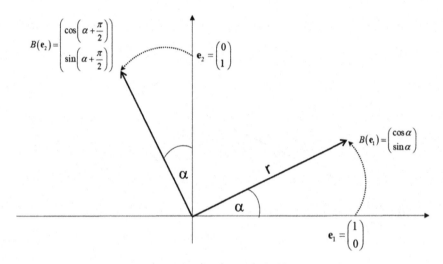

Figure 5.2. Illustration of a rotation

Therefore, the matrix representation of this transformation is

$$\mathbf{B}_\alpha = \begin{pmatrix} \cos\alpha & -\sin\alpha \\ \sin\alpha & \cos\alpha \end{pmatrix}.$$

Consider now the transformation C that gives the mirror image of each vector with respect to the vertical axis. Figure 5.3 shows that

$$C(\mathbf{e}_1) = \begin{pmatrix} -1 \\ 0 \end{pmatrix} \quad \text{and} \quad C(\mathbf{e}_2) = \begin{pmatrix} 0 \\ 1 \end{pmatrix}.$$

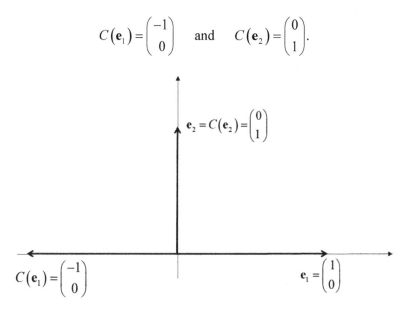

Figure 5.3. Illustration of transformation that gives mirror image
with respect to vertical axis

Therefore, the matrix representation is as follows:

$$C = \begin{pmatrix} -1 & 0 \\ 0 & 1 \end{pmatrix}.$$

Assume that transformation D gives the mirror image of each vector with respect to the horizontal axis. From Figure 5.4 we see that

$$D(\mathbf{e}_1) = \begin{pmatrix} 1 \\ 0 \end{pmatrix} \quad \text{and} \quad D(\mathbf{e}_2) = \begin{pmatrix} 0 \\ -1 \end{pmatrix},$$

therefore, the matrix representation is given as

$$D = \begin{pmatrix} 1 & 0 \\ 0 & -1 \end{pmatrix}.$$

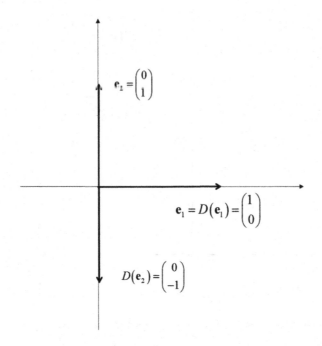

Figure 5.4. Illustration of transformation that gives mirror image with
respect to horizontal axis

The transformation E which gives the mirror image of each vector with
respect the origin is equivalent to a rotation by π. Therefore, the matrix
representation of E is the following:

$$E = \begin{pmatrix} -1 & 0 \\ 0 & -1 \end{pmatrix}.$$

Easy to see that these transformations satisfy the following relations:

$$\mathbf{A}_0 = \mathbf{O}, \ \mathbf{A}_1 = \mathbf{I},$$
$$\mathbf{B}_0 = \mathbf{I}, \ \mathbf{B}_\pi = -\mathbf{I} = \mathbf{E},$$

and

$$\mathbf{B}_\alpha \mathbf{B}_\beta = \mathbf{B}_{\alpha+\beta},$$

since repeated rotations by angles α and β are equivalent to a single rotation by angle $\alpha + \beta$. Notice that this relation can be rewritten as

$$\begin{pmatrix} \cos\alpha & -\sin\alpha \\ \sin\alpha & \cos\alpha \end{pmatrix}\begin{pmatrix} \cos\beta & -\sin\beta \\ \sin\beta & \cos\beta \end{pmatrix} = \begin{pmatrix} \cos(\alpha+\beta) & -\sin(\alpha+\beta) \\ \sin(\alpha+\beta) & \cos(\alpha+\beta) \end{pmatrix}.$$

Compare the corresponding elements on the left and right hand sides to see that

$$\cos\alpha\cos\beta - \sin\alpha\sin\beta = \cos(\alpha+\beta)$$
$$\sin\alpha\cos\beta + \cos\alpha\sin\beta = \sin(\alpha+\beta).$$

This simple derivation provides an elegant new proof of these well-known trigonometric identities.
Notice furthermore that

$$D = -C \quad \text{and} \quad D^2 = C^2 = I.$$

This last identity can be interpreted as

$$D^{-1} = D \quad \text{and} \quad C^{-1} = C.$$

2. Let $V = C^2[a, b]$ be the set of twice continously differentiable functions on interval $[a, b]$ and consider mapping T defined as

$$y(t) \mapsto y''(t) - p(t)y'(t) - q(t)y(t)$$

with continous functions $p(t)$ and $q(t)$. Clearly this mapping is linear, and functions satisfying the boundary conditions $y(a) = y(b) = 0$ is a subspace of V.
The solutions of the "homogenous" boundary value problem

$$y''(t) - p(t)y'(t) - q(t)y(t) = 0, \qquad y(a) = y(b) = 0$$

form a subspace of V. The general solution of the "inhomogenous" boundary value problem

$$y''(t) - p(t)y'(t) - q(t)y(t) = r(t), \qquad y(a) = A, \ y(b) = B$$

with a continous function $r(t)$ can be obtained as the sum of a given (particular) solution of the inhomogenous problem and the general solution of the homogenous problem.

3. Consider next the vector space $V = C[a,b]$ of continuous real functions, and assume that k is a continuous function $\left(k \in C\left([a,b] \times [a,b]\right)\right)$ of two variables. Consider the transformation $A : V \mapsto V$ defined as

$$A(x)(t) = \int_a^b k(t,s)x(s)ds \quad (t \in [a,b]) \tag{5.24}$$

for all $x \in C[a, b]$. Here $A(x)$ denotes the image function, the value of which at point t is defined by equation (5.24). Many problems of applied mathematics and engineering can be reduced to the solution of the *Fredholm integral equation:*

$$x(t) = \int_a^b k(t,s)x(s)ds + f(t) \quad (t \in [a,b]), \tag{5.25}$$

where $f \in C[a,b]$ is a given function. This equation can be rewritten as

$$x = A(x) + f,$$

or

$$(I - A)(x) = f.$$

If transformation $I - A$ is invertible, then the solution can be obtained in the following form:

$$x = (I - A)^{-1}(f). \tag{5.26}$$

A modified version of equation (5.25) can be written as

$$x(t) = \int_a^t k(t,s) x(s) ds + f(t) \qquad (t \in [a,b]) \tag{5.27}$$

which is called the *Volterra integral equation*. The difference between equations (5.25) and (5.27) is the different integration intervals. In the case equation (5.25) it is [a, b], while in the case of (5.27) it is a *t*-dependent interval [a, t].

Introduce the modified transformation

$$B(x)(t) = \int_a^t k(x,s) x(s) ds,$$

then equation (5.27) can be rewritten as

$$x = B(x) + f,$$

or

$$(I - B)(x) = f.$$

It can be proved that mapping B is invertible, therefore equation (5.27) always has a unique solution in $C[a, b]$. For more detail see, for example Szidarovszky and Yakowitz (1978, Chapter 4). Equation (5.25) (and similarly equation (5.27)) can be solved numerically in the following way.

Divide interval [a, b] into subintervals $[t_0, t_1]$, $[t_1, t_2]$, ..., $[t_{n-1}, t_n]$ where $t_0 = a$, $t_n = b$ and $t_{l+1} - t_l = h = \dfrac{b-a}{n}$ for $l = 0, 1, ..., n-1$.

If we use the trapezoidal rule to approximate the integral of the right-hand side at $t = t_l$, then

$$\int_a^b k(t_l, s) x(s) ds \approx \frac{h}{2} k(t_l, a) x(a) + h \sum_{m=1}^{n-1} k(t_l, t_m) x(t_m) + \frac{h}{2} k(t_l, b) x(b)$$

for $l = 1, \ldots, n$, therefore we get a system of n linear equations for the n unknown function values $x(t_l)$, $(l = 1, \ldots, n)$:

$$\frac{h}{2}k(t_l,a)x(a) + h\sum_{m=1}^{n-1}k(t_l,t_m)x(t_m) + \frac{h}{2}k(t_l,b)x(b) + f(t_l) - x(t_l) \quad (5.28)$$

Then the values of $t = t_l$, h, $k(t_l,a)$, $k(t_l,t_m)(1 \le m \le n-1)$, $k(t_l,b)$, $f(t_l)$ are all known, only the function values $x(t_m)$, $(1 \le m \le n-1)$ and $x(b)$ are unknown. Notice that $x(a) = f(a)$.

In the case of equation (5.27) the system of equations is modified as follows:

$$\frac{h}{2}k(t_l,a)x(a) + h\sum_{m=1}^{l-1}k(t_l,t_m)x(t_m) + \frac{h}{2}k(t_l,t_l)x(t_l) + f(t_l) = x(t_l) \quad (5.29)$$

where we have again n equations for the n unknown function values. This system can be solved easily by noticing that at $l = 1$ we have

$$\frac{h}{2}k(t_1,a)x(a) + \frac{h}{2}k(t_1,t_1)x(t_1) + f(t_1) = x(t_1)$$

from which $x(t_1)$ can be obtained. For $l = 2$ we have

$$\frac{h}{2}k(t_2,a)x(a) + hk(t_2,t_1)x(t_1) + \frac{h}{2}k(t_2,t_2)x(t_2) + f(t_2) = x(t_2)$$

and here only $x(t_2)$ is unknown, since $x(t_1)$ was already determined. Assume in general that $x(a)$, $x(t_1)$, $\ldots$, $x(t_{l-1})$ are already determined, then $x(t_l)$ can be easily obtained from (5.29), since all other quantities are given in the problem or already computed. In the case of system (5.28) the elimination method can be used to find the solutions.

5.9 Exercises

1. Show that the following transformations are not linear:

a) $\begin{pmatrix} x \\ y \end{pmatrix} \mapsto \begin{pmatrix} x \\ 2 \end{pmatrix}$

b) $\begin{pmatrix} x \\ y \end{pmatrix} \mapsto \begin{pmatrix} x^2 \\ x+y \end{pmatrix}$

c) $\begin{pmatrix} x \\ y \end{pmatrix} \mapsto \begin{pmatrix} x+1 \\ y+1 \end{pmatrix}$

d) $\begin{pmatrix} x \\ y \end{pmatrix} \mapsto \begin{pmatrix} \sin x \\ y \end{pmatrix}$.

2. Show that the following transformations are linear:

a) $\begin{pmatrix} x \\ y \end{pmatrix} \mapsto \begin{pmatrix} -3x \\ 2y \end{pmatrix}$

b) $\begin{pmatrix} x \\ y \end{pmatrix} \mapsto \begin{pmatrix} x+y \\ x-y \end{pmatrix}$

c) $\begin{pmatrix} x \\ y \end{pmatrix} \mapsto \begin{pmatrix} 2x+3y \\ x-y \end{pmatrix}$

d) $\begin{pmatrix} x \\ y \end{pmatrix} \mapsto \begin{pmatrix} -x+2y \\ x-2y \end{pmatrix}$.

3. Find the matrix representations of the transformations of the previous Exercise.

4. Denote the transformations

$$\begin{pmatrix} x \\ y \end{pmatrix} \mapsto \begin{pmatrix} -3x \\ 2y \end{pmatrix} \quad \text{and} \quad \begin{pmatrix} x \\ y \end{pmatrix} \mapsto \begin{pmatrix} x+y \\ x-y \end{pmatrix}$$

by A and B, respectively.

Find $A+B$, $A-B$, $2A$, $3B$, $2A+3B$, AB, A^2, A^2B, and BA.

5. Check which of the mappings of Exercise 2 are one-to-one.

6. Let V be the set of all single variable real polynomials of degree at most n. If $p(x) = a_0 + a_1 x + ... + a_n x^n$, then define $A(p)$ as

$$A(p)(x) = a_0 + \frac{a_1 x}{1!} + \frac{a_2 x^2}{2!} + ... + \frac{a_n x^n}{n!}. \text{ Is } A \text{ a linear transformation?}$$

7. Find the ranks of the resulting transformations of Exercise 4.

8. Prove that if $A \in L(V, V)$, then $A^2 = O$ if and only if $R(A) \subseteq N(A)$.

9. Let $A \in L(V, V)$. Is it possible that $R(A) \cap N(A) \neq \{0\}$?

10. Is there any linear mapping $A \in L(R^3, R^2)$ such that

$$A \begin{pmatrix} 1 \\ 1 \\ 1 \end{pmatrix} = \begin{pmatrix} 1 \\ 1 \end{pmatrix}, \quad A \begin{pmatrix} 1 \\ 2 \\ 3 \end{pmatrix} = \begin{pmatrix} 1 \\ 2 \end{pmatrix}, \quad \text{and} \quad A \begin{pmatrix} 0 \\ 1 \\ 1 \end{pmatrix} = \begin{pmatrix} 0 \\ 1 \end{pmatrix}?$$

11. Let V be the set of all continuously differentiable functions on $[0, 1]$, and let W be the set of all continuous functions on the same interval $[0, 1]$. Define $A \in L(V, V)$ as for all $f \in V$, $A(f) = f'$. Is A a linear mapping?

12. Let $\mathbf{a} \in R^n$ be a fixed vector. Define mapping A as $A(\mathbf{x}) = \mathbf{x} + \mathbf{a}$ for all $\mathbf{x} \in R^n$. Is A a linear mapping?

13. Let $A \in L\left(R^n, R^{n-1}\right)$ be defined as

$$A \begin{pmatrix} x_1 \\ x_2 \\ \vdots \\ x_n \end{pmatrix} = \begin{pmatrix} x_1 - x_2 \\ x_2 - x_3 \\ \vdots \\ x_{n-1} - x_n \end{pmatrix}.$$

a) Show that A is a linear mapping;
b) Give a matrix representation of A;
c) Find $N(A)$ and $R(A)$;
d) Find the rank of A.

14. Let A be a mapping of R^3 into itself such that $A(\mathbf{x})$ is the mirror image of $\mathbf{x}$ with respect to the origin. Find a matrix representation of A.

15. Find the coordinates of vector $\mathbf{x}$ in basis $B = \{\mathbf{x}_1, \mathbf{x}_2\}$, where

$$\mathbf{x} = \begin{pmatrix} 1 \\ 1 \end{pmatrix}, \ \mathbf{x}_1 = \begin{pmatrix} 2 \\ 1 \end{pmatrix}, \ \mathbf{x}_2 = \begin{pmatrix} 1 \\ 2 \end{pmatrix}.$$

16. Let $V = C[a, b]$, and let $g \in C[a, b]$ be a given function. Define the mapping $A : V \to R$ as

$$A(f) = \int_a^b f(t) g(t) \, dt.$$

Is this mapping linear?

17. Find the inverses of the linear transformations given in Exercise 2.

18. Repeat Example 5.13 for vectors

$$\mathbf{x} = \begin{pmatrix} 1 \\ 2 \\ 3 \end{pmatrix}, \ \mathbf{v}_1 = \begin{pmatrix} 1 \\ 1 \\ 1 \end{pmatrix}, \ \mathbf{v}_2 = \begin{pmatrix} 2 \\ 1 \\ 1 \end{pmatrix}, \ \mathbf{v}_3 = \begin{pmatrix} 1 \\ 1 \\ 2 \end{pmatrix}.$$

19. Repeat Example 5.14 for matrix

$$\mathbf{A}_{B,B} = \begin{pmatrix} 1 & 2 & 2 \\ 1 & 1 & 1 \\ 2 & 1 & 3 \end{pmatrix}$$

and basis elements as given in the previous Exercise.

20. Show how the matrix of the previous Exercise changes in the new basis

$$\mathbf{v}_1 = \begin{pmatrix} 1 \\ 1 \\ 0 \end{pmatrix}, \ \mathbf{v}_2 = \begin{pmatrix} 1 \\ 0 \\ 1 \end{pmatrix}, \ \mathbf{v}_3 = \begin{pmatrix} 0 \\ 1 \\ 1 \end{pmatrix}.$$

21. Assume that $k(t,s) = \alpha \neq \dfrac{1}{b-a}$ for all t and s, where α is a given constant. Show that equation (5.25) has a unique solution

$$x(t) = f(t) + c$$

with

$$c = \frac{\alpha \int_a^b f(s)\, ds}{1 - \alpha (b-a)}.$$

22. In addition to the assumptions of the previous Exercise assume that f is continuously differentiable. Show that the solution of the Volterra equation (5.27) solves the initial-value-problem

$$\dot{x}(t) = \alpha x(t) + \dot{f}(t), \quad x(a) = f(a).$$

23. (Continuation of Exercise 22). Show that under the conditions of the previous problem,

$$x(t) = e^{\alpha t}\left(f(a)e^{-\alpha a} + \int_{a}^{t} \dot{f}(s)e^{-\alpha s}\,ds \right).$$

24. Assume that $k(t,s) \equiv K(s)$, where K is a continuous function on $[a, b]$, and f is continuously differentiable. Show that the solution of the Volterra equation (5.27) solves the initial-value-problem

$$\dot{x}(t) = K(t)x(t) + \dot{f}(t), \quad x(a) = f(a).$$

25. Generalize the result of Exercise 23 for the more general case of the previous Exercise.

26. Apply the numerical method outlined in the last application with $n = 10$ for equation

$$\int_{0}^{1}(t+s)x(s)+1 = x(t).$$

27. Apply the numerical method of the last application with $n = 10$ for equation

$$\int_{0}^{t}(t+s)x(s)+1 = x(t) \qquad (0 \le t \le 1).$$

Chapter 6

Eigenvalues, Invariant Subspaces, Canonical Forms

6.1 Introduction

In this chapter the general theory of the space $L(V)$ of linear transformations on a vector space V will be discussed. It will be always assumed that V is finitely generated, and therefore in a given basis, each $A \in L(V)$ has an $n \times n$ matrix representation with some n. The main objective of this chapter is to introduce methodology that can be used to transform matrices into simple forms. Most of such methods are based on the idea of decomposing space V into special low dimensional subspaces with the property that a given transformation A maps each subspace into itself. Therefore, A can also be decomposed into more simple transformations of lower ranks, which will result in matrix representations with special structures.

6.2 Basic Concepts

Let V be a vector space and assume that $A \in L(V)$ is a linear transformation of V into itself. The methodology to be introduced and examined in this chapter will be based on the following concept.

Definition 6.1. A subspace V_1 of V is called an *A-invariant subspace* if for all $\mathbf{x} \in V_1, A(\mathbf{x}) \in V_1$.

Notice first that V itself is obviously an A-invariant subspace, and the subspace $V_1 = \{\mathbf{0}\}$ is also A-invariant, since $A(\mathbf{x}) = 0$. These subspaces are the trivial A-invariant subspaces. A nontrivial case is shown in the following example.

Example 6.1. Select $V = R^2$, and for all $\mathbf{x} \subset V$, define

$$A(\mathbf{x}) = \begin{pmatrix} 1 & 1 \\ 2 & 2 \end{pmatrix} \cdot \mathbf{x}.$$

Consider the subspace generated by vector $\mathbf{v} = \begin{pmatrix} 1 \\ 2 \end{pmatrix}$:

$$V_1 = \left\{ \mathbf{x} \,\middle|\, \mathbf{x} = \begin{pmatrix} a \\ 2a \end{pmatrix}, \quad a \in R \right\}.$$

We can easily prove that V_1 is A-invariant. Simple calculation shows that for all vectors of V_1,

$$A(\mathbf{x}) = \begin{pmatrix} 1 & 1 \\ 2 & 2 \end{pmatrix}\begin{pmatrix} a \\ 2a \end{pmatrix} = \begin{pmatrix} a & + & 2a \\ 2a & + & 4a \end{pmatrix} = \begin{pmatrix} 3a \\ 6a \end{pmatrix} \in V_1.$$

♦

We start our discussion with the simplest nontrivial A-invariant subspaces. Let V_1 be a one-dimensional subspace of V, then it is generated by a nonzero vector $\mathbf{v}$:

$$V_1 = \left\{ \mathbf{x} \,\middle|\, \mathbf{x} = a\mathbf{v}, \; a \text{ is a scalar} \right\}.$$

This subspace is A-invariant if and only if for all $\mathbf{x} \in V_1$, $A(\mathbf{x})$ is a constant multiple of $\mathbf{x}$. This condition can be reformulated by saying that

$$A(\mathbf{x}) = \lambda \mathbf{x} \tag{6.1}$$

with some scalar λ. The selection $\mathbf{x} = \mathbf{0}$ always satisfies this equation with arbitrary value of λ, therefore this trivial solution has no interest to us.

Definition 6.2. Let $A \in L(V)$ be a linear transformation. A scalar λ is called an *eigenvalue* of A, if there is an $\mathbf{x} \neq \mathbf{0}$ in V such that equation (6.1) is satisfied. The solutions $\mathbf{x} \neq \mathbf{0}$ are called the *eigenvectors* of A. Equation (6.1) is usually called the *eigenvector equation* of mapping A.

Notice that if V is a real vector space, then scalars are also real, therefore only real eigenvalues are considered. If V is complex, then eigenvalues can be both real and complex numbers.

First, we point out an important relation between the eigenvalues of a linear transformation and the transformation being one-to-one. From Theorem 5.5 we know that a linear transformations A is one-to-one if and only if its nullspace consists of only the zero element. The eigenvector equation implies that this is the case if and only if $\lambda = 0$ is not an eigenvalue of A. That is, A is one-to-one if and only if all eigenvalues of A are nonzero.

Theorem 6.1. Let $\lambda_1, \ldots, \lambda_k$ be distinct eigenvalues of a transformation $A \in L(V)$, and let $\mathbf{x}_1, \ldots, \mathbf{x}_k$ be associated eigenvectors. Then $\mathbf{x}_1, \ldots, \mathbf{x}_k$ are linearly independent.

Proof. Assume that a linear combination of these elements is zero,

$$a_1\mathbf{x}_1 + a_2\mathbf{x}_2 + \ldots + a_k\mathbf{x}_k = \mathbf{0}, \tag{6.2}$$

where at last one coefficient, say, a_k is nonzero. By renumbering the elements, we can always have nonzero coefficient in the last term. Multiply this equation by λ_1 to have

$$a_1\lambda_1\mathbf{x}_1 + a_2\lambda_1\mathbf{x}_2 + \ldots + a_k\lambda_1\mathbf{x}_k = \mathbf{0}. \tag{6.3}$$

Apply mapping A to both sides of equality (6.2),

$$a_1 A(\mathbf{x}_1) + a_2 A(\mathbf{x}_2) + \ldots + a_k A(\mathbf{x}_k) = \mathbf{0}.$$

Use the fact that for $i = 1, 2, \ldots, k$, $A(\mathbf{x}_k) = \lambda_k \mathbf{x}_k$ to rewrite this last relation as

$$a_1\lambda_1\mathbf{x}_1 + a_2\lambda_2\mathbf{x}_2 + ... + a_k\lambda_k\mathbf{x}_k = \mathbf{0}.$$

Subtract this equality from (6.3) to see that

$$a_2\left(\lambda_1 - \lambda_2\right)\mathbf{x}_2 + ... + a_k\left(\lambda_1 - \lambda_k\right)\mathbf{x}_k - \mathbf{0},$$

that is,

$$a_2^{(1)}\mathbf{x}_2 + ... + a_k^{(1)}\mathbf{x}_k = \mathbf{0}, \tag{6.4}$$

where $a_i^{(1)} = a_i\left(\lambda_1 - \lambda_i\right)$ for $i = 2, ..., k$. Notice that for all i, $a_i^{(1)} \neq 0$ if and only if $a_i \neq 0$, since the eigenvalues are distinct and therefore $\lambda_1 - \lambda_i \neq 0$. The above derivation shows that if a linear combination of $\mathbf{x}_1, \mathbf{x}_2, ..., \mathbf{x}_k$ is zero, then $\mathbf{x}_1$ can be eliminated from the zero linear combination, and all nonzero coefficients remain nonzero. Repeat the same procedure for equation (6.4) to eliminate $\mathbf{x}_2$, and repeat again for the resulting equation to eliminate $\mathbf{x}_3$, and so on. Finally, after eliminating all elements $\mathbf{x}_1, ..., \mathbf{x}_{k-1}$, we have

$$a_k^{(k-1)}\mathbf{x}_k = \mathbf{0}.$$

Since $\mathbf{x}_k \neq \mathbf{0}$, $a_k^{(k-1)} = 0$, which implies that $a_k = 0$. This conclusion contradicts to the initial assumption that a_k is nonzero. Thus, the proof is complete.

♣

In the special case when V is finitely generated, with some n, mapping A has an $n \times n$ matrix representation and each element $\mathbf{x} \in V$ has an n-element coordinate vector in any given basis B. For the sake of simplicity let $\mathbf{A}$ and $\mathbf{x}$ denote the matrix representation of A and the coordinate vector of $\mathbf{x}$. In this notation we do not indicate basis B in the subscripts of these symbols as we have done earlier in Chapter 5, since the basis is considered now fixed. Then equation (6.1) can be rewritten as

$$\mathbf{A}\mathbf{x} = \lambda\mathbf{x},$$

which is equivalent to the homogeneous equation

$$\left(\mathbf{A} - \lambda\mathbf{I}\right)\mathbf{x} = \mathbf{0}, \tag{6.5}$$

where $\mathbf{I}$ is the $n \times n$ identity matrix and $\mathbf{0}$ is the n-element zero vector. A scalar λ is an eigenvalue of $\mathbf{A}$ if and only if this equation has a nonzero solution $\mathbf{x}$. Since $\mathbf{0}$ is always a solution of this equation, nonzero solution exists if and only if multiple solutions exist. From Theorem 4.6 we know that a necessary and sufficient condition for the existence of multiple solutions is given by the condition that the determinant of the coefficient matrix is zero:

$$\det(\mathbf{A} - \lambda\mathbf{I}) = 0 \tag{6.6}$$

Equation (6.6) is called the *characteristic equation* of matrix $\mathbf{A}$. Notice that $\mathbf{A} - \lambda\mathbf{I}$ can be obtained easily from $\mathbf{A}$, we have to subtract λ from each of its diagonal elements. That is, equation (6.6) has the equivalent form:

$$\det\begin{pmatrix} a_{11} - \lambda & a_{12} & \cdots & a_{1,n-1} & a_{1n} \\ a_{21} & a_{22} - \lambda & \cdots & a_{2,n-1} & a_{2n} \\ \cdots & \cdots & \cdots & \cdots & \cdots \\ a_{n1} & a_{n2} & \cdots & a_{n,n-1} & a_{nn} - \lambda \end{pmatrix} = 0. \tag{6.7}$$

Expanding this determinant with respect to the first row it is easy to show by mathematical induction that the left hand side is always a polynomial of degree n and the coefficient of λ^n is $(-1)^n$. This polynomial is called the *characteristic polynomial* of matrix $\mathbf{A}$. It is easy to see that the characteristic polynomial does not depend on the basis being selected in V. The matrix representation of $\mathbf{A}$ in another basis is $\mathbf{T}^{-1}\mathbf{AT}$, where $\mathbf{T}$ is a nonsingular matrix. The characteristic polynomial of $\mathbf{T}^{-1}\mathbf{AT}$ is the following:

$$\det(\mathbf{T}^{-1}\mathbf{AT} - \lambda\mathbf{I}) = \det(\mathbf{T}^{-1}(\mathbf{A} - \lambda\mathbf{I})\mathbf{T})$$
$$= \det(\mathbf{T}^{-1}) \cdot \det(\mathbf{A} - \lambda\mathbf{I}) \cdot \det(\mathbf{T}).$$

From the Corollary of Theorem 4.4 we know that

$$\det(\mathbf{T}^{-1}) = \frac{1}{\det(\mathbf{T})},$$

therefore, the first and last factors cancel showing that the characteristic polynomial of $\mathbf{T}^{-1}\mathbf{AT}$ coincides with that of matrix $\mathbf{A}$.

Based on the above discussion, the eigenvalues and eigenvectors of an $n \times n$ real or complex matrix can be obtained by using the following procedure:

Step 1. Expand determinant (6.7) to obtain the characteristic polynomial of matrix $\mathbf{A}$;

Step 2. Find the (real or complex) roots of the characteristic polynomial to get the eigenvalues;

Step 3. For each eigenvalue, solve the homogeneous system (6.5) to recover the eigenvectors.

This procedure is illustrated in the following example.

Example 6.2. Consider the 3×3 matrix

$$\mathbf{A} = \begin{pmatrix} 1 & 4 & 1 \\ 1 & 2 & 3 \\ 1 & 3 & 2 \end{pmatrix}.$$

Equation (6.7) has now the form

$$\det \begin{pmatrix} 1-\lambda & 4 & 1 \\ 1 & 2-\lambda & 3 \\ 1 & 3 & 2-\lambda \end{pmatrix} = 0.$$

Expanding the determinant with respect to its first row, the left-hand side can be written as

$$(1-\lambda) \cdot \det\begin{pmatrix} 2-\lambda & 3 \\ 3 & 2-\lambda \end{pmatrix} - 4\det\begin{pmatrix} 1 & 3 \\ 1 & 2-\lambda \end{pmatrix} + 1 \cdot \det\begin{pmatrix} 1 & 2-\lambda \\ 1 & 3 \end{pmatrix}$$

$$= (1-\lambda) \cdot \left(4 - 4\lambda + \lambda^2 - 9\right) - 4 \cdot (2-\lambda-3) + 1 \cdot (3-2+\lambda)$$

$$= (1-\lambda)\left(-5 - 4\lambda + \lambda^2\right) + (4\lambda+4) + (\lambda+1) = -\lambda^3 + 5\lambda^2 + 6\lambda.$$

The roots of this polynomial are

$$\lambda_1 = 0, \quad \lambda_2 = -1, \quad \text{and} \quad \lambda_3 = 6.$$

For eigenvalue $\lambda_1 = 0$, equation (6.5) has the form

$$\begin{pmatrix} 1 & 4 & 1 \\ 1 & 2 & 3 \\ 1 & 3 & 2 \end{pmatrix}\begin{pmatrix} x_1 \\ x_2 \\ x_3 \end{pmatrix} = \begin{pmatrix} 0 \\ 0 \\ 0 \end{pmatrix},$$

that is,

$$x_1 + 4x_2 + x_3 = 0$$
$$x_1 + 2x_2 + 3x_3 = 0$$
$$x_1 + 3x_2 + 2x_3 = 0.$$

The elimination procedure is shown in Table 6.1, where the right-hand side numbers are not presented, since they are always equal to zero.

Table 6.1. Elimination for $\lambda_1 = 0$

x_1	x_2	x_3
1	4	1
1	2	3
1	3	2
①	4	1
0	−2	2
0	−1	1
1	4	1
0	①	−1
0	−1	1

The last equation is identical with the second, so we can ignore it.

Hence, x_3 is the free variable, and back substitution shows that $x_2 = x_3$ and $x_1 = -4x_2 - x_3 = -5x_3$. The general solution is therefore the following:

$$\mathbf{x} = \begin{pmatrix} -5x_3 \\ x_3 \\ x_3 \end{pmatrix}.$$

Thus, this vector is the eigenvector associated to $\lambda_1 = 0$ with arbitrary nonzero value of x_3.

For $\lambda_2 = -1$, equation (6.5) can be written as

$$\begin{pmatrix} 2 & 4 & 1 \\ 1 & 3 & 3 \\ 1 & 3 & 3 \end{pmatrix} \begin{pmatrix} x_1 \\ x_2 \\ x_3 \end{pmatrix} = \begin{pmatrix} 0 \\ 0 \\ 0 \end{pmatrix},$$

which is equivalent to the system

$$2x_1 + 4x_2 + x_3 = 0$$
$$x_1 + 3x_2 + 3x_3 = 0$$
$$x_1 + 3x_2 + 3x_3 = 0.$$

Similar to the first eigenvalue, one may verify that the general solution is

$$\mathbf{x} = \begin{pmatrix} \dfrac{9}{2}x_3 \\ -\dfrac{5}{2}x_3 \\ x_3 \end{pmatrix}.$$

For $\lambda_3 = 6$, equation (6.5) is the following:

$$\begin{pmatrix} -5 & 4 & 1 \\ 1 & -4 & 3 \\ 1 & 3 & -4 \end{pmatrix} \begin{pmatrix} x_1 \\ x_2 \\ x_3 \end{pmatrix} = \begin{pmatrix} 0 \\ 0 \\ 0 \end{pmatrix},$$

and the general solution is:

$$\mathbf{x} = \begin{pmatrix} x_3 \\ x_3 \\ x_3 \end{pmatrix}.$$

If we wish to determine a particular eigenvector associated to each eigenvalue, then we might select a particular value for x_3. If $x_3 = 1$ is the choice, then the eigenvectors for $\lambda = 0, -1$ and 6 are

$$\begin{pmatrix} -5 \\ 1 \\ 1 \end{pmatrix}, \quad \begin{pmatrix} \dfrac{9}{2} \\ -\dfrac{5}{2} \\ 1 \end{pmatrix}, \quad \text{and} \quad \begin{pmatrix} 1 \\ 1 \\ 1 \end{pmatrix},$$

respectively.

◆

The procedure discussed above is very useful to find one-dimensional invariant subspaces for any given $n \times n$ real or complex matrix. However, to find higher dimensional invariant subspaces, the straightforward extension of this algorithm becomes very complicated, and therefore has no practical value. The case of finding two dimensional invariant subspaces of a 3×3 matrix is illustrated in the following example.

Example 6.3. Consider again matrix

$$\mathbf{A} = \begin{pmatrix} 1 & 4 & 1 \\ 1 & 2 & 3 \\ 1 & 3 & 2 \end{pmatrix}.$$

Any two-dimensional subspace is characterized as

$$V_1 = \left\{ \lambda\mathbf{x} + \mu\mathbf{y} \,\middle|\, \{\mathbf{x},\mathbf{y}\} \text{ is a basis of } V_1, \text{ and } \lambda,\mu \text{ are scalars} \right\}.$$

This subspace is **A**-invariant if and only if both **Ax** and **Ay** are in V_1, since for any vector $\mathbf{z} \in V_1$,

$$\mathbf{Az} = \mathbf{A}(\lambda\mathbf{x} + \mu\mathbf{y}) = \lambda\mathbf{Ax} + \mu\mathbf{Ay} \in V_1.$$

This observation can be mathematically formulated as the system:

$$\mathbf{Ax} = \alpha\mathbf{x} + \beta\mathbf{y}$$
$$\mathbf{Ay} = \gamma\mathbf{x} + \delta\mathbf{y},$$

where vectors **x**, **y** and scalars α, β, γ, δ are the unknowns. These equations can be summarized as the homogeneous system

$$\begin{pmatrix} \mathbf{A} - \alpha\mathbf{I} & -\beta\mathbf{I} \\ -\gamma\mathbf{I} & \mathbf{A} - \delta\mathbf{I} \end{pmatrix} \begin{pmatrix} \mathbf{x} \\ \mathbf{y} \end{pmatrix} = \begin{pmatrix} \mathbf{0} \\ \mathbf{0} \end{pmatrix}.$$

Similar to the eigenvector equations the unknown scalars can be obtained as the solutions of the equation

$$\det\begin{pmatrix} \mathbf{A} - \alpha\mathbf{I} & -\beta\mathbf{I} \\ -\gamma\mathbf{I} & \mathbf{A} - \delta\mathbf{I} \end{pmatrix} = 0.$$

Notice that this determinant is 6×6 with four unknown parameters, therefore its expansion becomes complicated. Notice that if a k-dimensional invariant subspace of an $n \times n$ matrix is to be determined, then this equation has k^2 unknown scalars, and the size of the determinant is $kn \times kn$.

♦

Due to the difficulty arising from the increasing size and more complicated structure of the determinant being expanded, a different approach has to be developed to find invariant subspaces. This new methodology is based on the following simple result.

Theorem 6.2. Let $A \in L(V)$, and assume that for a linear transformation $B \in L(V)$, $AB = BA$. Then the nullspace of B is an A-invariant subspace.

Proof. Let $\mathbf{x} \in N(B)$ be an arbitrary element. Then $A(\mathbf{x}) \in N(B)$, since

$$B\big(A(\mathbf{x})\big) = (B \cdot A)(\mathbf{x}) = (A \cdot B)(\mathbf{x}) = A\big(B(\mathbf{x})\big) = A(\mathbf{0}) = \mathbf{0},$$

which proves the assertion.

♣

If we wish to use this theorem in constructing A-invariant subspaces, then we have to find a large class of mappings that commute with A. Let p be a real or complex polynomial depending on whether V is a real or complex vector space. Since A commute with A^k for all $k \geq 0$, A and $p(A)$ commute as well. This simple observation shows that $N(p(A))$ is an A-invariant subspace for all polynomials p. Our general methodology for constructing invariant subspaces will be mainly based on this idea. In developing further results we will use the fundamental properties of matrix polynomials, which are the subject of the next section.

6.3 Matrix Polynomials

We start this section with the summary of the most important properties of real and complex polynomials. As in some earlier examples, a single variable single valued function of t having the form

$$p(t) = a_0 + a_1 t + a_2 t^2 + \ldots + a_n t^n$$

is called a *polynomial of degree n* if the coefficients are real or complex numbers, and the leading coefficient, $a_n \neq 0$. If all coefficients are real, then the polynomial is called real, otherwise the polynomial is complex.

A real (or complex) number t^* is called a *root* of p if $p(t^*) = 0$. We will use the following facts (see, for example, Herstein, 1964) regarding the roots of an nth degree polynomial p:

1. For any complex polynomial of degree n, there are n (not necessarily distinct) real or complex roots.

2. If all coefficients are real and complex roots exist, they occur in conjugate pairs.

3. If n is odd and all coefficients are real, then there is at least one real root.

4. If t^* is a root of p, then necessarily

$$p(t) = (t - t^*)q(t),$$

where q is a polynomial of degree $n-1$.

The first fact is known as the *fundamental theorem of algebra,* the proof of which is based on the theory of functions of complex variables, therefore it is not included in this book. The second property can be proved by simple substitution. Let t^* be a complex root. If complex conjugate is denoted by overbar, then

$$\overline{p(t^*)} = \overline{a_0 + a_1 t^* + a_2 t^{*2} + \ldots + a_n t^{*n}}$$
$$= \overline{a_0} + \overline{a_1 t^*} + \overline{a_2 t^{*2}} + \ldots + \overline{a_n t^{*n}} = \overline{p(t^*)} = \overline{0} = 0,$$

that is, $\overline{t^*}$ is also a root of p. Fact 3 is a consequence of Fact 2, since the number of complex roots of a real polynomial is always even, but the total number of real and complex roots is odd. Therefore, the number of real roots is odd, hence nonzero. The last property is known as the *Factor Theorem*, and it can be proved by using the well-known algebraic identity

$$a^k - b^k = (a - b)\left(a^{k-1} + a^{k-2}b + \ldots + ab^{k-2} + b^{k-1}\right).$$

Simple algebra shows that

$$p(t) = p(t) - p(t^*) = a_1(t - t^*) + a_2(t^2 - t^{*2}) + \dots + a_n(t^n - t^{*n})$$
$$= a_1(t - t^*) + a_2(t - t^*)(t + t^*) + \dots + a_n(t - t^*)$$
$$\times \left(t^{n-1} + t^{n-2}t^* + \dots + t \cdot t^{*n-2} + t^{*n-1}\right)$$
$$= (t - t^*)q(t)$$

where

$$q(t) = a_1 + a_2(t + t^*) + \dots + a_n\left(t^{n-1} + t^{n-2}t^* + \dots + t \cdot t^{*n-2} + t^{*n-1}\right)$$

is a polynomial of degree $n - 1$.

Some additional properties of polynomials which will be used later in this chapter are discussed next.

5. If p and q are polynomials of degree n and $m \le n$, respectively, then there are unique polynomials h and r of degree $n - m$ and at most $m - 1$ such that

$$p(t) = q(t)h(t) + r(t). \tag{6.8}$$

Equation (6.8) is called the *Remainder Theorem,* and polynomials h and r are called the *quotient* and the *remainder polynomials*, respectively. Relation (6.8) can be proved in the following way. Denote

$$p(t) = a_n t^n + a_{n-1} t^{n-1} + \dots + a_1 t + a_0,$$
$$q(t) = b_m t^m + b_{m-1} t^{m-1} + \dots + b_1 t + b_0,$$

and look for polynomials h and r in the following form:

$$h(t) = c_{n-m} t^{n-m} + \dots + c_1 t + c_0$$

and

$$r(t) = d_{m-1} t^{m-1} + \dots + d_1 t + d_0.$$

Then relation (6.8) is equivalent to equality

$$a_n t^n + \ldots + a_1 t + a_0 = \left(b_m t^m + \ldots + b_1 t + b_0\right)\left(c_{n-m} t^{n-m} + \ldots + c_1 t + c_0\right)$$
$$+ \left(d_{m-1} t^{m-1} + \ldots + d_1 t + d_0\right).$$

Comparing the corresponding coefficients of the two sides we obtain the system of equations

$$a_n = b_m \cdot c_{n-m}$$
$$a_{n-1} = b_m c_{n-m-1} + b_{m-1} c_{n-m}$$
$$\vdots$$
$$a_m = b_m c_0 + b_{m-1} c_1 + \ldots + b_0 c_m,$$
$$a_{m-1} = b_{m-1} c_0 + b_{m-2} c_1 + \ldots + b_0 c_{m-1} + d_{m-1}$$
$$\vdots$$
$$a_1 = b_1 c_0 + b_0 c_1 + d_1$$
$$a_0 = b_0 c_0 + d_0,$$

where for $k > n - m$, $c_k = 0$. From the first equation the value of c_{n-m} can be determined, since $b_m \neq 0$. From the second equation we have c_{n-m-1}, and from the third equation we obtain c_{n-m-2}. Continuing the process the unknown coefficients are obtained in the order c_{n-m-2}, c_{n-m-3}, ..., c_0, d_{m-1}, d_0.

Example 6.4. Let $p(t) = t^5 + 2t^4 + t^3 + 2t^2 + t + 1$ and $q(t) = t^2 + t + 1$. Then $n = 5$ and $m = 2$, and equation (6.8) has the form

$$t^5 + 2t^4 + t^3 + 2t^2 + t + 1 = \left(t^2 + t + 1\right)\left(c_3 t^3 + c_2 t^2 + c_1 t + c_0\right) + \left(d_1 t + d_0\right),$$

and the comparison of the corresponding coefficients gives the system of equations

$$1 = c_3$$
$$2 = c_2 + c_3$$
$$1 = c_1 + c_2 + c_3$$
$$2 = c_0 + c_1 + c_2$$
$$1 = c_0 + c_1 + d_1$$
$$1 = c_0 + d_0.$$

The solution is obtained as follows:

$$c_3 = 1$$
$$c_2 = 2 - c_3 = 1$$
$$c_1 = 1 - c_2 - c_3 = -1$$
$$c_0 = 2 - c_1 - c_2 = 2$$
$$d_1 = 1 - c_0 - c_1 = 0$$
$$d_0 = 1 - c_0 = -1,$$

therefore

$$h(t) = t^3 + t^2 - t + 2 \quad \text{and} \quad r(t) = -1.$$

♦

A real (or complex) polynomial is called *prime* if it cannot be represented as the product of two real (or complex) polynomials of lower degree. From properties 1. and 4. we know that in the set of complex polynomials only the linear polynomials are primes. In the set of real polynomials all linear polynomials are primes, a quadratic polynomial is prime if and only if it has no real root, and no real polynomial of degree three or more is prime. This fact follows from the observation that if the degree of a real polynomial p is at least three, then it can be factored either as $(t - t^*)q(t)$ where t^* is a real root of p, or as $(t - t^*)(t - \overline{t}^*)q(t)$, where t^* and $\overline{t}^*$ are complex conjugate roots. In the second case, the quadratic polynomial

$$\left(t - t^*\right)\left(t - \overline{t^*}\right) = t^2 - t\left(t^* + \overline{t^*}\right) + t^*\,\overline{t^*}$$

is real, since both $t^* + \overline{t}^*$ and $t^*\,\overline{t}^*$ are real numbers.

6. Every polynomial can be decomposed as the product of powers of primes. For complex polynomials this decomposition has the form

$$p(t) = a_n \left(t - t_1^*\right)^{m_1} \left(t - t_2^*\right)^{m_2} \ldots \left(t - t_r^*\right)^{m_r}, \tag{6.9}$$

where $t_1{}^*, t_2{}^*, \ldots, t_r{}^*$ are the distinct roots of p with multiplicities m_1, m_2, ..., m_r. In the case of real polynomials this decomposition can be given as

$$p(t) = a_n \left(t - t_1^*\right)^{m_1} \ldots \left(t - t_l^*\right)^{m_l} q_{l+1}(t)^{m_{l+1}} \ldots q_r(t)^{m_r}, \tag{6.10}$$

where $t_1{}^*, \ldots, t_l{}^*$ are the distinct real roots with multiplicities $m_1, \ldots m_l$, and $q_{l+1}, \ldots, q_r$ are prime quadratic polynomials with unit leading coefficients. For $i = l + 1, \ldots, r$ let z_i and $\overline{z}_i$ denote the roots of q_i, then m_i is the common multiplicity of these complex roots. It can be proved that decompositions (6.9) and (6.10) are unique up to the order of the prime factors. This property is called the *Primary Decomposition Theorem*.

The *greatest common divisor* of polynomials is next introduced. Let p_1, ..., p_k be real (or complex) polynomials, and assume that for $i = 1, 2, \ldots,$ k, p_i is factored into powers of distinct primes:

$$p_i(t) = a_n^{(i)} q_1(t)^{m_1^{(i)}} \ldots q_r(t)^{m_r^{(i)}} q_{r+1}^{(i)}(t)^{m_{r+1}^{(i)}} \ldots q_{k_i}^{(i)}(t)^{m_{k_i}^{(i)}}.$$

Without loosing generality, we may assume that the prime factors $q_1(t), \ldots, q_r(t)$ are common in these polynomials, and there is no other common prime factor. For all $j = 1, 2, \ldots, r$, define

$$m_j = \min\left\{m_j^{(1)}; \ldots; m_j^{(k)}\right\}.$$

Then the greatest common divisor is defined as the polynomial

$$q_1(t)^{m_1} \ldots q_r(t)^{m_r}.$$

That is, the greatest common divisor is the product of the powers of the distinct common prime factors, and the power of each common prime factor is the smallest value among the powers of this factor in the different polynomials. This definition implies that any common factor of given polynomials is a divisor of their greatest common factor.

We say that polynomials $p_1, \ldots, p_k$ are *relative prime*, if they have no common divisor of degree at least one. Relative prime polynomials satisfy the following condition.

7. Assume that polynomials $p_1, \ldots, p_k$ are relative prime. Then there exist polynomials $q_1, q_2, \ldots, q_k$ such that

$$p_1(t)q_1(t) + p_2(t)q_2(t) + \ldots + p_k(t)q_k(t) = 1$$

for all t.

This result can be proved in the following way. Consider the set of polynomials

$$P = \left\{ \sum_{i=1}^{k} p_i q_i \mid q_i \text{ is a polynomial for all } i \right\}.$$

It is obvious that $p_1, p_2, \ldots, p_k \in P$, all linear combinations of $p_1, \ldots, p_k$ belong to P, and if $p \in P$, then for any polynomial q, $pq \in P$. Since the degrees of the nonzero elements of P are nonnegative integers, there is a nonzero polynomial $r \in P$ with smallest degree. Since $p \in P$, the degree of r is not larger than the degree of p. First, we prove that r is a divisor of all $p_i (i = 1, 2, \ldots, k)$. Using the remainder theorem, we see that there are polynomials q_i and r_i such that

$$p_i = r \cdot q_i + r_i,$$

where either $r_i = 0$ or the degree of r_i is smaller than the degree of r. In the first case r is a divisor of p_i, and in the second case

$$r_i = p_i - r \cdot q_i \in P,$$

which contradicts the selection of r (since the degree of r_i is smaller than the degree of r, r must not be the element of P with smallest degree). Therefore, only $r_i = 0$ is possible. Since polynomials $p_1, \ldots, p_k$ are relatively prime, r has to be a nonzero constant. Hence

$$r = p_1 q_1 + \ldots + p_k q_k,$$

and by selecting $q_i^*(t) = \dfrac{1}{r} q_i(t),$

$$1 = p_1(t) q_1^*(t) + \ldots + p_k(t) q_k^*(t),$$

which proves the assertion.

♣

The main properties of matrix polynomials are discussed next. Assume that V is finitely generated and $A \in L(V)$ is a linear transformation. Let $\mathbf{A}$ be an $n \times n$ matrix representation of A. In Example 2.19 we have proved that the dimension of the vector space of $n \times n$ matrices is n^2. Consider now matrices

$$\mathbf{I}, \mathbf{A}, \mathbf{A}^2, \ldots, \mathbf{A}^{n^2}.$$

Since we have $n^2 + 1$ matrices in an n^2-dimensional vector space, they are linearly dependent. Therefore

$$a_{n^2} \mathbf{A}^{n^2} + \ldots + a_2 \mathbf{A}^2 + a_1 \mathbf{A} + a_0 \mathbf{I} = \mathbf{O}$$

with some scalars $a_0, a_1, \ldots, a_{n^2}$. That is, there is a polynomial p of degree at most n^2 such that

$$p(\mathbf{A}) = \mathbf{O}.$$

Consider now the set of all nonzero polynomials p such that $p(\mathbf{A}) = \mathbf{O}$. Since their degrees are nonnegative integers, there is a polynomial of least degree with this property.

Definition 6.3. A nonzero polynomial r of least degree such that $r(\mathbf{A}) = \mathbf{O}$ is called a *minimal polynomial* of $\mathbf{A}$.

Minimal polynomials satisfy the following properties.

Theorem 6.3. Let r be a minimal polynomial of an $n \times n$ matrix $\mathbf{A}$. Then

a) It is unique up to a constant multiplier;

b) If for a nonzero polynomial p, $p(\mathbf{A}) = \mathbf{O}$, then r is a factor of p.

Proof. a) Assume that r_1 and r_2 are both minimal polynomials of $\mathbf{A}$ and their common degree is m. The remainder theorem implies that there is a scalar a and a polynomial r^* of degree at most $m-1$ such that

$$r_1 = a \cdot r_2 + r^*.$$

Then

$$r^*(\mathbf{A}) = r_1(\mathbf{A}) - a \cdot r_2(\mathbf{A}) = \mathbf{O} - a \cdot \mathbf{O} = \mathbf{O},$$

which is impossible for $r^* \neq 0$ because of the minimality of the degree of r_1 and r_2. Thus $r^* = 0$, and r_1 is a constant multiple of r_2.

b) The remainder theorem implies that

$$p = r \cdot q + r^*$$

where q is a polynomial and the degree of r^* is at most $m-1$. Then

$$r^*(\mathbf{A}) = p(\mathbf{A}) - r(\mathbf{A})q(\mathbf{A}) = \mathbf{O} - \mathbf{O} \cdot q(\mathbf{A}) = \mathbf{O}.$$

Similar to the previous part of the proof we see that this is possible only if r^* is the zero polynomial, that is, r is a divisor of p.

♣

Example 6.5. We will now find a minimal polynomial of the 2×2 matrix

$$\mathbf{A} = \begin{pmatrix} 1 & 1 \\ 2 & 2 \end{pmatrix}.$$

Constant nonzero polynomials have the general form $p_0(t) = a_0$ with some scalar $a_0 \neq 0$. Then $p_0(\mathbf{A}) = a_0\mathbf{I} \neq \mathbf{O}$. Linear polynomials have the general form $p_1(t) = a_1 t + a_0$. Then

$$p_1(\mathbf{A}) = a_1\mathbf{A} + a_0\mathbf{I} = a_1 \begin{pmatrix} 1 & 1 \\ 2 & 2 \end{pmatrix} + a_0 \begin{pmatrix} 1 & 0 \\ 0 & 1 \end{pmatrix},$$

which is zero if and only if

$$a_1 + a_0 = a_1 = 2a_1 = 2a_1 + a_0 = 0.$$

Since these equations are satisfied only if $u_1 = u_0 = 0$, the minimal polynomials are at least quadratic. The general form of a quadratic polynomial is $p_2(t) = a_2 t^2 + a_1 t + a_0$, so $p(\mathbf{A}) = \mathbf{O}$ if and only if with some scalars $a_2, a_1, a_0,$

$$a_2 \begin{pmatrix} 3 & 3 \\ 6 & 6 \end{pmatrix} + a_1 \begin{pmatrix} 1 & 1 \\ 2 & 2 \end{pmatrix} + a_0 \begin{pmatrix} 1 & 0 \\ 0 & 1 \end{pmatrix} = \begin{pmatrix} 0 & 0 \\ 0 & 0 \end{pmatrix}.$$

This matrix equation is equivalent to relations

$$\begin{aligned} 3a_2 + a_1 + a_0 &= 0 \\ 3a_2 + a_1 &= 0 \\ 6a_2 + 2a_1 &= 0 \\ 6a_2 + 2a_1 + a_0 &= 0. \end{aligned}$$

The general solution of these equations is the following: $a_0 = 0$, $a_1 = -3a_2$, where a_2 is arbitrary. Hence (by the selection of $a_2 = 1$),

$$p_2(t) = t^2 - 3t$$

is a minimal polynomial of $\mathbf{A}$.

◆

6.4 The Construction of Invariant Subspaces

We start this section with a simple result, which will play an important role in the construction of invariant subspaces and special matrix forms. Let V be a vector space and assume that for $i = 1, 2, \ldots, k$, transformations $B_i \in L(V)$ satisfy the following properties:
a) $B_1 + \ldots + B_k = I$
b) $B_i B_j = O$ for $i \neq j$, where O is the zero mapping such that $O(\mathbf{x}) = \mathbf{0}$ for all $\mathbf{x} \in V$.

We will first show that for all i, $B_i^2 = B_i$. Such linear transformations are called *idempotent*. Properties (a) and (b) imply that

$$B_i = B_i \cdot I = B_i \cdot \left(B_1 + B_2 + \dots + B_k \right) = B_i^2 + \sum_{j \neq i} B_i B_j = B_i^2,$$

since all terms in the summation are zeros. Define V_i as the range of B_i:

$$V_i = R(B_i) = \left\{ \mathbf{x} \mid \mathbf{x} = B_i(\mathbf{z}) \text{ with some } \mathbf{z} \in V \right\}$$

Theorem 6.4. Subspace V is the direct sum of $V_1, V_2, \dots, V_k$:

$$V = V_1 \oplus V_2 \oplus \dots \oplus V_k.$$

Proof. Using Theorem 2.16 we have to prove first that each $\mathbf{v} \in V$ can be written as the sum $\mathbf{v}_1 + \mathbf{v}_2 + \dots + \mathbf{v}_k$, where for $i = 1, 2, \dots, k$, $\mathbf{v}_k \in V_k$. And second, we have to show that if for some elements $\mathbf{v}_i \in V_i$ ($i = 1, 2, \dots, k$), $\mathbf{0} = \mathbf{v}_1 + \mathbf{v}_2 + \dots + \mathbf{v}_k$, then necessarily $\mathbf{v}_1 = \mathbf{v}_2 = \dots = \mathbf{v}_k = \mathbf{0}$. Let $\mathbf{v} \in V$ be arbitrary. Then

$$\mathbf{v} = I(\mathbf{v}) = \left(B_1 + B_2 + \dots + B_k \right)(\mathbf{v}) = B_1(\mathbf{v}) + B_2(\mathbf{v}) + \dots + B_k(\mathbf{v}).$$

Notice tha $B_i(\mathbf{v}) \in V_i$ for all i, therefore the first statement is verified. Assume next that for some $\mathbf{v}_i \in V_i$ ($i = 1, 2, \dots, k$)

$$\mathbf{0} = \mathbf{v}_1 + \mathbf{v}_2 + \dots + \mathbf{v}_k.$$

The definition of subspaces V_i implies that for all i, there is a $\mathbf{w}_i \in V$ such that $v_i = B_i(\mathbf{w}_i)$. Then

$$\mathbf{0} = B_i(\mathbf{0}) = B_i(\mathbf{v}_1) + B_i(\mathbf{v}_2) + \dots + B_i(\mathbf{v}_k),$$

and if $j \neq i$, then

$$B_i(\mathbf{v}_j) = B_i\left(B_j(\mathbf{w}_j) \right) = \left(B_i B_j \right)(\mathbf{w}_j) = 0$$

as the consequence of assumption (b), therefore $B_i(\mathbf{v}_i) = \mathbf{0}$, which implies that for all i,

$$\mathbf{v}_i = B_i(\mathbf{w}_i) = B_i^2(\mathbf{w}_i) = B_i(B_i(\mathbf{w}_i)) = B_i(\mathbf{v}_i) = \mathbf{0},$$

which completes the proof.

♣

Let now $A \in L(V)$ be a linear transformation and assume that V is finitely generated. Assume furthermore that the minimal polynomial r of A is factored into powers of distinct primes:

$$r(t) = a p_1(t)^{m_1} p_2(t)^{m_2} \ldots p_l(t)^{m_l},$$

where the leading coefficient is a. For $i = 1, 2, \ldots, l$, define

$$r_i(t) = \frac{r(t)}{p_i(t)^{m_i}}. \tag{6.11}$$

These polynomials are relative prime, and therefore Property 7 discussed in the previous section implies that there are polynomials $q_i(t)$ such that

$$1 = r_1(t) q_1(t) + r_2(t) q_2(t) + \ldots + r_l(t) q_l(t). \tag{6.12}$$

Introduce notation

$$f_i(t) = r_i(t) q_i(t) \qquad (i = 1, 2, \ldots, l),$$

and consider transformations $B_i = f_i(A)$. We will next show that these mappings satisfy all conditions of Theorem 6.4.

Theorem 6.5.

(i) $I = B_1 + B_2 + \ldots + B_l$;

(ii) $B_i B_j = O$ for $j \neq i$;

(iii) define $V_i = R(B_i)$ for all i, then $V = V_1 \oplus V_2 \oplus \ldots \oplus V_l$;

(iv) for all i, $V_i = N\left(p_i(A)^{m_i}\right)$.

Proof. The proof consists of several steps.

(i) From equation (6.12) and the definition of $f_i(t)$. we have

$$1 = f_1(t) + f_2(t) + \ldots + f_l(t).$$

Substitution of A into this equation shows that

$$I = f_1(A) + f_2(A) + \ldots + f_l(A),$$

and since $B_i = f_i(A)$, the assertion is valid.

(ii) If $i \neq j$, then $B_i B_j = f_i(A) f_j(A) = r_i(A) q_i(A) r_j(A) q_j(A) = 0$, since $r_i(t) r_j(t)$ is a multiple of $r(t)$, and $r(A) = 0$.

(iii) Properties (i), (ii) and Theorem 6.4 imply that V is the direct sum of $V_1, V_2, \ldots, V_l$.

(iv) First, we show that $V_i \subseteq N\left(p_i(A)^{m_i}\right)$. Let $\mathbf{v} \in V_i$ be an arbitrary element. Then $\mathbf{v} = B_i(\mathbf{w})$ with some $\mathbf{w} \in V$, and the definition of f_i implies that

$$p_i(A)^{m_i}(\mathbf{v}) = p_i(A)^{m_i}(B_i(\mathbf{w})) = p_i(A)^{m_i} f_i(A)(\mathbf{w})$$
$$= p_i(A)^{m_i} r_i(A) q_i(A)(\mathbf{w}) = r(A) q_i(A)(\mathbf{w}) = \mathbf{0},$$

since $r(A) = 0$. Therefore $\mathbf{v} \in N\left(p_i(A)^{m_i}\right)$.

Next, we prove that $V_i \supseteq N\left(p_i(A)^{m_i}\right)$. Assume that $\mathbf{v} \in N\left(p_i(A)^{m_i}\right)$ with some fixed i. From property (iii) of Theorem 6.5 we know that $\mathbf{v}$ can be expressed as

$$\mathbf{v} = \mathbf{v}_1 + \mathbf{v}_2 + \ldots + \mathbf{v}_l \tag{6.13}$$

where $\mathbf{v}_j \in V_j$ for all j. Since $V_j = R\left(B_j\right)$, there are elements $\mathbf{w}_j$ such that $\mathbf{v}_j = B_j(\mathbf{w}_j)$, therefore

$$0 = p_i(A)^{m_i}(\mathbf{v}) = p_i(A)^{m_i}\left(B_1(\mathbf{w}_1) + B_2(\mathbf{w}_2) + \ldots + B_l(\mathbf{w}_l)\right)$$

$$= \left(p_i(A)^{m_i} B_1\right)(\mathbf{w}_1) + \left(p_i(A)^{m_i} B_2\right)(\mathbf{w}_2) + \ldots + \left(p_i(A)^{m_i} B_l\right)(\mathbf{w}_l)$$

$$= \left(p_i(A)^{m_i} f_1(A)\right)(\mathbf{w}_1) + \left(p_i(A)^{m_i} f_2(A)\right)(\mathbf{w}_2) + \ldots + \left(p_i(A)^{m_i} f_l(A)\right)(\mathbf{w}_l).$$

Notice that for $j = 1, 2, \ldots, l$ the jth term can be written as

$$\left(f_j(A) p_i(A)^{m_i}\right)(\mathbf{w}_j) = \left(B_j p_i(A)^{m_i}\right)(\mathbf{w}_j) = B_j\left(p_i(A)^{m_i}(\mathbf{w}_j)\right)$$

which is in $R(B_j) = V_j$. Hence $\mathbf{0}$ is decomposed as the sum of l elements, where the jth element belongs to V_j for $j = 1, 2, \ldots, l$. Then Theorem 2.16 implies that each term is necessarily zero, so for all j,

$$0 = B_j\left(p_i(A)^{m_i}(\mathbf{w}_j)\right) = \left(B_j p_i(A)^{m_i}\right)(\mathbf{w}_j). \qquad (6.14)$$

For all $j \neq i$, $p_i(t)^{m_i}$ and $p_j(t)^{m_j}$ are relative primes, therefore Property 7 of the previous section implies that there exist polynomials $s_i(t)$ and $s_j(t)$ such that

$$1 = s_i(t) p_i(t)^{m_i} + s_j(t) p_j(t)^{m_j}.$$

Substituting A into this equation and applying both sides to the element $B_j(\mathbf{w}_j)$, with $j \neq i$ we have

$$B_j(\mathbf{w}_j) = \left(s_i(A) p_i(A)^{m_i}\right)\left(B_j(\mathbf{w}_j)\right) + \left(s_j(A) p_j(A)^{m_j}\right)\left(B_j(\mathbf{w}_j)\right)$$

$$= s_i(A)\left(\left(B_j p_i(A)^{m_i}\right)(\mathbf{w}_j)\right) + s_j(A)\left(\left(B_j p_j(A)^{m_j}\right)(\mathbf{w}_j)\right) = \mathbf{0},$$

since $B_i = f_i(A)$ commutes with $p_i(A)^{m_i}$ and $p_j(A)^{m_j}$, relation (6.14) holds, and

$$B_j p_j(A)^{m_j} = f_j(A) p_j(A)^{m_j} = r_j(A) q_j(A) p_j(A)^{m_j} = q_j(A) r(A) = O.$$

And finally, equation (6.13) implies that

$$\mathbf{v} = \mathbf{v}_1 + \mathbf{v}_2 + \ldots + \mathbf{v}_l = B_1\left(\mathbf{w}_1\right) + B_2\left(\mathbf{w}_2\right) + \ldots + B_l\left(\mathbf{w}_l\right) = B_i\left(\mathbf{w}_i\right)$$

since all other terms are equal to zero. Hence $\mathbf{v} \in R(B_i) = V_i$, that is, $N(p_i(A)^{m_i}) \subseteq V_i$, and thus, the proof is complete.

♣

6.5 Diagonal and Triangular Forms

In this section we will always assume that V is finitely generated. The main result of the previous section will be first applied to answer the question of when a basis of V can be chosen that consists of eigenvectors of A. In such cases the matrix representation of A in this basis becomes diagonal. This important fact can be proved as follows. Assume that $\dim(V) = n$, and $B = \{\mathbf{x}_1, \mathbf{x}_2, \ldots, \mathbf{x}_n\}$ is a basis of V consisting of eigenvectors of A. For all $i = 1, 2, \ldots, n$, the eigenvector equation

$$A\left(\mathbf{x}_i\right) = \lambda_i \mathbf{x}_i$$

is satisfied, where λ_i is the eigenvalue associated to eigenvector $\mathbf{x}_i$. That is,

$$A\left(\mathbf{x}_i\right) = 0 \cdot \mathbf{x}_1 + \ldots + 0 \cdot \mathbf{x}_{i-1} + \lambda_i \mathbf{x}_i + 0 \cdot \mathbf{x}_{i+1} + \ldots + 0 \cdot \mathbf{x}_n,$$

and equation (5.14) implies that the matrix representation of transformation A in basis B is diagonal:

$$\mathbf{A}_{B,B} = \begin{pmatrix} \lambda_1 & & & \\ & \lambda_2 & & \\ & & \ddots & \\ & & & \lambda_n \end{pmatrix}.$$

Definition 6.4. A linear transformation $A \in L(V)$ is called *diagonable* if there is a basis of V consisting of eigenvectors of A.

In the above discussion we have derived that the matrix representation of a diagonable transformation A in basis B is diagonal. Assume first that A has n distinct eigenvalues, then Theorem 6.1 implies that the associated eigenvectors form a basis in V, hence, A is diagonable. This condition is only sufficient for a transformation being diagonable. The case of the identity transformation shows that transformations may be diagonable even if they have multiple eigenvalues.

For the $n \times n$ identity transformation, $\lambda = 1$ is the only eigenvalue with multiplicity n, and any arbitrary vector $\mathbf{x}$ is an associated eigenvector, since $\mathbf{I}(\mathbf{x}) = \mathbf{x} = 1 \cdot \mathbf{x}$. Therefore, any basis of V consists of eigenvectors.

Example 6.6. Consider again the 3×3 matrix

$$\mathbf{A} = \begin{pmatrix} 1 & 4 & 1 \\ 1 & 2 & 3 \\ 1 & 3 & 2 \end{pmatrix},$$

which was the subject of Example 6.2. We have shown there that the eigenvalues of $\mathbf{A}$ are $\lambda_1 = 0$, $\lambda_2 = -1$, $\lambda_3 = 6$ with associated eigenvectors

$$\mathbf{x}_1 = \begin{pmatrix} -5 \\ 1 \\ 1 \end{pmatrix}, \quad \mathbf{x}_2 = \begin{pmatrix} 9 \\ -5 \\ 2 \end{pmatrix}, \quad \text{and } \mathbf{x}_3 = \begin{pmatrix} 1 \\ 1 \\ 1 \end{pmatrix}.$$

Since the eigenvalues are different, the eigenvectors form a basis in R^3. Introduce matrix

$$\mathbf{T} = \begin{pmatrix} -5 & 9 & 1 \\ 1 & -5 & 1 \\ 1 & 2 & 1 \end{pmatrix},$$

then in the new basis $\{\mathbf{x}_1, \mathbf{x}_2, \mathbf{x}_3\}$, matrix $\mathbf{A}$ will have the diagonal form

$$\mathbf{T}^{-1}\mathbf{A}\mathbf{T} = \begin{pmatrix} -\dfrac{1}{6} & -\dfrac{7}{42} & \dfrac{7}{21} \\ 0 & -\dfrac{6}{42} & \dfrac{3}{21} \\ \dfrac{1}{6} & \dfrac{19}{42} & \dfrac{8}{21} \end{pmatrix} \begin{pmatrix} 1 & 4 & 1 \\ 1 & 2 & 3 \\ 1 & 3 & 2 \end{pmatrix} \begin{pmatrix} -5 & 9 & 1 \\ 1 & -5 & 1 \\ 1 & 2 & 1 \end{pmatrix} = \begin{pmatrix} 0 & 0 & 0 \\ 0 & -1 & 0 \\ 0 & 0 & 6 \end{pmatrix}.$$

♦

A necessary and sufficient condition is next given for a matrix being diagonable.

Theorem 6.6. A linear transformation A is diagonable if and only if the minimal polynomial of A can be factored as

$$r(t) = (t - \lambda_1)(t - \lambda_2)...(t - \lambda_k),$$

where $\lambda_1, \lambda_2, ..., \lambda_k$ are distinct values.

Proof. Assume first that A is diagonable, and let $\{\mathbf{x}_1, \mathbf{x}_2, ..., \mathbf{x}_n\}$ be a basis of V consisting of eigenvectors of A. Let $\lambda_1, \lambda_2, ..., \lambda_k$ be the distinct eigenvalues of A, which means that any other eigenvalue of A coincides with some λ_i ($i = 1, 2, ..., k$). Define

$$p(t) = (t - \lambda_1)(t - \lambda_2)...(t - \lambda_k).$$

We will show that p is the minimal polynomial of A. Since $\mathbf{x}_i$ is an eigenvector associated to an eigenvalue λ_i ($i = 1, 2, ..., k$), the eigenvector equation implies that

$$\mathbf{0} = A(\mathbf{x}_i) - \lambda_i \mathbf{x}_i = (A - \lambda_i I)(\mathbf{x}_i).$$

Multiply both sides by polynomial $\prod_{l \neq i}(A - \lambda_l I)$ to see that $p(A)\,\mathbf{x}_i = \mathbf{0}$. If $\mathbf{x} \in V$ is arbitrary, then

$$\mathbf{x} = c_1 \mathbf{x}_1 + ... + c_n \mathbf{x}_n$$

with some scalars c_i $(i = 1, 2, \ldots, k)$, and

$$p(A)(\mathbf{x}) = p(A)(c_1\mathbf{x}_1 + \ldots + c_n\mathbf{x}_n) = c_1 \cdot p(A)(\mathbf{x}_1) + \ldots + c_n p(A)(\mathbf{x}_n) = \mathbf{0}.$$

Hence $p(A) = O$. Let $r(A)$ denote the minimal polynomial of A. If $p \neq r$, then Theorem 6.3 implies that r is a factor of p, therefore r is obtained from p by deleting at least one prime factor. Assume that factor $t - \lambda_i$ is among the deleted prime factors, then

$$r(t) = (t - \lambda_{i_1})\ldots(t - \lambda_{i_l}),$$

where the values $\lambda_{i_1}, \ldots, \lambda_{i_l}$ differ from λ_i. Let $\mathbf{x}_i$ be an eigenvector associated to λ_i, then

$$\begin{aligned}
r(A)(\mathbf{x}_i) &= (A - \lambda_{i_1}I)\ldots(A - \lambda_{i_l}I)(\mathbf{x}_i) \\
&= (A - \lambda_{i_1}I)\ldots(A - \lambda_{i_{l-1}}I)\left((\lambda_i - \lambda_{i_l})\mathbf{x}_i\right) \\
&= \ldots = (\lambda_i - \lambda_{i_1})\ldots(\lambda_i - \lambda_{i_{l-1}})(\lambda_i - \lambda_{i_l})\mathbf{x}_i \neq 0.
\end{aligned}$$

That is, $r(A)$ is nonzero contradicting the assumption that r is the minimal polynomial of A.

Assume next that the minimal polynomial of A has the form

$$r(t) = (t - \lambda_1)(t - \lambda_2)\ldots(t - \lambda_k)$$

with distinct values of $\lambda_1, \lambda_2, \ldots, \lambda_k$. From Theorem 6.5 we know that

$$V = V_1 \oplus V_2 \oplus \ldots \oplus V_k,$$

where for $i = 1, 2, \ldots, k$,

$$V_i = N\left((A - \lambda_i I)\right).$$

Let $\mathbf{x}_{i1}, \mathbf{x}_{i2}, \ldots, \mathbf{x}_{in_i}$ be a basis of V_i. First, we show that all basis elements are eigenvectors of A. The definition of V_i implies that for all j,

$$(A - \lambda_i I)(\mathbf{x}_{ij}) = \mathbf{0},$$

that is,

$$A\left(\mathbf{x}_{ij}\right) = \lambda_i \mathbf{x}_{ij},$$

which shows that $\mathbf{x}_{ij}$ is an eigenvector of A associated to λ_i. Since V is the direct sum of subspaces $V_1, V_2, \ldots, V_k$, the system

$$\left\{\mathbf{x}_{11}, \ldots, \mathbf{x}_{1n_1}, \mathbf{x}_{21}, \ldots, \mathbf{x}_{2n_2}, \ldots, \mathbf{x}_{k1}, \ldots, \mathbf{x}_{kn_k}\right\}$$

is a basis of V, which completes the proof.

♣

For an $n \times n$ matrix $\mathbf{A}$ the assertion of the theorem can be restated as follows. If $\mathbf{A}$ is real, then there exists a nonsingular real matrix $\mathbf{T}$ such that $\mathbf{T}^{-1}\mathbf{A}\mathbf{T}$ is diagonal if and only if all roots of the minimal polynomial of $\mathbf{A}$ are real and distinct. If $\mathbf{A}$ is complex, then there is a nonsingular (maybe complex) matrix $\mathbf{T}$ such that $\mathbf{T}^{-1}\mathbf{A}\mathbf{T}$ is diagonal if and only if the roots (which maybe real or complex) of the minimal polynomial of $\mathbf{A}$ are distinct.

The condition of the theorem can fail under two circumstances. One is that the minimal polynomial has multiple roots; the other is that the minimal polynomial cannot be factored into linear factors. For complex polynomials the second case never occurs, and for real polynomials it occurs if and only if the minimal polynomial has at least one complex root. In such cases we may consider the matrix representation as complex and diagonalize the matrix with complex arithmetic. The first case however cannot be handled so easily. The best way is to find a class of matrices of a special structure that is as close to the set of diagonal matrices as possible and for all $A \in L(V)$, there is a basis such that the matrix representation of A has that special structure.

We start our analysis with examining a special group of linear transformations.

Definition 6.5. Let $A \in L(V)$ be a linear transformation. It is called *nilpotent* if $A^k = O$ for some positive integer k. The smallest such integer k is called the *degree* of A.

Example 6.7. Consider the 3×3 matrix

$$\mathbf{A} = \begin{pmatrix} 0 & 1 & 0 \\ 0 & 0 & 1 \\ 0 & 0 & 0 \end{pmatrix}.$$

Simple calculation shows that

$$\mathbf{A}^2 = \begin{pmatrix} 0 & 0 & 1 \\ 0 & 0 & 0 \\ 0 & 0 & 0 \end{pmatrix} \quad \text{and} \quad \mathbf{A}^3 = \begin{pmatrix} 0 & 0 & 0 \\ 0 & 0 & 0 \\ 0 & 0 & 0 \end{pmatrix}.$$

Let A be a linear transformation the matrix representation of which is $\mathbf{A}$ in a given basis. Then A is nilpotent, and its degree is 3.

◆

We will first show that nilpotent matrices are singular. Let $\mathbf{A}$ be a nonsingular matrix. The repeated application of property (iv) of Theorem 5.8 implies that $\mathbf{A}^k$ is also nonsingular for all $k \geq 2$ and $\left(\mathbf{A}^k\right)^{-1} = \left(\mathbf{A}^{-1}\right)^k$. Therefore, nilpotent matrices must be singular, since $\mathbf{O}$ is a singular matrix.

The following property of nilpotent transformations will be useful later in this section.

Theorem 6.7. Let V be finite dimensional and let $A \in L(V)$ be nilpotent. Then there is a basis $B = \{\mathbf{x}_1, \mathbf{x}_2, \ldots, \mathbf{x}_n\}$ of V such that $A(\mathbf{x}_1) = \mathbf{0}$, and for all $i = 2, \ldots, n$, $A(\mathbf{x}_i)$ belongs to the subspace $V(\mathbf{x}_1, \ldots, \mathbf{x}_{i-1})$ generated by the previous basis elements $\mathbf{x}_1, \ldots, \mathbf{x}_{i-1}$.

Proof. The theorem is proved by induction. Let n denote the dimension if the vector space V. Notice first that A is singular, therefore there is an $\mathbf{x}_1 \neq 0$ such that $A(\mathbf{x}_1) = \mathbf{0}$. As the inductive hypothesis assume that for an $i < n$, we have found linearly independent elements $\mathbf{x}_1, \ldots, \mathbf{x}_i$ satisfying the conditions of the theorem. The subspace $V(\mathbf{x}_1, \ldots, \mathbf{x}_i)$ generated by these elements is a nontrivial subspace since it is generated by less than n elements.

If $R(A) \subset V(\mathbf{x}_1, ..., \mathbf{x}_i)$, then let $\mathbf{x}_{i+1}$ be any element that does not belong to $V(\mathbf{x}_1, ..., \mathbf{x}_i)$. If $R(A) \not\subset V(\mathbf{x}_1, ..., \mathbf{x}_i)$, then there exists a positive integer l such that $R(A^l) \not\subset V(\mathbf{x}_1, ..., \mathbf{x}_i)$, but $R(A^{l+1}) \not\subset V(\mathbf{x}_1, ..., \mathbf{x}_i)$, since with some k, $\mathbf{A}^k = \mathbf{O}$, and $R(A^k) = \{\mathbf{0}\} \subset V(\mathbf{x}_1, ..., \mathbf{x}_i)$.
Select now $\mathbf{x}_{i+1}$ such that $\mathbf{x}_{i+1} \in R(A^l)$ but $\mathbf{x}_{i+1} \notin V(\mathbf{x}_1, ..., \mathbf{x}_i)$. Then vectors $\mathbf{x}_1, ..., \mathbf{x}_i, \mathbf{x}_{i+1}$ are linearly independent and $A(\mathbf{x}_{i+1}) \in V(\mathbf{x}_1, ..., \mathbf{x}_i)$. Thus, the proof is complete.

♣

The construction of the basis $B = \{\mathbf{x}_1, ..., \mathbf{x}_n\}$ implies that the matrix representation of a nilpotent transformation A has the special form that all diagonal elements as well as all elements under the diagonal are equal to zero. This is a special upper triangular matrix. We will next prove that all linear transformations of finitely generated vector spaces can be represented by an upper triangular matrix in a special basis.

Theorem 6.8. Let $A \in L(V)$ be a linear transformation. Assume that V is finitely generated and the minimal polynomial of A is given as

$$r(t) = (t - \lambda_1)^{m_1} (t - \lambda_2)^{m_2} ... (t - \lambda_l)^{m_l}.$$

Then there is a basis B of V such that the matrix representation of A in this basis has the special form

$$\mathbf{A}_{B,B} = \begin{pmatrix} \mathbf{A}_1 & & & \mathbf{O} \\ & \mathbf{A}_2 & & \\ & & \ddots & \\ \mathbf{O} & & & \mathbf{A}_l \end{pmatrix}, \tag{6.15}$$

where for $i = 2, ..., l$, A_i is an upper triangular $n_i \times n_i$ matrix with some $n_i \geq m_i$, and all diagonal elements of A_i are equal to λ_i.

Proof. As in the earlier theorems, define $V_i = N\left((A - \lambda_i I)^{m_i}\right)$, and let $n_i = \dim(V_i)$. In Theorem 6.5 we have proved that $V = V_1 \oplus V_2 \oplus ... \oplus V_l$, and from Theorem 6.2 we know that all subspaces V_i are A-invariant. If for $i = 1, 2, ..., l$, $B_i = \{\mathbf{x}_{i1}, ..., \mathbf{x}_{in_i}\}$ is a basis of V_i, then $B = \left\{ \mathbf{x}_{11}, ..., \mathbf{x}_{1n_1}, \mathbf{x}_{21}, ..., \mathbf{x}_{2n_2}, ..., \mathbf{x}_{l1}, ..., \mathbf{x}_{ln_l} \right\}$ is a basis of V and the matrix representation of A in this basis has the block-diagonal form (6.15) where the size of each diagonal matrix $\mathbf{A}_i$ is $n_i \times n_i$. In order the complete the proof we have to show that each diagonal block $\mathbf{A}_i$ has a special upper triangular form with diagonal elements λ_i.

Define $A_i \in L(V_i)$ as a linear transformation on V_i such that for all $\mathbf{x} \in V_i$ $A_i(\mathbf{x}) = (A - \lambda_i I)(\mathbf{x})$.

The definition of subspace V_i implies that on V_i, $\left(A - \lambda_i I\right)^{m_i} = O$, that is, $A - \lambda_i I$ is nilpotent. Then Theorem 6.7 implies that there is a basis $B_i = \left\{ \mathbf{x}_{i1}, \mathbf{x}_{i2}, ..., \mathbf{x}_{in_i} \right\}$ of V_i such that

$$\left(A - \lambda_i I\right)\left(\mathbf{x}_{i1}\right) = \mathbf{0},$$

and for $j = 2, ..., n_i$,

$$\left(A - \lambda_i I\right)\left(\mathbf{x}_{ij}\right) \in V\left(\mathbf{x}_{i1}, ..., \mathbf{x}_{i,j-1}\right).$$

These relations can be rewritten as

$$A\left(\mathbf{x}_{i1}\right) = \lambda_i \underline{\mathbf{x}}_{i1}$$

and for $j = 2, ..., n_i$,

$$A\left(\mathbf{x}_{ij}\right) = \lambda_i \mathbf{x}_{ij} + a_{1j}\mathbf{x}_{i1} + ... + a_{j-1,j}\mathbf{x}_{i,j-1}.$$

Hence the matrix representation of A in subspace V_i has the special form

$$\mathbf{A}_i = \begin{pmatrix} \lambda_i & a_{12} & a_{13} & \cdots & & a_{1n} \\ & \lambda_i & a_{23} & \cdots & & a_{2n} \\ & & \ddots & & & \vdots \\ & & & & & a_{n_i-1,n_i} \\ O & & & & & \lambda_i \end{pmatrix},$$

which completes the proof.

♣

For $n \times n$ matrices this theorem can be restated in the following way. For an arbitrary $n \times n$ real or complex matrix there is a nonsingular matrix **T** such that $\mathbf{T}^{-1}\mathbf{A}\mathbf{T}$ has the special form (6.15). If the minimal polynomial has only real roots, then **T**, $\mathbf{T}^{-1}$, and all diagonal blocks $\mathbf{A}_i$ are real. In the case of complex roots the corresponding diagonal blocks $\mathbf{A}_i$ as well as matrices $\mathbf{T}^{-1}$ and **T** might have complex elements.

Corollary. Let $r(t)$ and $\varphi(t)$ be the minimal polynomial and characteristic polynomial of A, respectively. Then

    a) r is a factor of φ;

    b) all roots of φ are roots of r as well;

    c) $\varphi(A) = O$.

Proof. From the matrix representation (6.15) we see that the characteristic polynomial of A (which is independent of the selected basis) is the following:

$$\varphi(t) = (t - \lambda_1)^{n_1}(t - \lambda_2)^{n_2} \ldots (t - \lambda_l)^{n_l}$$
$$= (t - \lambda_1)^{n_1 - m_1}(t - \lambda_2)^{n_2 - m_2} \ldots (t - \lambda_l)^{n_l - m_l} r(t) = q(t)r(t),$$

from which assertions (a) and (b) follow immediately, furthermore

$$\varphi(A) = (A - \lambda_1 I)^{n_1 - m_1}(A - \lambda_2 I)^{n_2 - m_2} \ldots (A - \lambda_l I)^{n_l - m_l} r(A) = O,$$

since $r(A) = O$. Here we applied part b) of Theorem 6.3.

♣

Assertion c) is known as the famous *Cayley-Hamilton* theorem, which can be restated by saying that if we substitute any arbitrary $n \times n$ matrix into its characteristic polynomial we always obtain the zero matrix as the result.

6.6   The Jordan Canonical Form

In the previous section we have proved that any real or complex square matrix can be transformed into a block-diagonal form, where each diagonal block is an upper triangular matrix (that might have complex elements). In this section we will present a further refinement of this result by showing that the upper triangular blocks can be made very specially: the diagonal elements equal to an eigenvalue of the matrix, all elements just above the diagonal are equal to one, and all other elements equal zero. Our analysis will be based on the following concept.

Definition 6.6. Let $A \in L(V)$, and let $\mathbf{x}$ be a nonzero element of V. The subspace generated by the system $\{\mathbf{x}, A(\mathbf{x}), A^2(\mathbf{x}), ...\}$ is called a *cyclic subspace*, and is denoted by $V(\mathbf{x}, A)$.

We show first that subspace $V(\mathbf{x}, A)$ is A-invariant for all $\mathbf{x} \in V$. Let $\mathbf{z} \in V(\mathbf{x}, A)$ be an arbitrary element. Then Theorem 2.4 implies that

$$\mathbf{z} = a_1 A^{i_1}(\mathbf{x}) + a_2 A^{i_2}(\mathbf{x}) + ... + a_l A^{i_l}(\mathbf{x})$$

with some integers $l \geq 1$, $0 \leq i_1 < i_2 < ... < i_l$, and scalars $a_1, a_2, ..., a_l$. Then the linearity of A implies that

$$A(\mathbf{z}) = a_1 A^{i_1+1}(\mathbf{x}) + a_2 A^{i_2+1}(\mathbf{x}) + ... + a_l A^{i_l+1}(\mathbf{x}) \in V(\mathbf{x}, A)$$

which implies the assertion.

Notice next that if A is nilpotent with degree k, then $V(\mathbf{x}, A)$ is generated by $\mathbf{x}, A(\mathbf{x}), A^2(\mathbf{x}), ..., A^{k-1}(\mathbf{x})$. Since $A^{k-1} \neq O$, there exists an $\mathbf{x} \in V$ such that $A^{k-1}(\mathbf{x}) \neq \mathbf{0}$. We will first prove that in this case the elements $\mathbf{x}, A(\mathbf{x}), A^2(\mathbf{x}), ..., A^{k-1}(\mathbf{x})$ are linearly independent. Assume that a linear combination of these elements is zero;

$$a_0 \mathbf{x} + a_1 A(\mathbf{x}) + a_2 A^2(\mathbf{x}) + ... + a_{k-1} A^{k-1}(\mathbf{x}) = \mathbf{0}.$$

Assume that i is the smallest subscript such that $a_i \neq 0$. Apply A^{k-i-1} on both sides of this equation to have

$$a_i A^{k-1}(\mathbf{x}) = \mathbf{0}.$$

Since $A^{k-1}(\mathbf{x}) \neq \mathbf{0}$, this is possible only if $a_i = 0$, which contradicts the assumption that $a_i \neq 0$. Hence the linear independence of elements $\mathbf{x}, A(\mathbf{x}), A^2(\mathbf{x}), ..., A^{k-1}(\mathbf{x})$ is verified, and therefore they form a basis of the subspace $V(\mathbf{x}, A)$.

Assume again that A is nilpotent on V with degree k. Let $\mathbf{x} \in V_1 = R(A)$ and define the linear transformation A_1 on V_1 such that for all $\mathbf{x} \in V_1$, $A_1(\mathbf{x}) = A(\mathbf{x})$. This transformation is called the *restriction* of A to subspace V_1. We can easily prove that A_1 is also nilpotent, and its degree is $k - 1$. First, let $\mathbf{x} \in V_1$ be an arbitrary element, then $\mathbf{x} = A(\mathbf{z})$ with some $\mathbf{z} \in V$. Then

$$A_1^{k-1}(\mathbf{x}) = A_1^{k-1}(A(\mathbf{z})) = A^{k-1}(A(\mathbf{z})) = A^k(\mathbf{z}) = \mathbf{0},$$

that is, the degree of A_1 is not greater than $k - 1$. Notice next that there is an $\mathbf{x} \in V$ such that $A^{k-1}(\mathbf{x}) \neq \mathbf{0}$, then by choosing $\mathbf{y} = A(\mathbf{x}) \in V_1$,

$$A_1^{k-2}(\mathbf{y}) = A_1^{k-2}(A(\mathbf{x})) = A^{k-2}(A(\mathbf{x})) = A^{k-1}(\mathbf{x}) \neq \mathbf{0}$$

showing that the degree of A_1 is not less than $k - 1$. Hence the degree of A_1 equals $k - 1$.

The main result of this section that will be used in obtaining the Jordan form can be formulated as follows.

Theorem 6.9. Assume that $A \in L(V)$ is nilpotent with degree k. Assume that $A^{k-1}(\mathbf{u}) \neq \mathbf{0}$. Then there exists an A-invariant subspace $V_1 \subset V$ such that

$$V = V(\mathbf{u}, A) \oplus V_1 \qquad (6.16)$$

Proof. Finite induction will be used with respect to the degree of A. Assume first that $k = 1$. Then $A(\mathbf{x}) = \mathbf{0}$ for all $\mathbf{x} \in V$. Select an arbitrary element $\mathbf{u} \neq \mathbf{0}$. If $\dim(V) = n$, then there exist elements $\mathbf{u}_1, \mathbf{u}_2, ..., \mathbf{u}_{n-1}$ such that $\{\mathbf{u}, \mathbf{u}_1, \mathbf{u}_2, ..., \mathbf{u}_{n-1}\}$ is a basis of V. Consider the subspace $V_1 = V(\mathbf{u}_1, \mathbf{u}_2, ..., \mathbf{u}_{n-1})$ which is obviously an A-invariant subspace and the fact that $V(\mathbf{u}, A) = V(\mathbf{u})$ implies that (6.16) holds. Assume next that the assertion is true for nilpotent transformations with degrees up to $k - 1$, and consider an A with degree k. The subspace $R(A) \subset V$ is A-invariant, and the restriction of A to $R(A)$ is also nilpotent with degree $k - 1$. Let $\mathbf{u} \in R(A)$ be an element such that $A^{k-2}(\mathbf{u}) \neq \mathbf{0}$. Then by the inductive hypotheses there is an A-invariant subspace V_0 such that

$$R(A) = V\left(A(\mathbf{v}), A\right) \oplus V_0,$$

where $\mathbf{v} \in V$ is selected as $\mathbf{u} = A(\mathbf{v})$. Notice that the first term is generated by the elements $A(\mathbf{v}), A^2(\mathbf{v}), ..., A^{k-1}(\mathbf{v})$, which are linearly independent. Consider the set

$$\overline{V}_0 = \left\{ \mathbf{v} \in V \mid A(\mathbf{v}) \in V_0 \right\}.$$

We will next prove that it is an A-invariant subspace. First, we show that $\overline{V}_0$ is a subspace of V. Let a_1, a_2 be two scalars and $\mathbf{u}_1$, $\mathbf{u}_2 \in V_0$. Then

$$A\left(a_1 \mathbf{u}_1 + a_2 \mathbf{u}_2\right) = a_1 A(\mathbf{u}_1) + a_2 A(\mathbf{u}_2) \in V_0,$$

that is, $a_1 \mathbf{u}_1 + a_2 \mathbf{u}_2 \in \overline{V}_0$. To prove that $\overline{V}_0$ is A-invariant, consider an arbitrary element $\mathbf{u} \in \overline{V}_0$. Then $A(\mathbf{u}) \in V_0$, and since V_0 is A-invariant, $A(A(\mathbf{u})) \in V_0$ showing that $A(\mathbf{u}) \in V_0$. Hence, $\overline{V}_0$ is also A-invariant. We will next show that $V(\mathbf{v}, A) \cup \overline{V}_0$ generates the entire space V. Let $\mathbf{x} \in V$ be arbitrary. Since $A(\mathbf{x}) \in R(A)$, we have

$$A(\mathbf{x}) = \sum_{i=1}^{k-1} a_i A^i(\mathbf{v}) + \mathbf{w},$$

where $\mathbf{w} \in V_0$ and a_1, a_2, ..., a_{k-1} are scalars. This equation can be rewritten as

$$A\left(\mathbf{x} - \sum_{i=1}^{k-1} a_i A^{i-1}(\mathbf{v}) \right) = \mathbf{w},$$

therefore, the definition of $\overline{V}_0$ implies that

$$\mathbf{x} - \sum_{i=1}^{k-1} a_i A^{i-1}(\mathbf{v}) \in \overline{V}_0.$$

That is, $\mathbf{x}$ can be expressed as the sum of an element of $V(\mathbf{v}, A)$ and an element of $\overline{V}_0$.

Unfortunately, in completing the proof, we cannot select $\mathbf{u} = \mathbf{v}$ and $V_1 = \overline{V}_0$, since $V(\mathbf{v}, A)$ and $\overline{V}_0$ might have nonzero common elements. We can show however that $V(\mathbf{v}, A) \cap V_0 = \{\mathbf{0}\}$. If

$$\mathbf{x} \in V(\mathbf{v}, A) \cap V_0,$$

then

$$A(\mathbf{x}) \in V\left(A(\mathbf{v}), A\right) \cap V_0,$$

therefore $A(\mathbf{x}) = \mathbf{0}$. Since $\mathbf{x} \in V(\mathbf{v}, A)$, with some scalars $c_0, c_1, \ldots, c_{k-1}$,

$$\mathbf{x} = c_0 \mathbf{v} + c_1 A(\mathbf{v}) + \ldots + c_{k-1} A^{k-1}(\mathbf{v}).$$

Assume that for an index i $(i < k-1)$, $c_i \neq 0$. Let i denote the smallest such value. Apply A^{k-i-1} to both sides to this equality to have

$$A^{k-i-1}(\mathbf{x}) = c_i A^{k-1}(\mathbf{v}),$$

which is an obvious contradiction, since the left-hand side is zero, and the right-hand side is nonzero. Consequently, $\mathbf{x}$ is a scalar-multiple of $A^{k-1}(\mathbf{v})$, and therefore

$$\mathbf{x} \in V\left(A(\mathbf{v}), A\right) \cap V_0 = \{\mathbf{0}\},$$

hence $\mathbf{x} = \mathbf{0}$. Notice finally, that both $V(\mathbf{v}, A) \cap \overline{V}_0$ and V_0 are subspaces of $\overline{V}_0$, and their intersection is only the zero element. Therefore, a basis of $\overline{V}_0$ can be obtained as

$$\{B_1, B_2, \mathbf{v}_1, \mathbf{v}_2, \ldots, \mathbf{v}_l\},$$

where B_1 is a basis of $V(\mathbf{v}, A) \cap \overline{V}_0$, B_2 is a basis of V_0, and $\mathbf{v}_1, \mathbf{v}_2, \ldots, \mathbf{v}_l$ are the elements which complete the linearly independent system $\{B_1, B_2\}$ into a basis of $\overline{V}_0$. Define now

$$V_1 = V(B_2, \mathbf{v}_1, \ldots, \mathbf{v}_l),$$

then obviously

$$V = V(\mathbf{v}, A) \oplus V_1.$$

The definition of V_1 implies that $V_0 \subseteq V_1 \subseteq \overline{V}_0$ and for all $\mathbf{x} \in V_1$, $A(\mathbf{x}) \in V_0$. We have assumed that V_0 is an A-invariant subspace, therefore for all $\mathbf{x} \in V_1 \subseteq \overline{V}_0$, $A(\mathbf{x}) \in V_0 \subseteq V_1$ proving that V_1 is also an A-invariant subspace. Thus, the proof is complete.

♣

Corollary 1. The repeated application of the assertion of the theorem implies that if A is a nilpotent linear transformation with degree m then there exist elements $\mathbf{v}_0, \mathbf{v}_1, \ldots, \mathbf{v}_r \in V$ and positive integers $k = k_0 \geq k_1 \geq k_2 \geq \ldots \geq k_r$ such that vectors

$$\mathbf{v}_0, A(\mathbf{v}_0), \ldots, A^{k_0-1}(\mathbf{v}_0), \mathbf{v}_1, A(\mathbf{v}_1), \ldots, A^{k_1-1}(\mathbf{v}_1), \ldots, \mathbf{v}_r, A(\mathbf{v}_r), \ldots, A^{k_r-1}(\mathbf{v}_r)$$

form a basis of V, furthermore

$$V = V(\mathbf{v}_0, A) \oplus V(\mathbf{v}_1, A) \oplus \ldots \oplus V(\mathbf{v}_r, A).$$

Corollary 2. Let now $A \in L(V)$ be an arbitrary linear transformation. Assume that the minimal polynomial of A is given as

$$r(t) = (t - \lambda_1)^{m_1} (t - \lambda_2)^{m_2} \ldots (t - \lambda_r)^{m_r}.$$

From Theorem 6.5 we know that

$$V = N\left((A - \lambda_1 I)^{m_1}\right) \oplus N\left((A - \lambda_2 I)^{m_2}\right) \oplus \ldots \oplus N\left((A - \lambda_r I)^{m_r}\right).$$

The restriction of mapping $(A - \lambda_i I)$ into the subspace

$V_i = N\left(\left(A - \lambda_i I\right)^{m_i}\right)$ is nilpotent with degree m_i, since on V_i,

$\left(A - \lambda_i I\right)^{m_i} = O$, and the exponent m_i cannot be lowered.

Otherwise, the exponent m_i could also be lowered in the minimal polynomial r. Use the previous Corollary to see that for all $i = 1, 2, \ldots, r$, there exist integers $m_i = m_{i0} \geq m_{i1} \geq \ldots \geq m_{is_i}$ and elements $\mathbf{v}_{i0}, \mathbf{v}_{i1}, \ldots, \mathbf{v}_{is_i}$ that

$$B_i = \left\{ \begin{array}{l} \mathbf{v}_{i0}, \left(A - \lambda_i I\right)\left(\mathbf{v}_{i0}\right), \ldots, \left(A - \lambda_i I\right)^{m_{i0}-1}\left(\mathbf{v}_{i0}\right), \\ \ldots, \mathbf{v}_{is_i}, \left(A - \lambda_i I\right)\left(\mathbf{v}_{is_i}\right), \ldots, \left(A - \lambda_i I\right)^{m_{is_i}-1}\left(\mathbf{v}_{is_i}\right) \end{array} \right\}$$

is a basis of V_i and

$$V_i = V\left(\mathbf{v}_{i0}, \left(A - \lambda_i I\right)\right) \oplus V\left(\mathbf{v}_{i1}, \left(A - \lambda_i I\right)\right) \oplus \ldots \oplus V\left(\mathbf{v}_{is_i}, \left(A - \lambda_i I\right)\right).$$

It is easy to show that all terms of the right-hand side are A-invariant subspaces. Consider next the restriction A_i of mapping A into subspace V_i. The matrix representation of A_i in basis B_i has the special block diagonal form

$$A_{i,B_i,B_i} = \begin{pmatrix} A_{i1} & & & \\ & A_{i2} & & \\ & & \ddots & \\ & & & A_{is_i} \end{pmatrix}$$

where for $j = 1, 2, \ldots, s_i, A_{ij}$ is an $m_{ij} \times m_{ij}$ matrix:

$$A_{ij} = \begin{pmatrix} \lambda_i & & & & \\ 1 & \lambda_i & & & \\ & 1 & \ddots & & \\ & & \ddots & \lambda_i & \\ & & & 1 & \lambda_i \end{pmatrix}.$$

All diagonal elements of A_{ij} are equal to λ_i, and just under the diagonal, all elements are equal to one, and all other elements are equal to zero. This representation follows from the identities

$$
A\left(\left(A-\lambda_i I\right)^j\left(\mathbf{v}_{ip}\right)\right)
$$
$$
=\begin{cases}
\left(A-\lambda_i I\right)^{j+1}\left(\mathbf{v}_{ip}\right)+\lambda_i\left(A-\lambda_i I\right)^j\left(\mathbf{v}_{ip}\right) & \text{if } 0\le j<m_{ip}-1, \\
\lambda_i\left(A-\lambda_i I\right)^j\left(\mathbf{v}_{ip}\right) & \text{if } j=m_{ip}-1.
\end{cases}
$$

And finally, $B = \{B_1, B_2, \ldots, B_r\}$ gives a basis of V, and the matrix representation of mapping A in basis B is the following:

$$
A_{B,B}=\begin{pmatrix}
A_{1,B_1,B_1} & & & \\
& A_{2,B_2,B_2} & & \\
& & \ddots & \\
& & & A_{r,B_r,B_r}
\end{pmatrix}.
$$

This representation is called the *Jordan canonical-form* of transformation (or matrix) A. The number of diagonal blocks equals $s_1+s_2+\ldots+s_r$, each block has the same eigenvalue in the diagonal, it has the form of A_{ij}, and each eigenvalue of A can be found in at least one diagonal block.

6.7 Complexification

In examining the structure of linear mappings, in several cases we had to distinguish between real and complex vector spaces, since different results apply to these different cases. In this section we will introduce a simple method which will enable us to apply directly results originally proven for the complex case to real vector and inner product spaces, and to real mappings. The idea is known as *imbedding*, and it constructs a complex vector space or inner-product space, a certain subset of which is isomorphic to the given real vector space and inner-product space. In addition, the restriction of the complex extension of a real mapping to this subset is equivalent to the original real mapping.

Consider now a real vector space V, and denote $\hat{V} = V \times V$. Define addition and multiplication by complex numbers on $\hat{V}$ in the following ways:

$$(\mathbf{x}, \mathbf{y}) + (\mathbf{u}, \mathbf{v}) = (\mathbf{x} + \mathbf{u}, \ \mathbf{y} + \mathbf{v}),$$

and

$$(a + ib)(\mathbf{x}, \mathbf{y}) = (a\mathbf{x} - b\mathbf{y}, \ b\mathbf{x} + a\mathbf{y}).$$

It is easy to see that with these operations, $\hat{V}$ is a complex vector space with zero element $(\mathbf{0}, \mathbf{0})$, and if multiplication by only real numbers is considered, then V is isomorphic to the set

$$\hat{V}_1 = \left\{ (\mathbf{x}, \mathbf{0}) \mid \mathbf{x} \in V \right\}.$$

Therefore, $\dim(V) = \dim(\hat{V}_1)$, and if $\{\mathbf{x}_1, \mathbf{x}_2, \ldots, \mathbf{x}_n\}$ is a basis of V then $\{(\mathbf{x}_1, \mathbf{0}), (\mathbf{x}_2, \mathbf{0}), \ldots, (\mathbf{x}_n, \mathbf{0})\}$ forms a basis in $\hat{V}_1$. For easy reference we may use the notation $\mathbf{x} + i\mathbf{y}$ for $(\mathbf{x}, \mathbf{y})$.

If V is a real inner-product space, then a complex inner-product on $\hat{V}$ can be defined as

$$(\mathbf{x} + i\mathbf{y}, \mathbf{u} + i\mathbf{v}) = \big((\mathbf{x}, \mathbf{u}) - (\mathbf{y}, \mathbf{v})\big) + i\big((\mathbf{x}, \mathbf{v}) + (\mathbf{y}, \mathbf{u})\big).$$

The reader may easily verify that this inner-product satisfies the properties of complex inner-products. The length of the elements of $\hat{V}$ are defined as

$$|\mathbf{x} + i\mathbf{y}| = \sqrt{(\mathbf{x} + i\mathbf{y}, \ \mathbf{x} + i\mathbf{y})} = \sqrt{|\mathbf{x}|^2 + |\mathbf{y}|^2},$$

where $|\mathbf{x}|$ and $|\mathbf{y}|$ are the lengths of $\mathbf{x}$ and $\mathbf{y}$ in V.

Consider next a linear transformation $A \in L(V)$. The extension $\hat{A}$ of A to the complex vector space $\hat{V}$ is defined by equation

$$\hat{A}(\mathbf{x} + i\mathbf{y}) = A(\mathbf{x}) + iA(\mathbf{y}),$$

and it is easy to see that $\hat{A} \in L(\hat{V})$, and the restriction of $\hat{A}$ into the real vector space is the original transformation A.

Based on this complexification method, real vector spaces, real inner-product spaces and their mappings can be treated as complex spaces and mappings, respectively, hence all general results proved for complex vector spaces can be directly applied for real cases.

6.8 Applications

The diagonal and Jordan canonical forms of square matrices are very useful in many fields of the application of matrix theory. In this section we will first focus on dynamic systems described by difference or differential equations, and then some simple examples will be presented.

1. Consider first the *discrete time-invariant system* with state transition equation

$$\mathbf{x}(t+1) = \mathbf{A} \cdot \mathbf{x}(t) + \mathbf{b},$$

where $t = 0, 1, 2, \ldots$, and $\mathbf{A}$ is an $n \times n$ real matrix, $\mathbf{b}$ is an n-element real vector. Let $\mathbf{T}$ be an $n \times n$ nonsingular matrix and introduce the new state variable

$$\mathbf{z} = \mathbf{Tx},$$

then

$$\begin{aligned}
\mathbf{z}(t+1) &= \mathbf{Tx}(t+1) = \mathbf{T}\left(\mathbf{Ax}(t) + \mathbf{b}\right) \\
&= \mathbf{T}\left(\mathbf{A} \cdot \mathbf{T}^{-1}\mathbf{z}(t) + \mathbf{b}\right) = \left(\mathbf{TAT}^{-1}\right)\mathbf{z}(t) + \mathbf{T} \cdot \mathbf{b}
\end{aligned} \tag{6.17}$$

showing that the new state-transition equation has the same form as originally and the new coefficient matrix becomes $\mathbf{TAT}^{-1}$. If matrix $\mathbf{A}$ is diagonable, then there exists a nonsingular matrix such that $\mathbf{TAT}^{-1}$ is diagonal. In this case system (6.17) becomes the set of n independent difference equations, and in each of them the unknown function is real-valued. Assume that

$$\mathbf{TAT}^{-1} = \begin{pmatrix} \lambda_1 & & & \\ & \lambda_2 & & \\ & & \ddots & \\ & & & \lambda_n \end{pmatrix}, \quad \text{and} \quad \mathbf{Tb} = \begin{pmatrix} b'_1 \\ b'_2 \\ \vdots \\ b'_n \end{pmatrix},$$

then equations (6.17) reduce to the following system

$$z_1(t+1) = \lambda_1 z_1(t) + b'_1$$
$$z_2(t+1) = \lambda_2 z_2(t) + b'_2$$
$$\cdots\cdots\cdots\cdots$$
$$z_n(t+1) = \lambda_n z_n(t) + b'_n.$$

The state transition matrix of the system can also be determined in a simple way if $\mathbf{A}$ is diagonable. Since

$$\mathbf{A} = \mathbf{T}^{-1}\mathbf{DT}$$

with

$$\mathbf{D} = \begin{pmatrix} \lambda_1 & & & \\ & \lambda_2 & & \\ & & \ddots & \\ & & & \lambda_n \end{pmatrix},$$

for all $t \geq 1$,

$$\mathbf{A}^t = \left(\mathbf{T}^{-1}\mathbf{DT}\right)\left(\mathbf{T}^{-1}\mathbf{DT}\right)\ldots\left(\mathbf{T}^{-1}\mathbf{DT}\right)$$
$$= \mathbf{T}^{-1}\mathbf{D}\left(\mathbf{TT}^{-1}\right)\mathbf{D}\ldots\left(\mathbf{TT}^{-1}\right)\mathbf{DT} = \mathbf{T}^{-1}\mathbf{D}^t\mathbf{T} \qquad (6.18)$$
$$= \mathbf{T}^{-1}\begin{pmatrix} \lambda_1^t & & & \\ & \lambda_2^t & & \\ & & \ddots & \\ & & & \lambda_n^t \end{pmatrix}\mathbf{T}.$$

Example 6.8. Consider system

$$\mathbf{x}(t+1) = \begin{pmatrix} 1 & 4 & 1 \\ 1 & 2 & 3 \\ 1 & 3 & 2 \end{pmatrix} \mathbf{x}(t) + \begin{pmatrix} 1 \\ 1 \\ 1 \end{pmatrix}, \quad \mathbf{x}(0) = \begin{pmatrix} 0 \\ 0 \\ 0 \end{pmatrix}.$$

From Example 6.6 and equation (6.17) we know that by introducing the new variable

$$\mathbf{z}(t) = \begin{pmatrix} -\dfrac{1}{6} & -\dfrac{7}{42} & \dfrac{7}{21} \\ 0 & -\dfrac{6}{42} & \dfrac{3}{21} \\ \dfrac{1}{6} & \dfrac{19}{42} & \dfrac{8}{21} \end{pmatrix} \mathbf{x}(t),$$

a diagonal system is obtained

$$\mathbf{z}(t+1) = \begin{pmatrix} 0 & & \\ & -1 & \\ & & 6 \end{pmatrix} \mathbf{z}(t) + \begin{pmatrix} 0 \\ 0 \\ 1 \end{pmatrix}, \quad \mathbf{z}(0) = \mathbf{0}$$

It can be written as three independent equations

$$z_1(t+1) = 0,$$
$$z_2(t+1) = -z_2(t)$$
$$z_3(t+1) = 6z_3(t) + 1$$

with solutions

$$z_1(t) = 0, \ z_2(t) = 0, \ z_3(t) = 1 + 6 + 6^2 + \ldots + 6^{t-1} = \dfrac{6^t - 1}{5},$$

so

$$\mathbf{x}(t) = \begin{pmatrix} -5 & 9 & 1 \\ 1 & -5 & 1 \\ 1 & 2 & 1 \end{pmatrix} \begin{pmatrix} 0 \\ 0 \\ \dfrac{6^t - 1}{5} \end{pmatrix} = \begin{pmatrix} \dfrac{6^t - 1}{5} \\ \dfrac{6^t - 1}{5} \\ \dfrac{6^t - 1}{5} \end{pmatrix}$$

is the solution of the original system.

Example 6.9. Consider matrix

$$\mathbf{A} = \begin{pmatrix} 1 & 4 & 1 \\ 1 & 2 & 3 \\ 1 & 3 & 2 \end{pmatrix}.$$

In Example 6.6 we have shown that

$$\mathbf{A} = \begin{pmatrix} -5 & 9 & 1 \\ 1 & -5 & 1 \\ 1 & 2 & 1 \end{pmatrix} \begin{pmatrix} 0 & & \\ & -1 & \\ & & 6 \end{pmatrix} \begin{pmatrix} -\dfrac{1}{6} & -\dfrac{7}{42} & \dfrac{7}{21} \\ 0 & -\dfrac{6}{42} & \dfrac{3}{21} \\ \dfrac{1}{6} & \dfrac{19}{42} & \dfrac{8}{21} \end{pmatrix},$$

therefore equation (6.18) implies that for all $t \geq 1$,

$$\mathbf{A}' = \begin{pmatrix} -5 & 9 & 1 \\ 1 & -5 & 1 \\ 1 & 2 & 1 \end{pmatrix} \begin{pmatrix} 0 & & \\ & (-1)^t & \\ & & 6^t \end{pmatrix} \begin{pmatrix} -\dfrac{1}{6} & -\dfrac{7}{42} & \dfrac{7}{21} \\ 0 & -\dfrac{6}{42} & \dfrac{3}{21} \\ \dfrac{1}{6} & \dfrac{19}{42} & \dfrac{8}{21} \end{pmatrix}.$$

♦

If **A** cannot be diagonalized, then there is a nonsingular matrix such that $\mathbf{TAT}^{-1}$ has the Jordan canonical form. Then system (6.17) breaks up to small dimensional systems of the form

$$\mathbf{z}_{ij}\left(t+1\right) = \begin{pmatrix} \lambda_i & & & & \\ 1 & \lambda_i & & & \\ & 1 & \ddots & & \\ & & & \lambda_i & \\ & & & 1 & \lambda_i \end{pmatrix} \mathbf{z}_{ij}\left(t\right) + \begin{pmatrix} b'_{ij1} \\ b'_{ij2} \\ \vdots \\ b'_{ijm_{ij}} \end{pmatrix},$$

where we use the notation of the previous section. Notice that this system can be rewritten as follows:

$$z_{ij1}\left(t+1\right) = \lambda_i z_{ij1}\left(t\right) + b'_{ij1}$$
$$z_{ij2}\left(t+1\right) = \lambda_i z_{ij2}\left(t\right) + z_{ij1}\left(t\right) + b'_{ij2}$$
$$\vdots$$
$$z_{ijm_{ij}}\left(t+1\right) = \lambda_i z_{ijm_{ij}}\left(t\right) + z_{ij,\,m_{ij}-1}\left(t\right) + b'_{ijm_{ij}}.$$

From the first equation we can determine function $z_{ij1}\left(t\right)$, and then, $z_{ij2}\left(t\right)$ can be obtained from the second equation. After $z_{ij2}\left(t\right)$ is determined, function $z_{ij3}\left(t\right)$ is obtained from the third equation, and so on. Notice, that at each step a real-valued function has to be computed. In summary, in both cases, the solution of an $n \times n$ system has been reduced to the solution of n single equations, which can be done efficiently.

The state-transition matrix of the system also can be obtained in a simple way. Notice first that if

$$\mathbf{A}_{ij} = \begin{pmatrix} \lambda_i & & & & \\ 1 & \lambda_i & & & \\ & 1 & \ddots & & \\ & & & \lambda_i & \\ & & & 1 & \lambda_i \end{pmatrix}$$

is a Jordan-block, then $\mathbf{A}_{ij} = \lambda_i \mathbf{I} + \mathbf{E}$, where $\mathbf{I}$ is the $m_{ij} \times m_{ij}$ identity matrix, and $\mathbf{E}$ is nilpotent with degree m_{ij}. By applying the binomial theorem (which can be used here, since $\mathbf{I}$ and $\mathbf{E}$ commute) we have

$$\mathbf{A}_{ij}^t = \left(\lambda_i \mathbf{I} + \mathbf{E} \right)^t = \sum_{l=0}^t \binom{t}{l} \lambda_i^{t-l} \mathbf{E}^l$$

$$= \begin{pmatrix} \lambda_i^t & & & & \\ \binom{t}{1}\lambda_i^{t-1} & \lambda_i^t & & & \\ \binom{t}{2}\lambda_i^{t-2} & \binom{t}{1}\lambda_i^{t-1} & \ddots & & \\ \vdots & \vdots & & \lambda_i^t & \\ \binom{t}{m_{ij}-1}\lambda_i^{t-m_{ij}+1} & \binom{t}{m_{ij}-2}\lambda_i^{t-m_{ij}+2} & \cdots & \binom{t}{1}\lambda_i^{t-1} & \lambda_i^t \end{pmatrix}.$$

And finally

$$\mathbf{A}^t = \mathbf{T}^{-1} \begin{pmatrix} \mathbf{A}_{11}^t & & & & & & \\ & \ddots & & & & & \\ & & \mathbf{A}_{1s_1}^t & & & & \\ & & & \ddots & & & \\ & & & & \mathbf{A}_{r1}^t & & \\ & & & & & \ddots & \\ & & & & & & \mathbf{A}_{rs_r}^t \end{pmatrix} \mathbf{T}. \qquad (6.19)$$

2. Consider next the *continuous time-invariant system* with state transition equation

$$\dot{\mathbf{x}}(t) = \mathbf{A} \cdot \mathbf{x}(t) + \mathbf{b},$$

where $t \geq 0$ is the continuous time variable, and $\mathbf{A}$ is an $n \times n$ real matrix, $\mathbf{b}$ is an n-element real vector. Let $\mathbf{T}$ be again a nonsingular matrix, and by introducing the new state variable $\mathbf{z} = \mathbf{Tx}$ we have

$$
\begin{aligned}
\dot{\mathbf{z}}(t) = \mathbf{T}\dot{\mathbf{x}}(t) &= \mathbf{T}\big(\mathbf{Ax}(t) + \mathbf{b}\big) \\
&= \mathbf{T}\big(\mathbf{AT}^{-1}\mathbf{z}(t) + \mathbf{b}\big) = \big(\mathbf{TAT}^{-1}\big)\mathbf{z}(t) + \mathbf{T}\cdot\mathbf{b}.
\end{aligned} \tag{6.20}
$$

That is, similarly to the discrete case, the state-transition equation for $\mathbf{z}(t)$ has the same form as for the original state variable $\mathbf{x}(t)$ with the new coefficient matrix $\mathbf{TAT}^{-1}$. If matrix $\mathbf{A}$ is diagonable, then with an appropriate $\mathbf{T}$, $\mathbf{TAT}^{-1}$ is diagonal, and the state-transition equation (6.20) reduces to the following:

$$
\begin{aligned}
\dot{z}_1(t) &= \lambda_1 z_1(t) + b_1' \\
\dot{z}_2(t) &= \lambda_2 z_2(t) + b_2' \\
&\cdots\cdots\cdots\cdots \\
\dot{z}_n(t) &= \lambda_n z_n(t) + b_n',
\end{aligned}
$$

which can be solved much more efficiently than the original n-dimensional problem. This idea is illustrated in the next example.

Example 6.10. We will now solve the system

$$
\dot{\mathbf{x}}(t) = \begin{pmatrix} 1 & 4 & 1 \\ 1 & 2 & 3 \\ 1 & 3 & 2 \end{pmatrix} \mathbf{x}(t) + \begin{pmatrix} 1 \\ 1 \\ 1 \end{pmatrix}.
$$

From equation (6.20) and the results of Example 6.6 we know that $\mathbf{T}^{-1}\mathbf{AT}$ is diagonal and by introducing the new variable

$$\mathbf{z}(t) = \begin{pmatrix} -\dfrac{1}{6} & -\dfrac{7}{42} & \dfrac{7}{21} \\[2mm] 0 & -\dfrac{6}{42} & \dfrac{3}{21} \\[2mm] \dfrac{1}{6} & \dfrac{19}{42} & \dfrac{8}{21} \end{pmatrix} \mathbf{x}(t),$$

a diagonal system is obtained:

$$\dot{\mathbf{z}}(t) = \begin{pmatrix} 0 & & \\ & -1 & \\ & & 6 \end{pmatrix} \mathbf{z}(t) + \begin{pmatrix} 0 \\ 0 \\ 1 \end{pmatrix},$$

which can be rewritten as

$$\dot{z}_1 = 0,$$
$$\dot{z}_2 = -z_2,$$
$$\dot{z}_3 = 6z_3 + 1.$$

It is easy to see that the solutions are

$$z_1(t) = c_1,$$
$$z_2(t) = c_2 e^{-t},$$
$$z_3(t) = -\frac{1}{6} + c_3 e^{6t},$$

where c_1, c_2, c_3 are arbitrary constants. The solution $\mathbf{x}(t)$ can be then obtained as

$$\mathbf{x}(t) = \mathbf{T}^{-1}\mathbf{z}(t) = \begin{pmatrix} -5 & 9 & 1 \\ 1 & -5 & 1 \\ 1 & 2 & 1 \end{pmatrix} \begin{pmatrix} c_1 \\ c_2 e^{-t} \\ -\dfrac{1}{6} + c_3 e^{6t} \end{pmatrix}.$$

♦

The state transition matrix of the system can also be determined efficiently if $\mathbf{A}$ is diagonable. Using equation (6.18) we have

$$e^{\mathbf{A}t} = \sum_{l=0}^{\infty} \frac{t^l}{l!} \mathbf{A}^l = \sum_{l=0}^{\infty} \frac{t^l}{l!} \mathbf{T}^{-1} \begin{pmatrix} \lambda_1^l & & & \\ & \lambda_2^l & & \\ & & \ddots & \\ & & & \lambda_n^l \end{pmatrix} \mathbf{T}$$

$$= \mathbf{T}^{-1} \left\{ \sum_{l=0}^{\infty} \frac{t^l}{l!} \begin{pmatrix} \lambda_1^l & & & \\ & \lambda_2^l & & \\ & & \ddots & \\ & & & \lambda_n^l \end{pmatrix} \right\} \mathbf{T}$$

$$= \mathbf{T}^{-1} \begin{pmatrix} \sum_{l=0}^{\infty} \frac{t^l \lambda_1^l}{l!} & & & \\ & \sum_{l=0}^{\infty} \frac{t^l \lambda_2^l}{l!} & & \\ & & \ddots & \\ & & & \sum_{l=0}^{\infty} \frac{t^l \lambda_n^l}{l!} \end{pmatrix} \mathbf{T}$$

$$= \mathbf{T}^{-1} \begin{pmatrix} e^{\lambda_1 t} & & & \\ & e^{\lambda_2 t} & & \\ & & \ddots & \\ & & & e^{\lambda_n t} \end{pmatrix} \underline{T}.$$

Example 6.11. Consider again matrix

$$\mathbf{A} = \begin{pmatrix} 1 & 4 & 1 \\ 1 & 2 & 3 \\ 1 & 3 & 2 \end{pmatrix},$$

then similarly to Example 6.9 we have

$$
e^{\mathbf{A}t} = \begin{pmatrix} -5 & 9 & 1 \\ 1 & -5 & 1 \\ 1 & 2 & 1 \end{pmatrix} \begin{pmatrix} 1 & & \\ & e^{-t} & \\ & & e^{6t} \end{pmatrix} \begin{pmatrix} -\dfrac{1}{6} & -\dfrac{7}{42} & \dfrac{7}{21} \\ 0 & -\dfrac{6}{42} & \dfrac{3}{21} \\ \dfrac{1}{6} & \dfrac{19}{42} & \dfrac{8}{21} \end{pmatrix}.
$$

◆

If **A** cannot be diagonalized, then there is a nonsingular matrix **T** such that $\mathbf{TAT}^{-1}$ has the Jordan canonical-form. Then system (6.20) is a set of small dimensional systems

$$
\dot{\mathbf{z}}_{ij}(t) = \begin{pmatrix} \lambda_i & & & & \\ 1 & \lambda_i & & & \\ & 1 & \ddots & & \\ & & & \lambda_i & \\ & & & 1 & \lambda_i \end{pmatrix} \mathbf{z}_{ij}(t) + \begin{pmatrix} b'_{ij1} \\ b'_{ij2} \\ \vdots \\ b'_{ijm_{ij}} \end{pmatrix}.
$$

This system can be rewritten as follows:

$$
\dot{z}_{ij1}(t) = \lambda_i z_{ij1}(t) + b'_{ij1}
$$
$$
\dot{z}_{ij2}(t) = \lambda_i z_{ij2}(t) + z_{ij1}(t) + b'_{ij2}
$$
$$
\cdots\cdots\cdots\cdots
$$
$$
\dot{z}_{ijm_{ij}}(t) = \lambda_i z_{ijm_{ij}}(t) + z_{ij, m_{ij}-1}(t) + b'_{ijm_{ij}}.
$$

From the first equation we can determine function $z_{ij1}(t)$, and then, $z_{ij2}(t)$ can be obtained from the second equation. After $z_{ij2}(t)$ is determined, function $z_{ij3}(t)$ is obtained from the third equation, and so on. At each step a single dimensional linear differential equation has to be solved. The solution of an $n \times n$ system has been therefore reduced to the solution of n single dimensional equations, which can be done more efficiently.

Example 6.12. We will now solve the system

$$\dot{\mathbf{z}} = \begin{pmatrix} 2 & & \\ 1 & 2 & \\ & 1 & 2 \end{pmatrix} \mathbf{z} + \begin{pmatrix} 1 \\ 1 \\ 1 \end{pmatrix}, \quad \mathbf{z}(0) = \mathbf{0}$$

where the coefficient matrix has the Jordan block-form. This system can be rewritten as

$$\dot{z}_1 = 2z_1 + 1$$
$$\dot{z}_2 = 2z_2 + z_1 + 1$$
$$\dot{z}_3 = 2z_3 + z_2 + 1.$$

From the first equations we have

$$z_1(t) = -\frac{1}{2} + c_1 e^{2t}$$

where c_1 is a constant. Substituting this solution into the second equation we obtain a single equation for z_2 :

$$\dot{z}_2 = 2z_2 + c_1 e^{2t} + \frac{1}{2}.$$

The solution of this equation is the following:

$$z_2(t) = (c_1 t + c_2) e^{2t} - \frac{1}{4}.$$

Substituting this solution into the third equation we have

$$\dot{z}_3 = 2z_3 + (c_1 t + c_2) e^{2t} + \frac{3}{4},$$

the solution of which is the following:

$$z_3(t) = \left(\frac{c_1 t^2}{2} + c_2 t + c_3 \right) e^{2t} - \frac{3}{8}.$$

◆

The state transition matrix of the original system can also be obtained efficiently. Assume again that the Jordan canonical form of matrix **A** has the blocks

$$\mathbf{A}_{ij} = \begin{pmatrix} \lambda_i & & & & \\ 1 & \lambda_i & & & \\ & 1 & \ddots & & \\ & & & \lambda_i & \\ & & & 1 & \lambda_i \end{pmatrix},$$

then similarly to the case of diagonable matrices we can easily show that

$$e^{\mathbf{A}t} = \mathbf{T}^{-1} \begin{pmatrix} e^{\mathbf{A}_{11}t} & & & & & \\ & \ddots & & & & \\ & & e^{\mathbf{A}_{1\eta_1}t} & & & \\ & & & \ddots & & \\ & & & & e^{\mathbf{A}_{r1}t} & \\ & & & & & \ddots \\ & & & & & & e^{\mathbf{A}_{r s_r}t} \end{pmatrix} \mathbf{T}.$$

For each Jordan block, $\mathbf{A}_{ij} = \lambda_i \mathbf{I} + \mathbf{E}$, and therefore

$$e^{\mathbf{A}_{ij}t} = \sum_{k=0}^{\infty} \frac{t^k}{k!} \left(\lambda_i \mathbf{I} + \mathbf{E} \right)^k = \sum_{k=0}^{\infty} \frac{t^k}{k!} \sum_{l=0}^{k} \binom{k}{l} \mathbf{E}^l \lambda_i^{k-l}.$$

Using the fact that $\mathbf{E}^l = \mathbf{0}$ for $l \geq m_{ij}$ and $\binom{k}{l} = 0$ as $k < l$ we see that

$$e^{\mathbf{A}_{ij}t} = \sum_{k=0}^{\infty} \frac{t^k}{k!} \sum_{l=0}^{m_{ij}-1} \binom{k}{l} \mathbf{E}^l \lambda_i^{k-l} = \sum_{l=0}^{m_{ij}-1} \mathbf{E}^l \left(\sum_{k=0}^{\infty} \frac{t^k}{k!} \binom{k}{l} \lambda_i^{k-l} \right)$$

$$= \sum_{l=0}^{m_{ij}-1} \mathbf{E}^l \sum_{k=0}^{\infty} \frac{1}{l!} \frac{d^l}{d\lambda^l} \left[\frac{t^k}{k!} \lambda^k \right]_{\lambda=\lambda_i}$$

$$= \sum_{l=0}^{m_{ij}-1} \frac{1}{l!} \mathbf{E}^l \frac{d^l}{d\lambda^l} \left[e^{\lambda t} \right]_{\lambda=\lambda_i} = \sum_{l=0}^{m_{ij}-1} \mathbf{E}^l \frac{t^l}{l!} e^{\lambda_i t}$$

$$= e^{\lambda_i t} \begin{pmatrix} 1 & & & & \\ \dfrac{t}{1!} & 1 & & & \\ \dfrac{t^2}{2!} & \dfrac{t}{1!} & \ddots & & \\ \vdots & \vdots & & 1 & \\ \dfrac{t^{m_{ij}-1}}{(m_{ij}-1)!} & \dfrac{t^{m_{ij}-2}}{(m_{ij}-2)!} & \cdots & \dfrac{t}{1!} & 1 \end{pmatrix}.$$

3. Our third application is concerned with the *geometric series* of square real matrices. Let $\mathbf{A}$ be an $n \times n$ real matrix. The geometric series of $\mathbf{A}$ is defined as the infinite series

$$\mathbf{I} + \mathbf{A} + \mathbf{A}^2 + \mathbf{A}^3 + \ldots + \mathbf{A}^N + \ldots$$

From calculus we know that an infinite series is convergent only if the terms converge to zero. We will show that this condition is now sufficient and necessary, and in the case of convergence, the sum of the geometric series equals $(\mathbf{I} - \mathbf{A})^{-1}$. This result generalizes the sum of scalar geometric series.

First we notice that equation (6.19) implies that $\mathbf{A}^t \to \mathbf{O}$ as $t \to \infty$ if and only if for all blocks, $\mathbf{A}_{ij}^t \to \mathbf{O}$. The closed form representation of $\mathbf{A}_{ij}^t$ derived before (6.19) implies that this limit relation holds if and only if for all eigenvalue λ_i of $\mathbf{A}$, $|\lambda_i| < 1$. Assume next that this condition holds. Simple calculation shows that for any integer $N \geq 1$,

$$(\mathbf{I} - \mathbf{A})(\mathbf{I} + \mathbf{A} + \mathbf{A}^2 + ... + \mathbf{A}^N) = \mathbf{I} - \mathbf{A}^{N+1}.$$

It is easy to see that the eigenvalues of $\mathbf{I} - \mathbf{A}$ are the values $1 - \lambda_i$, which are nonzero, therefore $\mathbf{I} - \mathbf{A}$ is nonsingular. Therefore

$$\mathbf{I} + \mathbf{A} + \mathbf{A}^2 + ... + \mathbf{A}^N = (\mathbf{I} - \mathbf{A})^{-1}(\mathbf{I} - \mathbf{A}^{N+1}),$$

and since $\mathbf{A}^{N+1} \to \mathbf{O}$ as $N \to \infty$,

$$\mathbf{I} + \mathbf{A} + \mathbf{A}^2 + ... + \mathbf{A}^N + ... = (\mathbf{I} - \mathbf{A})^{-1}.$$

This identity is often used as an iteration method for inverting matrices. Assume that $\mathbf{B}^{-1}$ has to be determined. Assume that $\mathbf{A} = \mathbf{I} - \mathbf{B}$ satisfies the condition that all of its eigenvalues are inside the unit circle (that is, $|\lambda_i| < 1$ for all eigenvalues of $\mathbf{A}$). Then

$$\mathbf{I} + \mathbf{A} + \mathbf{A}^2 + ... + \mathbf{A}^N + ... = (\mathbf{I} - (\mathbf{I} - \mathbf{B}))^{-1} = \mathbf{B}^{-1}.$$

The infinite series is the limit of the sequence:

$$\mathbf{A}_0 = \mathbf{I}$$
$$\mathbf{A}_{k+1} = \mathbf{I} + \mathbf{A}_k \cdot \mathbf{A} \quad (k \geq 0),$$

since

$$\mathbf{A}_1 = \mathbf{I} + \mathbf{A}$$
$$\mathbf{A}_2 = \mathbf{I} + (\mathbf{I} + \mathbf{A})\mathbf{A} = \mathbf{I} + \mathbf{A} + \mathbf{A}^2,$$

and in general

$$\mathbf{A}_{k+1} = \mathbf{I} + \left(\mathbf{I} + \mathbf{A} + ... + \mathbf{A}^k \right) \mathbf{A} = \mathbf{I} + \mathbf{A} + \mathbf{A}^2 + ... + \mathbf{A}^{k+1}.$$

4. In equation (6.19) we have seen a closed-form representation of the matrix $\mathbf{A}^t$ for all $t \geq 1$. This formula was developed for solving discrete dynamic systems. Matrix powers however should be applied in many other cases. One example of such another application is concerned with the vertex matrix of directed graphs. In Application 7 of section 1.5 we have seen that if $\mathbf{A}$ is the vertex-matrix of a directed graph, then the elements of the matrix $\mathbf{A}^r$ $(r \geq 1)$ give the r-step connections from each vertex to all other vertices and to itself. The fast computation of $\mathbf{A}^r$ is therefore essential in determining the number of higher-step connections.

5. In section 3.8 we have briefly analysed Markov chains, and have seen that future state probabilities can be directly obtained by equation (3.33), which has the form

$$\mathbf{x}(t+n) = \mathbf{P}^n \mathbf{x}(t).$$

Here the entries of $\mathbf{x}(t)$ and $\mathbf{x}(t + n)$ give the state probabilities at time periods t and $t + n$, respectively, furthermore $\mathbf{P}$ is the transition matrix. For efficient computations we have to determine the matrix powers $\mathbf{P}^n$ for large values of n. This task can be performed for example, by using the method of the first application of this section (equations (6.18) and (6.19)).

6.9 Exercises

1. For an $n \times n$ square matrix $\mathbf{A} = (a_{ij})$, define the trace of $\mathbf{A}$ as
$$tr(\mathbf{A}) = a_{11} + a_{22} + ... + a_{nn}.$$
Prove that $tr(\mathbf{A}) = tr(\mathbf{T}^{-1} \mathbf{A} \mathbf{T})$ for all nonsingular $n \times n$ matrices.

2. Show that for any $n \times n$ matrix $\mathbf{A}$,
   a) $tr(\mathbf{A}^T) = tr(\mathbf{A})$;
   b) $tr(\mathbf{A} + \mathbf{B}) = tr(\mathbf{A}) + tr(\mathbf{B})$, if $\mathbf{B}$ is also $n \times n$;
   c) $tr(a\mathbf{A}) = a\,tr(\mathbf{A})$, if a is a scalar;

d) $tr(\mathbf{AB}) = tr(\mathbf{BA})a$, if $\mathbf{B}$ is also $n \times n$;

e) $tr(\mathbf{A}) = \lambda_1 + \lambda_2 + ... + \lambda_n$, if $\lambda_1, ..., \lambda_n$ are the eigenvalues of $\mathbf{A}$.

3. Prove that $tr(\mathbf{AA}^T) \geq 0$ and $tr(\mathbf{AA}^T) = 0$ if and only if $\mathbf{A} = \mathbf{O}$.

4. Find the eigenvalues of the complex matrix

$$\mathbf{A} = \begin{pmatrix} 2 & i \\ -i & 1 \end{pmatrix}.$$

5. Find the characteristic equations of the following matrices:

a) $\begin{pmatrix} 1 & 1 & 1 \\ 1 & 2 & 1 \\ 1 & 3 & 2 \end{pmatrix}$;

b) $\begin{pmatrix} 0 & 1 & 1 \\ 1 & 1 & 0 \\ 1 & 0 & 1 \end{pmatrix}$;

c) $\begin{pmatrix} 1 & 1 & 1 & 1 \\ 1 & 2 & 1 & 1 \\ 2 & 1 & 1 & 1 \\ 1 & 1 & 1 & 2 \end{pmatrix}$.

6. Find the eigenvalues and associated eigenvectors for the following matrices

a) $\begin{pmatrix} 1 & 1 \\ 1 & 1 \end{pmatrix}$;

b) $\begin{pmatrix} 0 & w \\ -w & 0 \end{pmatrix}$, where $w > 0$ is a given number;

c) $\begin{pmatrix} 1 & 2 \\ 2 & 4 \end{pmatrix}$;

d) $\begin{pmatrix} 1 & 2 & 3 \\ 0 & 2 & 1 \\ 0 & 0 & 3 \end{pmatrix}$.

7. Repeat Example 6.1 with matrix

$$A = \begin{pmatrix} 1 & 2 \\ 2 & 4 \end{pmatrix}.$$

8. Illustrate Theorem 6.1 for matrices of Exercise 6.

9. Repeat Example 6.3 with matrix

$$A = \begin{pmatrix} 1 & 1 & 1 \\ 0 & 1 & 1 \\ 1 & 1 & 0 \end{pmatrix}.$$

10. Find $p(A)$, where

$$p(t) = t^3 + t^2 + 2t + 1$$

and

$$A = \begin{pmatrix} 1 & 1 \\ 1 & 1 \end{pmatrix}.$$

11. Find the minimal polynomial of the matrix of the previous problem. (Repeat Example 6.5).

12. Diagonalize the matrices of Exercise 6.

13. Characterize all 2×2 real nilpotent matrices of degree 2.

14. Assume that the only eigenvalue of an $n \times n$ real matrix is zero. Prove that the matrix is nilpotent.

15. (Continuation of Exercise 14). An $n \times n$ real matrix $\mathbf{A}$ is called *unipotent*, if $\mathbf{A} - \mathbf{I}$ is nilpotent. Prove that $\mathbf{A}$ is unipotent if and only if 1 is the only eigenvalue of $\mathbf{A}$.

16. Illustrate the Cayley-Hamilton Theorem with the matrices of Exercise 5.

17. Illustrate the Cayley-Hamilton Theorem with the matrices of Exercise 6.

18. Find $\mathbf{A}^t$ for matrices

a) $\begin{pmatrix} 1 & 1 \\ 3 & 3 \end{pmatrix}$;

b) $\begin{pmatrix} 2 & 1 \\ 0 & 3 \end{pmatrix}$.

19. Find $e^{\mathbf{A}t}$ for matrices

a) $\begin{pmatrix} 1 & 1 \\ 3 & 3 \end{pmatrix}$;

b) $\begin{pmatrix} 2 & 1 \\ 0 & 3 \end{pmatrix}$.

20. Find $\mathbf{x}(t)$ for the diagonal discrete system

$$\mathbf{x}(t+1) = \begin{pmatrix} 1 & 0 \\ 0 & 2 \end{pmatrix} \mathbf{x}(t) + \begin{pmatrix} 1 \\ 1 \end{pmatrix}, \quad \mathbf{x}(0) = \begin{pmatrix} 1 \\ 0 \end{pmatrix}.$$

21. Find $\mathbf{x}(t)$ for the diagonal continuous system

$$\dot{\mathbf{x}}(t) = \begin{pmatrix} 1 & 0 \\ 0 & 2 \end{pmatrix} \mathbf{x}(t) + \begin{pmatrix} 1 \\ 1 \end{pmatrix}, \quad \mathbf{x}(0) = \begin{pmatrix} 1 \\ 0 \end{pmatrix}.$$

22. Find $\mathbf{x}(t)$ for the triangular discrete system

$$\mathbf{x}(t+1) = \begin{pmatrix} 1 & 1 \\ 0 & 2 \end{pmatrix} \mathbf{x}(t) + \begin{pmatrix} 1 \\ 1 \end{pmatrix}, \quad \mathbf{x}(0) = \begin{pmatrix} 0 \\ 0 \end{pmatrix}.$$

23. Find $\mathbf{x}(t)$ for the triangular continuous system

$$\dot{\mathbf{x}}(t) = \begin{pmatrix} 1 & 1 \\ 0 & 2 \end{pmatrix} \mathbf{x}(t) + \begin{pmatrix} 1 \\ 1 \end{pmatrix}, \quad \mathbf{x}(0) = \begin{pmatrix} 0 \\ 0 \end{pmatrix}.$$

24. Find the inverse of matrix

$$\mathbf{A} = \begin{pmatrix} \dfrac{3}{8} & -\dfrac{1}{8} \\ -\dfrac{1}{8} & \dfrac{3}{8} \end{pmatrix}$$

using the method given in Application 3 of Section 6.8.

25. Find the eigenvalues and eigenvectors of matrix $\mathbf{A} = \begin{pmatrix} a & b \\ -b & a \end{pmatrix}$.

26. Transform matrix $\mathbf{A}$ of the previous Exercise to diagonal form.

27. Based on the result of the previous Exercise find $\mathbf{A}^t$ for $t \geq 0$.

28. Based on the result of Exercise 26, find $e^{\mathbf{A}t}$.

Chapter 7

Special Matrices

7.1 Introduction

In this chapter the most important classes of special matrices will be examined. In the first part the main properties of diagonal, tridiagonal, and triangular matrices will be discussed. In the second part we will focus on idempotent and nilpotent matrices, and the third part will be devoted to special matrices defined in inner-product spaces.

7.2 Diagonal, Tridiagonal, and Triangular Matrices

From Chapter 1 we know that an $n \times n$ square matrix $\mathbf{A} = (a_{ij})$ is *diagonal* if $a_{ij} = 0$ for all $i \neq j$. That is, a diagonal matrix has the special form

$$\mathbf{A} = \begin{pmatrix} a_{11} & & & O \\ & a_{22} & & \\ & & \ddots & \\ O & & & a_{nn} \end{pmatrix},$$

where all element under and above the diagonal are equal to zero. The rank of $\mathbf{A}$ equals the number of nonzero diagonal elements, since the columns with nonzero diagonal elements are the nonzero-multiples of the corresponding natural basis vectors $\mathbf{e}_k$.

The characteristic polynomial of **A** is the determinant of the diagonal matrix

$$\mathbf{A} - \lambda\mathbf{I} = \begin{pmatrix} a_{11} - \lambda & & & O \\ & a_{22} - \lambda & & \\ & & \ddots & \\ O & & & a_{nn} - \lambda \end{pmatrix},$$

which is the product of the diagonal elements:

$$\varphi(\lambda) = (a_{11} - \lambda) \cdot (a_{22} - \lambda) \cdot \ldots \cdot (a_{nn} - \lambda),$$

therefore the eigenvalues are the diagonal elements of **A**. Consider now an eigenvalue λ, and assume that

$$\lambda = a_{i_1 i_1} = a_{i_2 i_2} = \ldots = a_{i_r i_r}$$

and

$$\lambda \neq a_{jj} \quad \text{for} \quad j \notin \{i_1, i_2, \ldots, i_r\}.$$

That is, λ can be found as the common diagonal element in positions i_1, $i_2, \ldots, i_r$. Let $\mathbf{x} = (x_i)$ be an associated eigenvector. Then the eigenvector equation $\mathbf{Ax} = \lambda\mathbf{x}$ implies that

$$a_{ii} x_i = \lambda x_i \quad (i = 1, 2, \ldots, n).$$

If $i \in \{i_1, i_2, \ldots, i_r\}$, then this equation is satisfied for all x_i, since $\lambda = a_{ii}$, and if $i \notin \{i_1, i_2, \ldots, i_r\}$ then $x_i = 0$. Hence the eigenvectors associated to λ have the special form:

$$x_i = \begin{cases} \text{arbitrary} & \text{if } i \in \{i_1, i_2, \ldots, i_r\} \\ 0 & \text{if } i \notin \{i_1, i_2, \ldots, i_r\} \end{cases}. \tag{7.1}$$

Notice that system $\{\mathbf{e}_{i_1}, \mathbf{e}_{i_2}, ..., \mathbf{e}_{i_r}\}$ forms a basis of the subspace of the eigenvectors associated to λ.

If $\mathbf{A} = \mathbf{O}$, then all diagonal elements are equal to zero, therefore $\lambda = 0$ is the only eigenvalue with multiplicity n, and any arbitrary n-element vector is an eigenvector of $\mathbf{A}$. Similarly, if $\mathbf{A} = \mathbf{I}$, then all diagonal elements are equal to one, therefore $\lambda = 1$ is the only eigenvalue with multiplicity n, and all n-element vectors are eigenvectors of $\mathbf{A}$.

Definition 7.1. A matrix $\mathbf{A} = (a_{ij})$ is called *triadiagonal*, if $a_{ij} = 0$ for $|i-j| \geq 2$.

This definition implies that $\mathbf{A}$ is tridiagonal if and only if it has the special form

$$\mathbf{A} = \begin{pmatrix} a_{11} & a_{12} & 0 & ... & 0 & 0 & 0 \\ a_{21} & a_{22} & a_{23} & ... & 0 & 0 & 0 \\ 0 & a_{32} & a_{33} & ... & 0 & 0 & 0 \\ ... & ... & ... & ... & ... & ... & ... \\ 0 & 0 & 0 & ... & a_{n-2,n-2} & a_{n-2,n-1} & 0 \\ 0 & 0 & 0 & ... & a_{n-1,n-2} & a_{n-1,n-1} & a_{n-1,n} \\ 0 & 0 & 0 & ... & 0 & a_{n,n-1} & a_{nn} \end{pmatrix}.$$

Two specialties of tridiagonal matrices will be discussed here: the solution of linear equations with tridiagonal matrices, and an easy way to compute the characteristic polynomials of such matrices.

Table 7.1. Initial elimination table for a tridiagonal system

x_1	x_2	x_3	$\cdots$	x_{n-2}	x_{n-1}	x_n	b
a_{11}	a_{12}	0	$\cdots$	0	0	0	b_1
a_{21}	a_{22}	a_{23}	$\cdots$	0	0	0	b_2
0	a_{32}	a_{33}	$\cdots$	0	0	0	b_3
$\cdots$	$\cdots$	$\cdots$	$\cdots$	$\cdots$	$\cdots$	$\cdots$	$\cdots$
0	0	0	$\cdots$	$a_{n-2,n-2}$	$a_{n-2,n-1}$	0	b_{n-2}
0	0	0	$\cdots$	$a_{n-1,n-2}$	$a_{n-1,n-1}$	$a_{n-1,n}$	b_{n-1}
0	0	0	$\cdots$	0	$a_{n,n-1}$	a_{nn}	b_n

In applying the elimination method for solving linear equations with tridiagonal matrices, the procedure can be largely simplified, if at each step, the next diagonal element is selected as the pivot. To illustrate this point, assume that a system of linear equations has a tridiagonal coefficient matrix. The initial elimination table is shown in Table 7.1. Select the pivot element a_{12} by assuming that it is nonzero. Subtract the a_{21}/a_{11} multiple of the first equation from the second equation, then the second row of the first derived system becomes the following:

0	$a_{22}^{(1)}$	a_{23}	$\cdots$	0	0	0	$b_2^{(2)}$

with $a_{22}^{(1)} = a_{22} - a_{12} \cdot \dfrac{a_{21}}{a_{11}}$, and $b_2^{(1)} = b_2 - b_1 \cdot \dfrac{a_{21}}{a_{11}}$. The other rows remain unchanged.

Notice that by omitting the first row, the remaining system with only $n-1$ equations for the unknowns $x_2, x_3, \ldots, x_n$ remains tridiagonal. Select next the pivot element $a_{22}^{(1)}$, then only the element a_{32} has to be eliminated by subtracting the $a_{32}/a_{22}^{(1)}$ multiple of the new second equation from the third equation. In the new table only the third row changes, and by omitting the second equation, the remaining $n - 2$

equations for the $n - 2$ remaining unknowns $x_3, \ldots, x_n$ is tridiagonal again. At each elimination step only one diagonal element and the right hand side number has to be modified, which makes the entire elimination procedure very inexpensive. After elimination is finished, the resulting table is also special as it is shown in Table 7.2, where all elements under the diagonal equal zero. Notice that back substitution also becomes simple in this case, since

$$x_n = \frac{b_n^{(n-1)}}{a_{nn}^{(n-1)}}, \tag{7.2}$$

and for $i = n - 1, n - 2, \ldots, 2, 1.$

$$x_i = \frac{b_i^{(i-1)} - a_{i,i+1} x_{i+1}}{a_{ii}^{(i-1)}}. \tag{7.3}$$

Table 7.2. Final elimination table for tridiagonal system

x_1	x_2	x_3	$\cdots$	x_{n-2}	x_{n-1}	x_n	b
a_{11}	a_{12}	0	$\cdots$	0	0	0	b_1
	$a_{22}^{(1)}$						$b_2^{(1)}$
		$a_{33}^{(2)}$					$b_3^{(2)}$
			$\ddots$	$\cdots$	$\cdots$	$\cdots$	$\cdots$
				$a_{n-2,n-2}^{(n-3)}$	$a_{n-2,n-1}$	0	$b_{n-2}^{(n-3)}$
					$a_{n-1,n-1}^{(n-2)}$	$a_{n-1,n}$	$b_{n-1}^{(n-2)}$
						$a_{n,n}^{(n-1)}$	$b_n^{(n-1)}$

The characteristic polynomial φ_n of an $n \times n$ tridiagonal matrix **A** can be determined by using a special recursion as it is shown next. The definition of characteristic polynomials implies that

$$\varphi_n(\lambda) = \det \begin{pmatrix} a_{11} - \lambda & a_{12} & 0 & \dots & 0 & 0 & 0 \\ a_{21} & a_{22} - \lambda & a_{23} & \dots & 0 & 0 & 0 \\ 0 & a_{32} & a_{33} - \lambda & \dots & 0 & 0 & 0 \\ \dots & \dots & \dots & \dots & \dots & \dots & \dots \\ 0 & 0 & 0 & & a_{n-2,n-2} - \lambda & a_{n-2,n-1} & 0 \\ 0 & 0 & 0 & & a_{n-1,n-2} & a_{n-1,n-1} - \lambda & a_{n-1,n} \\ 0 & 0 & 0 & & 0 & a_{n,n-1} & a_{nn} - \lambda \end{pmatrix}.$$

Expand this determinant with respect to its last column. The cofactor φ_{n-1} of $a_{nn} - \lambda$ has the same structure as the entire determinant but instead of n, its size is $n - 1$. That is, it is the characteristic polynomial of the $(n - 1) \times (n - 1)$ tridiagonal matrix with elements a_{ii} ($i = 1, 2, \dots, n-1$), $a_{i,i+1}$ and $a_{i+1,i}$ ($i = 1, 2, \dots, n-2$). The cofactor of the element $a_{n-1,n}$ is the following:

$$-\det \begin{pmatrix} a_{11} - \lambda & a_{12} & 0 & \dots & 0 & 0 & 0 \\ a_{21} & a_{22} - \lambda & a_{23} & \dots & 0 & 0 & 0 \\ 0 & a_{32} & a_{33} - \lambda & \dots & 0 & 0 & 0 \\ \dots & \dots & \dots & \dots & \dots & \dots & \dots \\ 0 & 0 & 0 & \dots & a_{n-2,n-3} & a_{n-2,n-2} - \lambda & a_{n-2,n-1} \\ 0 & 0 & 0 & \dots & 0 & 0 & a_{n,n-1} \end{pmatrix}.$$

Expanding this determinant with respect to the last row we get the product $a_{n,n-1}\, \varphi_{n-2}(\lambda)$, where $\varphi_{n-2}(\lambda)$ has the same structure as $\varphi_n(\lambda)$, its size is $(n - 2) \times (n - 2)$ with elements a_{ii} ($i = 1, 2, \dots, n-2$) and $a_{i,i+1}$ and $a_{i+1,i}$ ($i = 1, 2, \dots, n-3$). Hence, we derived a recursion to find φ_n:

$$\varphi_0(\lambda) = 1$$
$$\varphi_1(\lambda) = a_{11} - \lambda$$

and for $k = 2, 3, \ldots, n,$

$$\varphi_k(\lambda) = (a_{kk} - \lambda)\varphi_{k-1}(\lambda) - a_{k-1,k} \cdot a_{k,k-1}\, \varphi_{k-2}(\lambda). \qquad (7.4)$$

Example 7.1. We will first solve the system of linear equations

$$
\begin{aligned}
x_1 + x_2 &= 2, \\
x_1 + 2x_2 + x_3 &= 4, \\
x_2 + 2x_3 + x_4 &= 4, \\
x_3 + 2x_4 &= 3,
\end{aligned}
$$

where the coefficient matrix is tridiagonal. The elimination process is shown in Table 7.3.

Introduction to Matrix Theory

Table 7.3. Elimination of Example 7.1

	x_1	x_2	x_3	x_4	
	①	1	0	0	2
	1	2	1	0	4
	0	1	2	1	4
	0	0	1	2	3
	1	1	0	0	2
row 2 – pivot row	0	①	1	0	2
	0	1	2	1	4
	0	0	1	2	3
	1	1	0	0	2
	0	1	1	0	2
row 3 – pivot row	0	0	①	1	2
	0	0	1	2	3
	1	1	0	0	2
	0	1	1	0	2
	0	0	1	1	2
row 4 – pivot row	0	0	0	1	1

Back substitution shows that the solution is the following:

$$x_4 = 1,$$
$$x_3 = 2 - x_4 = 1,$$
$$x_2 = 2 - x_3 = 1,$$
$$x_1 = 2 - x_2 = 1.$$

The characteristic polynomial of the coefficient matrix

$$\mathbf{A} = \begin{pmatrix} 1 & 1 & 0 & 0 \\ 1 & 2 & 1 & 0 \\ 0 & 1 & 2 & 1 \\ 0 & 0 & 1 & 2 \end{pmatrix}$$

will be next determined by using recursion (7.4):

$$\varphi_0(\lambda) = 1;$$
$$\varphi_1(\lambda) = 1 - \lambda;$$
$$\varphi_2(\lambda) = (2 - \lambda)\varphi_1(\lambda) - 1 \cdot 1 \cdot \varphi_0(\lambda) = (2 - \lambda)(1 - \lambda) - 1 = \lambda^2 - 3\lambda + 1;$$
$$\varphi_3(\lambda) = (2 - \lambda)\varphi_2(\lambda) - 1 \cdot 1 \cdot \varphi_1(\lambda) = (2 - \lambda)(\lambda^2 - 3\lambda + 1) - (1 - \lambda)$$
$$= -\lambda^3 + 5\lambda^2 - 6\lambda + 1;$$

and finally,

$$\varphi_4(\lambda) = (2 - \lambda)\varphi_3(\lambda) - 1 \cdot 1 \cdot \varphi_2(\lambda)$$
$$= (2 - \lambda)(-\lambda^3 + 5\lambda^2 - 6\lambda + 1) - (\lambda^2 - 3\lambda + 1)$$
$$= \lambda^4 - 7\lambda^3 + 15\lambda^2 - 10\lambda + 1.$$

◆

Consider next an $n \times n$ *upper triangular* matrix

$$\mathbf{A} = \begin{pmatrix} a_{11} & a_{12} & a_{13} & \cdots & a_{1,n-1} & a_{1n} \\ 0 & a_{22} & a_{23} & \cdots & a_{2,n-1} & a_{2n} \\ \cdots & \cdots & \cdots & \cdots & \cdots & \cdots \\ 0 & 0 & 0 & \cdots & a_{n-1,n-1} & a_{n-1,n} \\ 0 & 0 & 0 & \cdots & 0 & a_{nn} \end{pmatrix},$$

where all elements under the diagonal are equal to zero. If a system of linear equations has an upper triangular coefficient matrix, then there is no need for elimination, the solution can be easily obtained by back substitution. This matrix $\mathbf{A}$ is nonsingular if and only if all diagonal elements differ from zero, since

$$\det(\mathbf{A}) = a_{11} \cdot a_{22} \cdot ... \cdot a_{nn},$$

and this product is nonzero if and only if all factors are nonzero. The characteristic polynomial of $\mathbf{A}$ equals

$$\varphi(\lambda) = (a_{11} - \lambda)(a_{22} - \lambda) ... (a_{nn} - \lambda),$$

that is, the eigenvalues of $\mathbf{A}$ are the diagonal elements.

Lower triangular matrices have similar properties. A lower triangular matrix $\mathbf{A}$ has the form

$$\mathbf{A} = \begin{pmatrix} a_{11} & 0 & 0 & & 0 & 0 \\ a_{21} & a_{22} & 0 & & 0 & 0 \\ ... & ... & ... & ... & ... & ... \\ a_{n-1,1} & a_{n-1,2} & a_{n-1,3} & ... & a_{n-1,n-1} & 0 \\ a_{n1} & a_{n2} & a_{n3} & ... & a_{n,n-1} & a_{nn} \end{pmatrix},$$

where all elements above the diagonal equal zero. If the coefficient matrix of a system of equations is lower triangular, then there is no need for elimination. The value of x_1 can be obtained from the first equation, and then, x_2 can be determined from the second equation, and so on. Finally, x_n is obtained from the last equation. This procedure is called the *forward substitution*. This matrix $\mathbf{A}$ is nonsingular if and only if all diagonal elements differ from zero. The characteristic polynomial of $\mathbf{A}$ is the following:

$$\varphi(\lambda) = (a_{11} - \lambda)(a_{22} - \lambda) ... (a_{nn} - \lambda),$$

and the eigenvalues are again the diagonal elements of $\mathbf{A}$.

7.3 Idempotent and Nilpotent Matrices

In Section 6.4 we called a square matrix $\mathbf{A}$ *idempotent*, if $\mathbf{A}^2 = \mathbf{A}$. We will first prove that the only nonsingular idempotent matrix is the identity matrix. Notice first that the definition implies that

$$\mathbf{A}(\mathbf{I} - \mathbf{A}) = \mathbf{O}. \tag{7.5}$$

If $\mathbf{A}$ is nonsingular, then $\mathbf{A}^{-1}$ exists. Multiply both sides of this equation by $\mathbf{A}^{-1}$ from the left. Then we have

$$\mathbf{I} - \mathbf{A} = \mathbf{O},$$

that is,

$$\mathbf{A} = \mathbf{I}$$

Consider next the eigenvalue equation of $\mathbf{A}$:

$$\mathbf{A}\mathbf{x} = \lambda\mathbf{x}.$$

Since $\mathbf{A}^2 = \mathbf{A}$, any eigenvalue λ and associated eigenvector satisfy relation

$$\lambda\mathbf{x} = \mathbf{A}\mathbf{x} = \mathbf{A}^2\mathbf{x} = \mathbf{A}(\mathbf{A}\mathbf{x}) = \mathbf{A}(\lambda\mathbf{x}) = \lambda\mathbf{A}\mathbf{x} = \lambda^2\mathbf{x},$$

and the fact that the eigenvector $\mathbf{x}$ differs from zero implies that $\lambda^2 = \lambda$. Therefore, the eigenvalues of $\mathbf{A}$ are 0 and 1. Equation (7.5) implies that for any column $\mathbf{u}$ of matrix $\mathbf{I} - \mathbf{A}$, $\mathbf{A}\mathbf{u} = 0\mathbf{u}$ showing that the columns of $\mathbf{I} - \mathbf{A}$ are eigenvectors of $\mathbf{A}$ associated to the zero eigenvalue. Equation (7.5) can be rewritten as $(\mathbf{A} - \mathbf{I})\mathbf{A} = \mathbf{O}$ which shows that any column $\mathbf{v}$ of $\mathbf{A}$ satisfies equation $\mathbf{A}\mathbf{v} = 1\cdot\mathbf{v}$. That is, the columns of $\mathbf{A}$ are eigenvectors associated to the eigenvalue $\lambda = 1$. Next we show that there are n linearly independent eigenvectors which implies that $\mathbf{A}$ is diagonable. Notice first that $\mathbf{A} + (\mathbf{I} - \mathbf{A}) = \mathbf{I}$, and therefore

$$n = \text{rank}(\mathbf{I}) \leq \text{rank}(\mathbf{A}) + \text{rank}(\mathbf{I} - \mathbf{A}).$$

Let $\mathbf{u}_1, \mathbf{u}_2, ..., \mathbf{u}_k$ be a basis of the column space of $\mathbf{A}$, and let $\mathbf{v}_1, \mathbf{v}_2, ..., \mathbf{v}_l$ be a basis of the column space of $\mathbf{I} - \mathbf{A}$. Theorem 6.1 implies that system $\{\mathbf{u}_1, \mathbf{u}_2, ..., \mathbf{u}_k, \mathbf{v}_1, \mathbf{v}_2, ..., \mathbf{v}_l\}$ is linearly independent. Since they are in an n-dimensional space and $k + l \geq n$, $k + l$ must be equal to n. Thus, this set of eigenvectors forms a basis of the n-dimensional vector space.

A square matrix $\mathbf{A}$ is called *nilpotent* if $\mathbf{A}^k = \mathbf{O}$ for some positive integer k. The smallest such integer k is called the *degree* of $\mathbf{A}$. In Section 6.5 we have seen that nilpotent matrices are always singular. We can easily show that the only eigenvalue of $\mathbf{A}$ is zero. Consider the eigenvector equation of $\mathbf{A}$,

$$\mathbf{A}\mathbf{x} = \lambda \mathbf{x}.$$

Let l be the smallest positive integer such that $\mathbf{A}^l \mathbf{x} = \mathbf{0}$. Such an l exists, since if l is selected as the degree of $\mathbf{A}$, $\mathbf{A}^l = \mathbf{O}$ as well as $\mathbf{A}^l \mathbf{x} = \mathbf{0}$. The choice of $l = 0$ is not appropriate, since $\mathbf{A}^0 \mathbf{x} = \mathbf{I}\mathbf{x} = \mathbf{x} \neq \mathbf{0}$. Multiply both sides of the eigenvector equation by $\mathbf{A}^{l-1}$ from the left to see that

$$\mathbf{A}^l \mathbf{x} = \lambda \mathbf{A}^{l-1} \mathbf{x}.$$

Since $\mathbf{A}^{l-1}\mathbf{x} \neq \mathbf{0}$ and $\mathbf{A}^l \mathbf{x} = \mathbf{0}$, necessarily $\lambda = 0$. This observation implies that all diagonal elements of the Jordan form of nilpotent matrices equal zero.

7.4 Matrices in Inner Product Spaces

In this section it is assumed that V is a finitely generated real vector space, and there is a real inner-product defined on V. Let $A \in L(V)$ be a linear transformation on V. From Theorem 2.12 we know that there is an orthonormal basis $B = \{\mathbf{x}_1, \mathbf{x}_2, ..., \mathbf{x}_n\}$ in V, where n is the dimension of V. Assume that

$$\mathbf{A} = \begin{pmatrix} a_{11} & a_{12} & \cdots & a_{1n} \\ a_{21} & a_{22} & \cdots & a_{2n} \\ \cdots & \cdots & \cdots & \cdots \\ a_{n1} & a_{n2} & \cdots & a_{nn} \end{pmatrix}$$

is the matrix-representation of A in basis B. Notice first that equality (5.14) implies that for all $i = 1, 2, \ldots, n$ and $j = 1, 2, \ldots, n$,

$$\left(A(\mathbf{x}_i), \mathbf{x}_j \right) = \left(\mathbf{c}_i, \mathbf{x}_j \right) = \left(a_{1i}\mathbf{x}_1 + a_{2i}\mathbf{x}_2 + \ldots + a_{ni}\mathbf{x}_n, \mathbf{x}_j \right)$$

$$= \sum_{l=1}^{n} a_{li} \left(\mathbf{x}_l, \mathbf{x}_j \right) = a_{ji},$$

where $\mathbf{c}_i$ denotes the ith column of matrix $\mathbf{A}$. Similarly,

$$\left(\mathbf{x}_i, A(\mathbf{x}_j) \right) = \left(\mathbf{x}_i, \mathbf{c}_j \right) = \left(\mathbf{x}_i, a_{1j}\mathbf{x}_1 + a_{2j}\mathbf{x}_2 + \ldots + a_{nj}\mathbf{x}_n \right)$$

$$= \sum_{l=1}^{n} a_{lj} \left(\mathbf{x}_i, \mathbf{x}_l \right) = a_{ij}.$$

Definition 7.2. The *adjoint* of an $A \in L(V)$ is defined as the linear transformation $A' \in L(V)$ such that for all $\mathbf{x}, \mathbf{y} \in V$,

$$\left(A(\mathbf{x}), \mathbf{y} \right) = \left(\mathbf{x}, A'(\mathbf{y}) \right). \tag{7.6}$$

Our first result guarantees the existence of the unique adjoint for all A, and gives the matrix representation of A'.

Theorem 7.1. For all $A \in L(V)$ there is a unique adjoint, and its matrix representation in basis B is given as the transpose $\mathbf{A}^T$ of matrix $\mathbf{A}$.

Proof. We have shown above that $\left(A(\mathbf{x}_i), \mathbf{x}_j \right) = a_{ji}$ and $\left(\mathbf{x}_i, A'(\mathbf{x}_j) \right) = a'_{ij}$, therefore the elements a'_{ij} of the matrix representation A' of the adjoint must satisfy the equation $a'_{ij} = a_{ji}$ for all i and j, hence A', if exists, must be the transpose $\mathbf{A}^T$ of $\mathbf{A}$. Next we show that this transformation satisfies relation (7.6) for all $\mathbf{x}$, $\mathbf{y} \in V$, therefore it is the adjoint of A. Assume that

$$\mathbf{x} = \alpha_1 \mathbf{x}_1 + \alpha_2 \mathbf{x}_2 + \ldots + \alpha_n \mathbf{x}_n$$

and

$$\mathbf{y} = \beta_1 \mathbf{x}_1 + \beta_2 \mathbf{x}_2 + \ldots + \beta_n \mathbf{x}_n,$$

then

$$(A(\mathbf{x}), \mathbf{y}) = \left(\sum_{i=1}^{n} \alpha_i A(\mathbf{x}_i), \sum_{j=1}^{n} \beta_j x_j \right)$$

$$= \sum_{i=1}^{n} \sum_{j=1}^{n} \alpha_i \beta_j \left(A(\mathbf{x}_i), \mathbf{x}_j \right) = \sum_{i=1}^{n} \sum_{j=1}^{n} \alpha_i \beta_j a_{ji},$$

and similarly,

$$(\mathbf{x}, A'(\mathbf{y})) = \left(\sum_{i=1}^{n} \alpha_i \mathbf{x}_i, \sum_{j=1}^{n} \beta_j A'(\mathbf{x}_j) \right)$$

$$= \sum_{i=1}^{n} \sum_{j=1}^{n} \alpha_i \beta_j \left(\mathbf{x}_i, A'(\mathbf{x}_j) \right) = \sum_{i=1}^{n} \sum_{j=1}^{n} \alpha_i \beta_j a'_{ij}$$

$$= \sum_{i=1}^{n} \sum_{j=1}^{n} \alpha_i \beta_j a_{ji} = (A(\mathbf{x}), \mathbf{y}).$$

♣

Remark. If V is a complex vector space and the inner product is complex, then a similar proof shows that $\mathbf{A}' = \overline{\mathbf{A}}^T$, where overbar denotes complex conjugate.

From Definition 7.2 it is easy to see that the adjoint of A' is A, furthermore

(i) $(aA)' = aA'$ for all scalars a and mappings A;

(ii) $(A + C)' = A' + C'$ for all A, $C \in L(V)$;

(iii) $(AC)' = C' A'$ for all A, $C \in L(V)$.

We mention also that adjoint mappings can be defined in the more general case when $A \in L(V, W)$ where V and W are two inner product spaces and either both are real or both are complex. For our purposes it is sufficient to discuss only the case of $V = W$. The general case is examined in a similar way as the one presented in this book.

In this section three special class of mappings (or matrices) will be examined. They are first defined.

Definition 7.3. Let $A \in L(V)$ be a linear tranformation on V.
(i) A is called *symmetric* (or *self-adjoint* in the complex case) if $A' = A$;
(ii) A is called *orthogonal* (or *unitary* in the complex case), if A^{-1} exists and $A^{-1} = A'$;
(iii) A is called *normal* if $AA' = A'A$.

First, we will examine the main properties of symmetric (or self-adjoint) matrices. If the matrix is $n \times n$ and is considered as complex, then from Chapter 6, we know that it always has n real or complex eigenvalues. However, if the matrix is self adjoint, then the eigenvalues are real.

Theorem 7.2. Let $\mathbf{A}$ be a self-adjoint matrix. Then
(a) All eigenvalues of $\mathbf{A}$ are real;
(b) Eigenvectors associated to different eigenvalues are orthogonal;
(c) There is an orthonormal basis in V that consists of eigenvectors of $\mathbf{A}$.

Proof. (a) Let λ be an eigenvalue of $\mathbf{A}$ with associated eigenvector $\mathbf{u}$. Then the eigenvector equation $\mathbf{Au} = \lambda\mathbf{u}$ implies that

$$\lambda(\mathbf{u}, \mathbf{u}) = (\lambda\mathbf{u}, \mathbf{u}) = (\mathbf{Au}, \mathbf{u}) = (\mathbf{u}, \mathbf{Au}) = (\mathbf{u}, \lambda\mathbf{u}) = \overline{\lambda}(\mathbf{u}, \mathbf{u})$$

since $\mathbf{A}$ is self adjoint. Since $\mathbf{u} \neq \mathbf{0}$, $(\mathbf{u}, \mathbf{u}) \neq 0$ as well and so, $\lambda = \overline{\lambda}$ showing that λ is real.

(b) Let λ_1 and λ_2 be distinct eigenvalues of $\mathbf{A}$ with associated eigenvectors $\mathbf{u}_1$ and $\mathbf{u}_2$. Then

$$\lambda_1(\mathbf{u}_1,\mathbf{u}_2) = (\lambda_1\mathbf{u}_1,\mathbf{u}_2) = (A\mathbf{u}_1,\mathbf{u}_2) = (\mathbf{u}_1, A\mathbf{u}_2)$$
$$= (\mathbf{u}_1,\lambda_2\mathbf{u}_2) = \overline{\lambda}_2(\mathbf{u}_1,\mathbf{u}_2) = \lambda_2(\mathbf{u}_1,\mathbf{u}_2),$$

since A is self-adjoint and λ_2 is real. Therefore

$$(\lambda_1 - \lambda_2)\cdot(\mathbf{u}_1,\mathbf{u}_2) = 0$$

which implies that

$$(\mathbf{u}_1,\mathbf{u}_2) = 0.$$

(c) We will next show by finite induction that for all $k \le \dim(V)$, there is an orthonormal system $\{\mathbf{u}_1,\mathbf{u}_2, ..., \mathbf{u}_k\}$ consisting of eigenvectors of A. For $k = 1$ the assertion is obvious, since A has at least one eigenvalue with associated eigenvector $\mathbf{v}$. Then $\mathbf{u}_1 = \dfrac{1}{(\mathbf{v},\mathbf{v})}\mathbf{v}$ satisfies the assertion.

Assume next that there is an orthonormal system $\{\mathbf{u}_1, \mathbf{u}_2, ..., \mathbf{u}_{k-1}\}$ of eigenvectors of A. Let V_1 denote the subspace generated by this orthonormal system, and let W be the orthogonal complementary subspace of V_1. For $i = 1, 2, ..., k-1$ and arbitrary $\mathbf{x} \in W$,

$$(A\mathbf{x},\mathbf{u}_i) = (\mathbf{x}, A\mathbf{u}_i) = (\mathbf{x},\lambda_i\mathbf{u}_i) = \lambda_i(\mathbf{x},\mathbf{u}_i) = 0,$$

where λ_i is the real eigenvalue associated to $\mathbf{u}_i$. Therefore, $A\mathbf{x}$ is orthogonal to the basis of V_1, which implies that $A\mathbf{x} \in W$. That is, W is an A-invariant subspace of V. Consider now the restriction of transformation A into subspace W. It is also self-adjoint, and has at least one eigenvalue with associated eigenvector $\mathbf{u}_k$ of unit length. Then system $\{\mathbf{u}_1, \mathbf{u}_2, ..., \mathbf{u}_{k-1}, \mathbf{u}_k\}$ is on orthonormal systems of eigenvectors. The proof is now completed.

♣

Corollary. If matrix A is self-adjoint, then its matrix representation in basis $\{\mathbf{u}_1, \mathbf{u}_2, ..., \mathbf{u}_n\}$ (where $n = \dim(V)$) is diagonal. The diagonal elements are the real eigenvalues, and the columns of the transformation matrix T form an orthonormal system. If matrix A has complex elements, then **T** also has complex elements. However, if **A** is a real symmetric matrix, then all eigenvectors can be given as real vectors since the eigenvalues are real, and if λ is an eigenvalue then the solutions of the homogeneous linear system $(\mathbf{A} - \lambda\mathbf{I})\mathbf{u} = \mathbf{0}$ determining the associated eigenvectors have real solutions. Hence the transformation matrix can be selected as a real matrix.

Example 7.2. Consider the 2×2 symmetric matrix

$$\mathbf{A} = \begin{pmatrix} 1 & 1 \\ 1 & 1 \end{pmatrix}.$$

The characteristic polynomial of **A** is the following:

$$\varphi(\lambda) = \det(\mathbf{A} - \lambda\mathbf{I}) = \det\begin{pmatrix} 1-\lambda & 1 \\ 1 & 1-\lambda \end{pmatrix} = \lambda^2 - 2\lambda,$$

therefore the eigenvalues are $\lambda_1 = 0$ and $\lambda_2 = 2$, both are real numbers. If $\mathbf{v}_1 = (v_{1i})$ and $\mathbf{v}_2 = (v_{2i})$ are eigenvectors associated to λ_1 and λ_2, then they satisfy equations

$$\begin{pmatrix} 1 & 1 \\ 1 & 1 \end{pmatrix}\begin{pmatrix} v_{11} \\ v_{12} \end{pmatrix} = \begin{pmatrix} 0 \\ 0 \end{pmatrix} \quad \text{and} \quad \begin{pmatrix} -1 & 1 \\ 1 & -1 \end{pmatrix}\begin{pmatrix} v_{21} \\ v_{22} \end{pmatrix} = \begin{pmatrix} 0 \\ 0 \end{pmatrix}.$$

We may therefore select

$$\mathbf{v}_1 = \begin{pmatrix} 1 \\ -1 \end{pmatrix} \quad \text{and} \quad \mathbf{v}_2 = \begin{pmatrix} 1 \\ 1 \end{pmatrix}.$$

The orthonormal system of eigenvectors can be obtained by normalizing the eigenvectors to unit lengths. Notice that $|\mathbf{v}_1| = |\mathbf{v}_2| = \sqrt{2}$, therefore we may select

$$\mathbf{u}_1 = \begin{pmatrix} 1/\sqrt{2} \\ -1/\sqrt{2} \end{pmatrix} \quad \text{and} \quad \mathbf{u}_2 = \begin{pmatrix} 1/\sqrt{2} \\ 1/\sqrt{2} \end{pmatrix}.$$

The transformation matrix $\mathbf{T}$ is given as the matrix with columns $\mathbf{u}_1$ and $\mathbf{u}_2$:

$$\mathbf{T} = \begin{pmatrix} \dfrac{1}{\sqrt{2}} & \dfrac{1}{\sqrt{2}} \\ -\dfrac{1}{\sqrt{2}} & \dfrac{1}{\sqrt{2}} \end{pmatrix} \quad \text{with inverse} \quad \mathbf{T}^{-1} = \begin{pmatrix} \dfrac{1}{\sqrt{2}} & -\dfrac{1}{\sqrt{2}} \\ \dfrac{1}{\sqrt{2}} & \dfrac{1}{\sqrt{2}} \end{pmatrix}.$$

The matrix representation of $\mathbf{A}$ in basis $\{\mathbf{u}_1, \mathbf{u}_2\}$ is diagonal:

$$\mathbf{T}^{-1}\mathbf{A}\mathbf{T} = \begin{pmatrix} \dfrac{1}{\sqrt{2}} & -\dfrac{1}{\sqrt{2}} \\ \dfrac{1}{\sqrt{2}} & \dfrac{1}{\sqrt{2}} \end{pmatrix} \begin{pmatrix} 1 & 1 \\ 1 & 1 \end{pmatrix} \begin{pmatrix} \dfrac{1}{\sqrt{2}} & \dfrac{1}{\sqrt{2}} \\ -\dfrac{1}{\sqrt{2}} & \dfrac{1}{\sqrt{2}} \end{pmatrix} = \begin{pmatrix} 0 & 0 \\ 0 & 2 \end{pmatrix}.$$

◆

Consider next a unitary matrix $\mathbf{A}$. Since for all $\mathbf{x}, \mathbf{y} \in V$,

$$(\mathbf{Ax}, \mathbf{Ay}) = (\mathbf{x}, \mathbf{A'Ay}) = (\mathbf{x}, \mathbf{Iy}) = (\mathbf{x}, \mathbf{y}), \tag{7.7}$$

unitary transformations preserve the values of the inner products. If we select $\mathbf{y} = \mathbf{x}$, then this equation shows that $|\mathbf{Ax}| = |\mathbf{x}|$, that is, unitary transformations preserve the lengths of the elements of V. These two simple observations also imply that the image of an orthonormal system is also an orthonormal system in V. Notice also that the columns of the matrix representation $\mathbf{A}$ of the unitary transformation A form an orthonormal system, since equation (7.7) implies that for all i and j, $(A(\mathbf{x}_i), A(\mathbf{x}_j)) = (\mathbf{x}_i, \mathbf{x}_j)$. We may also show that this property holds only for unitary transformations. If (7.7) holds, then for all $\mathbf{y} \in V$, $\mathbf{A'Ay} = \mathbf{y}$ showing that $\mathbf{A'A} = \mathbf{I}$.

Theorem 7.3. Let **A** be a unitary matrix. Then
(a) The absolute values of the eigenvalues of **A** are equal to one;
(b) There is an orthonormal basis in V that consists of eigenvectors of **A**.

Proof. (a) Let λ and **v** be a pair of an eigenvalue and associated eigenvector of **A**. Then

$$(\mathbf{v},\mathbf{v}) = (\mathbf{Av},\mathbf{Av}) = (\lambda\mathbf{v},\lambda\mathbf{v}) = \lambda(\mathbf{v},\lambda\mathbf{v}) = \lambda\bar{\lambda}(\mathbf{v},\mathbf{v}).$$

Since $(\mathbf{v},\mathbf{v}) \neq 0$, $\lambda\bar{\lambda} = |\lambda|^2 = 1$.

(b) The same proof can be repeated here as the one used in part (c) of Theorem 7.2. We will verify again that for all $k \leq \dim(V)$ there is an orthonormal system $\{\mathbf{u}_1, \mathbf{u}_2, \ldots, \mathbf{u}_k\}$ consisting of eigenvectors of **A**.

For $k = 1$, the assertion is obvious, since $\mathbf{v}_1 = \dfrac{1}{(\mathbf{v},\mathbf{v})}\mathbf{v}$ is again an appropriate selection, where **v** is an eigenvector of **A**.

Assume next that $\{\mathbf{u}_1, \mathbf{u}_2, \ldots, \mathbf{u}_{k-1}\}$ is an orthonormal system consisting of eigenvectors of **A**. Let V_1 be the subspace generated by this orthonormal system, and let W be the orthogonal complementary subspace of V_1. We can show that W is A-invariant, since for all $\mathbf{x} \in W$,

$$(\mathbf{Ax},\mathbf{u}_i) = (\mathbf{x},\mathbf{A}'\mathbf{u}_i) = (\mathbf{x},\mathbf{A}^{-1}\mathbf{u}_i) = \left(\mathbf{x},\frac{1}{\lambda_i}\mathbf{u}_i\right) = \frac{1}{\bar{\lambda_i}}(\mathbf{x},\mathbf{u}_i) = 0,$$

since if λ_i is an eigenvalue of **A** with associated eigenvector $\mathbf{u}_i$ then $1/\lambda_i$ is an eigenvalue of $\mathbf{A}^{-1}$ with the same associated eigenvector. Therefore, **Ax** is orthogonal to all basis elements of V_1, which implies that $\mathbf{Ax} \in W$. The restriction of transformation A into W has at least one eigenvector $\mathbf{u}_k$ of unit length, and system $\{\mathbf{u}_1, \mathbf{u}_2, \ldots, \mathbf{u}_k\}$ is orthonormal and all $\mathbf{u}_i$ ($i = 1, 2, \ldots, k$) are eigenvectors of **A**. Hence the proof is complete.

♣

Remark. If **A** is a real orthogonal matrix, then it can be transformed into diagonal form, however the diagonal matrix as well as the transformation matrix might have complex elements.

Example 7.3. Consider now the matrix

$$\mathbf{A} = \begin{pmatrix} \dfrac{1}{\sqrt{2}} & -\dfrac{1}{\sqrt{2}} \\ \dfrac{1}{\sqrt{2}} & \dfrac{1}{\sqrt{2}} \end{pmatrix}$$

which is obviously orthogonal. The characteristic polynomial of **A** is the following:

$$\varphi(\mathbf{A}) = \det(\mathbf{A} - \lambda\mathbf{I}) = \det\begin{pmatrix} \dfrac{1}{\sqrt{2}} - \lambda & -\dfrac{1}{\sqrt{2}} \\ \dfrac{1}{\sqrt{2}} & \dfrac{1}{\sqrt{2}} - \lambda \end{pmatrix} = \lambda^2 - \dfrac{2}{\sqrt{2}}\lambda + 1,$$

and therefore, the eigenvalues are

$$\lambda_1 = \frac{1}{\sqrt{2}} + i\frac{1}{\sqrt{2}} \quad \text{and} \quad \lambda_2 = \frac{1}{\sqrt{2}} - i\frac{1}{\sqrt{2}}.$$

It is clear that $|\lambda_1| = |\lambda_2| = 1$. The associated eigenvectors $\mathbf{v}_1 = (v_{1i})$ and $\mathbf{v}_2 = (v_{2i})$ satisfy the homogeneous equations

$$\begin{pmatrix} -\dfrac{i}{\sqrt{2}} & -\dfrac{1}{\sqrt{2}} \\ \dfrac{1}{\sqrt{2}} & -\dfrac{i}{\sqrt{2}} \end{pmatrix}\begin{pmatrix} v_{11} \\ v_{12} \end{pmatrix} = \begin{pmatrix} 0 \\ 0 \end{pmatrix} \quad \text{and} \quad \begin{pmatrix} \dfrac{i}{\sqrt{2}} & -\dfrac{1}{\sqrt{2}} \\ \dfrac{1}{\sqrt{2}} & \dfrac{i}{\sqrt{2}} \end{pmatrix}\begin{pmatrix} v_{21} \\ v_{22} \end{pmatrix} = \begin{pmatrix} 0 \\ 0 \end{pmatrix},$$

and therefore, we may have

$$\mathbf{v}_1 = \begin{pmatrix} i \\ 1 \end{pmatrix} \quad \text{and} \quad \mathbf{v}_2 = \begin{pmatrix} -i \\ 1 \end{pmatrix}.$$

Since $|\mathbf{v}_1| = |\mathbf{v}_2| = \sqrt{2}$, we may select $\mathbf{u}_1 = \dfrac{1}{\sqrt{2}}\begin{pmatrix} i \\ 1 \end{pmatrix}$ and $\mathbf{u}_2 = \dfrac{1}{\sqrt{2}}\begin{pmatrix} -i \\ 1 \end{pmatrix}$.

The complex inner product of these normalized eigenvectors is as follows:

$$\left(\mathbf{u}_1, \mathbf{u}_2\right) = \frac{1}{2}\left(\mathbf{v}_1, \mathbf{v}_2\right) = \frac{1}{2}\mathbf{v}_1^T \overline{\mathbf{v}}_2 = \frac{1}{2}\left(i \cdot (i) + 1 \cdot (1)\right) = 0,$$

that is $\{\mathbf{u}_1, \mathbf{u}_2\}$ is an orthonormal basis.

◆

Theorems 7.2 and 7.3 imply that self-adjoint and unitary matrices can be transformed into diagonal form by unitary transformation matrices. A common generalization of these very important matrix types is obtained by assuming that there is an orthonormal basis B in V such that the matrix representation $\mathbf{A}$ of mapping A in basis B is diagonal. Then $\mathbf{A}'$ is also diagonal, and its diagonal elements are the complex conjugates of the corresponding diagonal elements of $\mathbf{A}$. Since the product of diagonal matrices is commutative, we conclude that $\mathbf{A}'\mathbf{A} = \mathbf{A}\mathbf{A}'$, that is, $\mathbf{A}$ must be a normal matrix. The next theorem shows that the converse of this simple result also holds.

Theorem 7.4. Let $\mathbf{A}$ be a complex normal matrix. Then there exists an orthonormal basis $\{\mathbf{u}_1, \mathbf{u}_2, ..., \mathbf{u}_n\}$ of V such that all elements $\mathbf{u}_i$ are common eigenvectors of $\mathbf{A}$ and $\mathbf{A}'$.

Proof. Similarly to the proofs of Theorems 7.2 and 7.3 we will prove that for all $k \leq \dim(V)$, there is an orthonormal system $\{\mathbf{u}_1, ..., \mathbf{u}_k\}$ such that all elements $\mathbf{u}_i$ are common eigenvectors of $\mathbf{A}$ and $\mathbf{A}'$. We will use finite induction. Assume first that $k = 1$. Matrix $\mathbf{A}$ has an eigenvalue λ with associated eigenvector $\mathbf{u}$. First we show that $\mathbf{A}'\mathbf{u}$ is also an eigenvector associated to λ:

$$\left(\mathbf{A}-\lambda\mathbf{I}\right)\left(\mathbf{A'u}\right)=\mathbf{AA'u}-\lambda\mathbf{A'u}=\mathbf{A'Au}-\lambda\mathbf{A'u}$$
$$=\mathbf{A'}\left(\mathbf{Au}-\lambda\mathbf{u}\right)=0.$$

Consider next the subspace generated by the eigenvectors of **A** associated to λ. The restriction of transformation A' into this subspace is a linear transformation, which has at least one eigenvector $\mathbf{u}_1$, which is a common eigenvector of **A** and **A'** (since all vectors of this subspace are eigenvectors of **A**). Let μ be the eigenvalue of **A'** associated to $\mathbf{u}_1$. Then

$$\left(\mathbf{Au}_1,\mathbf{u}_1\right)=\left(\lambda\mathbf{u}_1,\mathbf{u}_1\right)=\lambda\left(\mathbf{u}_1,\mathbf{u}_1\right),$$

and similarly

$$\left(\mathbf{Au}_1,\mathbf{u}_1\right)=\left(\mathbf{u}_1,\mathbf{A'}\,\mathbf{u}_1\right)=\left(\mathbf{u}_1,\mu\mathbf{u}_1\right)=\overline{\mu}\left(\mathbf{u}_1,\mathbf{u}_1\right).$$

Since $\left(\mathbf{u}_1,\mathbf{u}_1\right)\neq0$, $\lambda=\overline{\mu}$, the eigenvalues are complex conjugates of each other.

Assume next that we have an orthonormal system $\{\mathbf{u}_1, \mathbf{u}_2, \ldots, \mathbf{u}_{k-1}\}$ such that each element $\mathbf{u}_i$ is a common eigenvector of **A** and **A'**. Let V_1 denote the subspace generated by this system, and let W denote the orthogonal complementary subspace of V_1. Let $\mathbf{x} \in W$ be an arbitrary element. Then for $I = 1, 2, \ldots, k-1$,

$$\left(\mathbf{Ax},\mathbf{u}_i\right)=\left(\mathbf{x},\mathbf{A'}\,\mathbf{u}_i\right)=\left(\mathbf{x},\overline{\lambda}_i\mathbf{u}_i\right)=\lambda_i\left(\mathbf{x},\mathbf{u}_i\right)=0,$$

and

$$\left(\mathbf{A'x},\,\mathbf{u}_i\right)=\left(\mathbf{x},\mathbf{Au}_i\right)=\left(\mathbf{x},\lambda_i\mathbf{u}_i\right)=\overline{\lambda}_i\left(\mathbf{x},\mathbf{u}_i\right)=0.$$

Therefore subspace W is A-invariant as well as A'-invariant. Consider next the restriction of transformations **A** and **A'** into W, which have at least one common eigenvector $\mathbf{u}_k$. The case of $k = 1$ discussed above provides the proof of the existence of $\mathbf{u}_k$. Then $\{\mathbf{u}_1, \mathbf{u}_2, \ldots, \mathbf{u}_{k-1}, \mathbf{u}_k\}$ is an orthonormal system. Thus the proof is complete.

♣

Remark. A linear transformation A on a complex vector space V is normal if and only if V has an orthonormal basis consisting of common eigenvectors of $\mathbf{A}$ and $\mathbf{A}'$. If a normal mapping is real, then both the transformation matrix and the diagonal form may contain complex elements similarly to the case of real orthogonal matrices.

Example 7.4. In this example we will find all 2×2 real normal matrices. Assume that

$$\mathbf{A} = \begin{pmatrix} a & b \\ c & d \end{pmatrix},$$

then

$$\mathbf{A}\mathbf{A}^T = \begin{pmatrix} a & b \\ c & d \end{pmatrix}\begin{pmatrix} a & c \\ b & d \end{pmatrix} = \begin{pmatrix} a^2 + b^2 & ac + bd \\ ac + bd & c^2 + d^2 \end{pmatrix}$$

and

$$\mathbf{A}^T\mathbf{A} = \begin{pmatrix} a & c \\ b & d \end{pmatrix}\begin{pmatrix} a & b \\ c & d \end{pmatrix} = \begin{pmatrix} a^2 + c^2 & ab + cd \\ ab + cd & b^2 + d^2 \end{pmatrix}.$$

Matrix $\mathbf{A}$ is normal if and only if

$$a^2 + b^2 = a^2 + c^2,$$
$$ac + db = ab + cd,$$
$$c^2 + d^2 = b^2 + d^2.$$

This equations are equivalent to the following two relations:

$$b^2 = c^2,$$
$$(a - d)(b - c) = 0. \tag{7.8}$$

If $b = c$, then **A** is symmetric, and if $b = -c$, then $a = d$. In the second case **A** is the sum of $a\mathbf{I}$ and a skew symmetric matrix, and it is a constant multiple of an orthogonal matrix.

♠

7.5 Definite Matrices

In this section special self-adjoint matrices will be introduced and examined. Let **A** be the $n \times n$ matrix representation of a self-adjoint mapping in an orthonormal basis of a complex vector space.

Definition 7.4. Introduce the following terminology:
(i) **A** is called *positive definite* if $(\mathbf{Ax}, \mathbf{x}) > 0$ for all $\mathbf{x} \neq \mathbf{0}$, $\mathbf{x} \in C^n$;
(ii) **A** is called *positive semidefinite* if $(\mathbf{Ax}, \mathbf{x}) \geq 0$ for all $\mathbf{x} \in C^n$;
(iii) **A** is called *negative definite* if $(\mathbf{Ax}, \mathbf{x}) < 0$ for all $\mathbf{x} \neq \mathbf{0}$, $\mathbf{x} \in C^n$;
(iv) **A** is called *negative semidefinite* if $(\mathbf{Ax}, \mathbf{x}) \leq 0$ for all $\mathbf{x} \in C^n$.

We can show that in the case of a real symmetric matrix **A** it is sufficient to check if the conditions of the above definition hold for only real vectors **x**. If **x** is complex, then it can be written as $\mathbf{x} = \mathbf{y} + i\mathbf{z}$ with real **y** and **z**, and

$$(\mathbf{Ax},\mathbf{x}) = \big(\mathbf{A}(\mathbf{y}+i\mathbf{z}),\mathbf{y}+i\mathbf{z}\big) = (\mathbf{Ay},\mathbf{y}) + (i\mathbf{Az},\mathbf{y}) + (\mathbf{Ay},i\mathbf{z}) + (i\mathbf{Az},i\mathbf{z})$$
$$= (\mathbf{Ay},\mathbf{y}) + i(\mathbf{Az},\mathbf{y}) + \overline{i}\,(\mathbf{Ay},\mathbf{z}) + i\overline{i}\,(\mathbf{Az},\mathbf{z}) = (\mathbf{Ay},\mathbf{y}) + (\mathbf{Az},\mathbf{z}),$$

since $\overline{i} = -i$ and $(\mathbf{Az},\mathbf{y}) = (\mathbf{z},\mathbf{A}^T\mathbf{y}) = (\mathbf{z},\mathbf{Ay}) = (\mathbf{Ay},\mathbf{z})$.

The assertion follows from the facts that both vectors **y** and **z** are real, and $\mathbf{x} \neq \mathbf{0}$ if and only if at least one of vectors **y** and **z** is nonzero.
From the definition it is clear that **A** is negative definite if and only if $-\mathbf{A}$ is positive definite, and **A** is negative semidefinite if and only if $-\mathbf{A}$ is positive semidefinite. Therefore we will discuss only the properties of positive definite and positive semidefinite matrices.

Theorem 7.5. a) A self-adjoint matrix **A** is positive definite if and only if all eigenvalues of **A** are positive;

b) A self-adjoint matrix **A** is positive semidefinite if and only if all eigenvalues of **A** are nonnegative.

Proof. a) Assume first that **A** is positive definite. Let λ be an eigenvalue of **A** with associated eigenvector **u**. Then

$$0 < (\mathbf{Au}, \mathbf{u}) = (\lambda \mathbf{u}, \mathbf{u}) = \lambda(\mathbf{u}, \mathbf{u}),$$

and since $(\mathbf{u},\ \mathbf{u}) > 0$, λ has to be positive. Assume next that all eigenvalues of an $n \times n$ matrix **A** are positive. Then

$$\mathbf{A} = \mathbf{T'DT},$$

where **T** is a unitary matrix, and **D** is diagonal with positive diagonal elements $\lambda_1, \lambda_2, \ldots, \lambda_n$. Let **x** be an arbitrary vector. Then

$$(\mathbf{Ax}, \mathbf{x}) = (\mathbf{T'DTx}, \mathbf{x}) = (\mathbf{DTx}, \mathbf{Tx}).$$

Introduce the notation $\mathbf{y} = \mathbf{Tx}$, and denote the elements of **y** by $y_1, y_2, \ldots, y_n$. Then

$$(\mathbf{Ax}, \mathbf{x}) = (\mathbf{Dy}, \mathbf{y}) = \sum_{i=1}^{n} \lambda_i y_i \overline{y}_i = \sum_{i=1}^{n} \lambda_i |y_i|^2 > 0,$$

since $\mathbf{y} \neq \mathbf{0}$, all terms are nonnegative and at least one term is positive.

b) The proof is the same as given for part a) with the slight modification that inequalities < and > have to replaced by ≤ and ≥, respectively.

♣

Corollary. A self adjoint matrix **A** is negative definite if and only if all eigenvalues of **A** are negative, and is negative semidefinite if and only if all eigenvalues of **A** are nonpositive.

Let now matrix **B** be an $m \times n$ real matrix. First we show that $\mathbf{A} = \mathbf{B}^T \mathbf{B}$ is positive semidefinite. Let **x** be an arbitrary n-element vector. Then

$$\left(\mathbf{Ax},\mathbf{x}\right)=\left(\mathbf{B}^T\mathbf{Bx},\mathbf{x}\right)=\left(\mathbf{Bx},\mathbf{Bx}\right)\geq 0.$$

Assume in addition that the columns of $\mathbf{B}$ are linearly independent, then $\mathbf{x}\neq\mathbf{0}$ implies that $\mathbf{Bx}\neq\mathbf{0}$, therefore

$$\left(\mathbf{Ax},\mathbf{x}\right)=\left(\mathbf{Bx},\mathbf{Bx}\right)>0,$$

which implies that $\mathbf{A}$ is positive definite. We will next prove that if $\mathbf{A}$ is a real $n\times n$ positive semidefinite matrix then there exists an $n\times n$ real matrix $\mathbf{B}$ such that $\mathbf{A}=\mathbf{B}^T\mathbf{B}$, and if in addition $\mathbf{A}$ is positive definite, then $\mathbf{B}$ is nonsingular. From Theorem 7.2 we know that

$$\mathbf{A}=\mathbf{T}^T\mathbf{DT}$$

where $\mathbf{T}$ is an orthogonal matrix and $\mathbf{D}$ is diagonal with diagonal elements $\lambda_1,\lambda_2,...,\lambda_n$. Select

$$\mathbf{B}=\begin{pmatrix}\sqrt{\lambda_1} & & & \\ & \sqrt{\lambda_2} & & \\ & & \ddots & \\ & & & \sqrt{\lambda_n}\end{pmatrix}\mathbf{T},$$

then obviously, $\mathbf{A}=\mathbf{B}^T\mathbf{B}$, and if $\mathbf{A}$ is positive definite, then $\lambda_i>0$ for all i, consequently both factors of $\mathbf{B}$ are nonsingular, hence $\mathbf{B}$ is also nonsingular. We mention here that in the decomposition $\mathbf{A}=\mathbf{B}^T\mathbf{B}$, matrix $\mathbf{B}$ can be selected as an upper triangular matrix. For more details on this decomposition see for example, Szidarovszky and Yakowitz (1978).

Definition 7.5. Let $\mathbf{A}$ be an $n\times n$ matrix, and let $1\leq i_1<i_2<...<i_r\leq n$ be arbitrary integers. The determinant of the principal submatrix

$$\begin{pmatrix} a_{i_1 i_1} & a_{i_1 i_2} & \cdots & a_{i_1 i_r} \\ a_{i_2 i_1} & a_{i_2 i_2} & \cdots & a_{i_2 i_r} \\ \cdots & \cdots & \cdots & \cdots \\ a_{i_r i_1} & a_{i_r i_2} & \cdots & a_{i_r i_r} \end{pmatrix} \tag{7.9}$$

is called the $(i_1, i_2, ..., i_r)$-*principal minor* of $\mathbf{A}$.

An important property of definite and semidefinite matrices is given next.

Theorem 7.6. a) All principal minors of a real positive definite matrix are positive;
b) All principal minors of a real positive semidefinite matrix are nonnegative.

Proof. a) First we prove that if $\mathbf{A}$ is positive definite, then $\det(\mathbf{A}) > 0$. From Theorem 7.2 we know that there exists a real orthogonal matrix $\mathbf{T}$ such that

$$\mathbf{A} = \mathbf{T} \begin{pmatrix} \lambda_1 & & & \\ & \lambda_2 & & \\ & & \ddots & \\ & & & \lambda_n \end{pmatrix} \mathbf{T}^{-1},$$

where all eigenvalues $\lambda_1, \lambda_2, ..., \lambda_n$ are positive. Then

$$\det(\mathbf{A}) = \det(\mathbf{T}) \lambda_1 \lambda_2 ... \lambda_n \det(\mathbf{T}^{-1}) = \lambda_1 \lambda_2 ... \lambda_n > 0,$$

since

$$\det(\mathbf{T}) \det(\mathbf{T}^{-1}) = \det(\mathbf{T}\mathbf{T}^{-1}) = \det(\mathbf{I}) = 1.$$

Next we show that if $\mathbf{A}$ is positive definite, then all principal submatrices of $\mathbf{A}$ are also positive definite, which implies the assertion. Notice first that for all $\mathbf{x}$, $(\mathbf{Ax},\mathbf{x}) = \mathbf{x}^T\mathbf{Ax}$. Let $\mathbf{A}_1$ be the principal submatrix (7.9) and let $\mathbf{z}$ be an r-element vector. Define the n-element vector as

$$x_j = \begin{cases} z_l & \text{if } j = i_l, \\ 0 & \text{otherwise,} \end{cases}$$

then it is easy to see that

$$\mathbf{z}^T \mathbf{A}_1 \mathbf{z} = \mathbf{x}^T \mathbf{Ax} > 0$$

since $\mathbf{x} \neq \mathbf{0}$ if and only if $\mathbf{z} \neq \mathbf{0}$.

b) The case of positive semidefinite matrices can be proved in the same way as shown above with the slight differences that the eigenvalues $\lambda_1, \lambda_2, ..., \lambda_n$ are nonnegative, and $\mathbf{x}^T\mathbf{Ax} \geq 0$ for all $\mathbf{x}$.

♣

Corollary. If $\mathbf{A}$ is real negative definite, then the principal minors of even order are all positive and the principal minors of odd order are all negative. Similarly, if $\mathbf{A}$ is real and negative semidefinite, then the even principal minors are all nonnegative, and those of odd order are all nonpositive.

Some important extensions of definite and semidefinite matrices will be discussed next. First special matrices with certain sign pattern on the principal minors are introduced.

Definition 7.6. Let $\mathbf{A}$ be an $n \times n$ real matrix.
(i) $\mathbf{A}$ is called a *P-matrix* if all principal minors of $\mathbf{A}$ are positive;
(ii) $\mathbf{A}$ is called an *N-matrix* if all principal minors of $\mathbf{A}$ are negative;
(iii) $\mathbf{A}$ is called an *N-P-matrix* if all principal minors of $\mathbf{A}$ of odd orders are negative and those of even orders are positive;
(iv) $\mathbf{A}$ is called a *P-N-matrix* if all principal minors of $\mathbf{A}$ of odd orders are positive and those of even orders are negative.

In the economic literature, N-P-matrices are sometimes called *Hicksian*. It is obvious that $\mathbf{A}$ is an N-P-matrix if and only if $-\mathbf{A}$ is a P-matrix, and $\mathbf{A}$ is a P-N-matrix if and only if $-\mathbf{A}$ is an N-matrix. These facts are the consequences of the simple properties of determinants that for any $m \times m$ real matrix $\mathbf{B}$, $\det(-\mathbf{B}) = (-1)^m \det(\mathbf{B})$. In addition, Definition 7.6 implies that every principal submatrix of a matrix $\mathbf{A}$ is a P-matrix, an N-matrix, an N-P-matrix, or a P-N-matrix, according as $\mathbf{A}$ is a P-matrix, an N-matrix, an N-P-matrix, or a P-N-matrix, since the principal minors of the principal submatrices of $\mathbf{A}$ are principal minors of $\mathbf{A}$.

In the definition of definite and semidefinite matrices we have always assumed that the matrix is symmetric. By dropping this assumption, more general matrix classes can be introduced.

Definition 7.7. Let $\mathbf{A}$ be an $n \times n$ real matrix

(i) $\mathbf{A}$ is called *positive quasidefinite*, if $(\mathbf{Ax}, \mathbf{x}) > 0$ for all $\mathbf{x} \neq \mathbf{0}$; $\mathbf{x} \in R^n$;

(ii) $\mathbf{A}$ is called *positive semi-quasidefinite*, if $(\mathbf{Ax}, \mathbf{x}) \geq 0$ for all $\mathbf{x} \in R^n$;

(iii) $\mathbf{A}$ is called *negative quasidefinite*, if $(\mathbf{Ax}, \mathbf{x}) < 0$ for all $\mathbf{x} \neq \mathbf{0}$; $\mathbf{x} \in R^n$;

(iv) $\mathbf{A}$ is called *negative semi-quasidefinite*, if $(\mathbf{Ax}, \mathbf{x}) \leq 0$ for all $\mathbf{x} \in R^n$.

It is clear from the definition that $\mathbf{A}$ is negative quasidefinite if and only if $-\mathbf{A}$ is positive quasidefinite, and $\mathbf{A}$ is negative semi-quasidefinite if and only if $-\mathbf{A}$ is positive semi-quasidefinite. If $\mathbf{A}$ is an $n \times n$ real matrix, then define

$$\mathbf{A}^{(1)} = \frac{1}{2}\left(\mathbf{A} + \mathbf{A}^T\right)$$

and

$$\mathbf{A}^{(2)} = \frac{1}{2}\left(\mathbf{A} - \mathbf{A}^T\right).$$

Obviously, $\mathbf{A}^{(1)}$ is symmetric and $\mathbf{A}^{(2)}$ is skew-symmetric. Matrices $\mathbf{A}^{(1)}$ and $\mathbf{A}^{(2)}$ are therefore called the *symmetric* and *skew-symmetric parts* of $\mathbf{A}$. Since for all $\mathbf{x}$,

$$\left(\mathbf{A}^{(2)}\mathbf{x},\mathbf{x}\right)=\left(\mathbf{x},\mathbf{A}^{(2)T}\mathbf{x}\right)=\left(\mathbf{x},-\mathbf{A}^{(2)}\mathbf{x}\right)=\left(-\mathbf{A}^{(2)}\mathbf{x},\mathbf{x}\right)=-\left(\mathbf{A}^{(2)}\mathbf{x},\mathbf{x}\right),$$

we conclude that $(\mathbf{A}^{(2)}\mathbf{x}, \mathbf{x}) = 0$. This relation implies that

$$\left(\mathbf{A}\mathbf{x},\mathbf{x}\right)=\left(\left(\mathbf{A}^{(1)}+\mathbf{A}^{(2)}\right)\mathbf{x},\mathbf{x}\right)=\left(\mathbf{A}^{(1)}\mathbf{x},\mathbf{x}\right)+\left(\mathbf{A}^{(2)}\mathbf{x},\mathbf{x}\right)=\left(\mathbf{A}^{(1)}\mathbf{x},\mathbf{x}\right).$$

From this observation we conclude that an $n \times n$ real matrix is positive quasidefinite, negative quasidefinite, positive semi-quasidefinite, or negative semi-quasidefinite if and only if the symmetric part of $\mathbf{A}$ is positive definite, negative definite, positive semidefinite, or negative semidefinite, respectively.

7.6 Nonnegative Matrices

In many linear models the nonnegativity of certain matrices has an important role. For example, in linear input-output models the input $\mathbf{x}$ and output $\mathbf{y}$ satisfy equation

$$\mathbf{x}=\left(\mathbf{I}-\mathbf{A}\right)^{-1}\mathbf{y},$$

where $\mathbf{A}$ is the input coefficient matrix. If $(\mathbf{I}-\mathbf{A})^{-1}$ is nonnegative (which means that all elements of the matrix are nonnegative), then for arbitrary nonnegative output vector $\mathbf{y}$ there exists nonnegative input which produces that output. In this section the most important properties of nonnegative matrices will be discussed. First, the fundamentals of the Perron-Frobenius theory will be outlined, and then, matrices with nonnegative inverses will be discussed.

Let $\mathbf{A}$ be an $n \times n$ matrix such that $\mathbf{A} > \mathbf{O}$. This assumption means that $a_{ij} > 0$ for all i and j. Let $\lambda_1, \lambda_2, ..., \lambda_r$ denote the distinct eigenvalues of $\mathbf{A}$ and assume that they are numbered as

$$\left|\lambda_1\right|\geq\left|\lambda_2\right|\geq...\geq\left|\lambda_r\right|.$$

Our first theorem below presents the most important properties of the dominant eigenvalue λ_1. In the matrix theory literature this result is known as the famous Perron-Frobenius theorem.

Theorem 7.7. (i) $\lambda_1 > 0$;
(ii) λ_1 has a positive associated eigenvector x_1;
(iii) The eigenvector associated to λ_1 is unique up to a scalar multiplier;
(iv) For all $l \geq 2$, $|\lambda_l| < \lambda_1$.

Proof. Consider the set

$$\Lambda = \left\{ \lambda \in R \mid \mathbf{A}\mathbf{x} \geq \lambda\mathbf{x} \text{ with some nonzero vector } \mathbf{x} \geq \mathbf{0} \right\} \quad (7.10)$$

Note first that $\mathbf{0} \in \Lambda$, since $\mathbf{A}\mathbf{x} \geq \mathbf{0}$ with all $\mathbf{x} \geq \mathbf{0}$. We prove next, that Λ is bounded from above. If $\lambda \in \Lambda$, then for all i,

$$\lambda x_i \leq \sum_{j=1}^{n} a_{ij} x_j \quad (7.11)$$

with some nonzero vector $\mathbf{x} \geq \mathbf{0}$.

Let $x_{i_0} = \max_j \left\{ x_j \right\}$, then inequality (7.11) with $i = i_0$ implies that

$$\lambda \leq \sum_{j=1}^{n} a_{i_0 j} \frac{x_j}{x_{i_0}} \leq \sum_{j=1}^{n} a_{i_0 j}.$$

Since the defining inequality of Λ is closed, nonempty and bounded, set Λ has a maximal element λ_0. We prove that this maximal element is the dominant eigenvalue of matrix $\mathbf{A}$, and it satisfies properties (i)-(iv).
(a) First we show that $\lambda_0 > 0$. Note first that if $\mathbf{x} \geq \mathbf{0}$ is nonzero, then it has at least one positive component, and therefore $\mathbf{A}\mathbf{x} > \mathbf{0} = 0\mathbf{x}$. Therefore selecting a sufficiently small positive λ, $\mathbf{A}\mathbf{x}$ is still greater than $\lambda\mathbf{x}$. Since λ_0 is the largest of such λ values, $\lambda_0 > 0$.

(b) Next we prove that λ_0 is an eigenvalue of $\mathbf{A}$. Let $\mathbf{x}_0$ be a nonnegative, nonzero vector such that $\mathbf{A}\mathbf{x}_0 \geq \lambda_0\mathbf{x}_0$. Assume that $\mathbf{A}\mathbf{x}_0 \neq \lambda_0\mathbf{x}_0$ then $\mathbf{A}\mathbf{x}_0 - \lambda_0\mathbf{x}_0$ is nonnegative and nonzero. Therefore

$$0 < \mathbf{A}\left(\mathbf{A}\mathbf{x}_0 - \lambda_0\mathbf{x}_0\right) = \mathbf{A}\left(\mathbf{A}\mathbf{x}_0\right) - \lambda_0\left(\mathbf{A}\mathbf{x}_0\right).$$

Denote $\mathbf{y} = \mathbf{A}\mathbf{x}_0$, then $\mathbf{y} > \mathbf{0}$ and $\mathbf{A}\mathbf{y} > \lambda_0\mathbf{y}$. Therefore with sufficiently small $\varepsilon > 0$, $\mathbf{A}\mathbf{y} > (\lambda_0+\varepsilon)\mathbf{y}$, which contradicts to the selection of λ_0. Therefore $\mathbf{A}\mathbf{x}_0 = \lambda_0\mathbf{x}_0$, that is, λ_0 is an eigenvalue with associated eigenvector $\mathbf{x}_0$.

(c) In the above argument we have shown that $\mathbf{A}\mathbf{x}_0 > \mathbf{0}$, $\lambda_0 > 0$ and $\mathbf{A}\mathbf{x}_0 = \lambda_0\mathbf{x}_0$, which imply that

$$\mathbf{x}_0 = \frac{1}{\lambda_0}\mathbf{A}\mathbf{x}_0 > \mathbf{0}.$$

(d) Let λ_l be now an eigenvalue of $\mathbf{A}$. Then

$$\mathbf{A}\mathbf{y} = \lambda_l\mathbf{y}$$

with some $\mathbf{y} \neq \mathbf{0}$. Define $\mathbf{z} = \mathbf{A}\mathbf{y}$, $\mathbf{z}^* = (|z_i|)$ and $\mathbf{y}^* = (|y_i|)$, where z_i and y_i are the ith elements of vectors $\mathbf{z}$ and $\mathbf{y}$, respectively. Then

$$\mathbf{A}\cdot\mathbf{y}^* = \left(\sum_{j=1}^{n}a_{ij}|y_j|\right)_{i=1}^{n} = \left(\sum_{j=1}^{n}|a_{ij}y_j|\right)_{i=1}^{n}$$

$$\geq \left(\left|\sum_{j=1}^{n}a_{ij}y_j\right|\right)_{i=1}^{n} = \left(|z_i|\right)_{i=1}^{n} = \left(|\lambda_l y_i|\right)_{i=1}^{n} = |\lambda_l|\cdot\mathbf{y}^*,$$

which implies that $|\lambda_l| \in \Lambda$. Since λ_0 is the maximal element in Λ, $|\lambda_l| \leq \lambda_0$.

(e) We prove next that $|\lambda_l| < \lambda_0$. If $\delta > 0$ is sufficiently small, then $\mathbf{A}_\delta = \mathbf{A} - \delta\mathbf{I} > \mathbf{O}$. The previous part of the proof implies that

$$\left|\lambda_l - \delta\right| \leq \lambda_0 - \delta, \tag{7.12}$$

since the eigenvalues of $\mathbf{A} - \delta\mathbf{I}$ are $\lambda_1 - \delta, ..., \lambda_r - \delta$. Contrary to the assertion assume that $|\lambda_l| = \lambda_0$. Let $\lambda_l = a + ib$ where a and b are the real and imaginary parts of λ_l, respectively. Since $|\lambda_l| = \sqrt{a^2 + b^2}$, $a^2 + b^2 = \lambda_0^2$, therefore, if $\lambda_l \neq \lambda_0$, then $a < \lambda_0$. This inequality implies that

$$
\begin{aligned}
\left|\lambda_l - \delta\right|^2 &= \left|(a - \delta) + ib\right|^2 = (a - \delta)^2 + b^2 = a^2 + b^2 + \delta^2 - 2a\delta \\
&> a^2 + b^2 + \delta^2 - 2\delta\lambda_0 = \lambda_0^2 + \delta^2 - 2\delta\lambda_0 = (\lambda_0 - \delta)^2,
\end{aligned}
$$

which contradicts to relation (7.12).

(f) Assume finally that $\mathbf{x}_0$ and $\mathbf{x}$ are linearly independent eigenvectors associated to λ_0. Since $\mathbf{x}_0 > 0$, there exists a scalar α such that vector $\mathbf{z} = \alpha\mathbf{x}_0 + \mathbf{x}$ is nonnegative, nonzero, and at least one element z_i of $\mathbf{z}$ is zero.

Observe that selection $\alpha = \max\left\{-\dfrac{x_i}{x_{0i}}\right\}$ is satisfactory, where $\mathbf{x} = (x_i)$ and $\mathbf{x}_0 = (x_{0i})$. Then

$$
\begin{aligned}
\lambda_0 \mathbf{z} &= \alpha\lambda_0\mathbf{x}_0 + \lambda_0\mathbf{x} = \alpha\mathbf{A}\mathbf{x}_0 + \mathbf{A}\mathbf{x} \\
&= \mathbf{A}(\alpha\mathbf{x}_0 + \mathbf{x}) = \mathbf{A}\mathbf{z} > \mathbf{0},
\end{aligned}
$$

which implies that $\mathbf{z} > \mathbf{0}$, which contradicts the assumption that at least one element of $\mathbf{z}$ is zero.

♣

Hence the proof is completed.

Remark. λ_1 is usually called the *Perron eigenvalue* of $\mathbf{A}$.

Corollary. Assume that $\mathbf{A}$ is an $n \times n$ real matrix with distinct eigenvalues $\lambda_1, \lambda_2, ..., \lambda_r$ such that $|\lambda_1| \geq |\lambda_2| \geq ... \geq |\lambda_r|$ and that $\mathbf{A}^{-1}$ exists and is positive. Since the distinct eigenvalues of $\mathbf{A}^{-1}$ are $\dfrac{1}{\lambda_1}, \dfrac{1}{\lambda_2}, ..., \dfrac{1}{\lambda_r}$ and they satisfy conditions (i)-(iv) of the theorem, we conclude that

(i') $\lambda_r > 0$;

(ii') λ_r has a positive associated eigenvector $\mathbf{x}_r$;

(iii') The eigenvector associated to λ_r is unique up to a scalar multiplier.

Assume now that $\mathbf{A} \geq \mathbf{O}$. Since the eigenvalues of real matrices continuously depend on the matrix elements, Theorem 7.7 implies the following result.

Theorem 7.8. Let $\mathbf{A}$ be a nonnegative matrix with distinct eigenvalues $|\lambda_1| \geq |\lambda_2| \geq \ldots \geq |\lambda_r|$. Then

(i) $\lambda_1 \geq 0$;

(ii) λ_1 has a nonnegative associated eigenvector $\mathbf{x}_1$;

(iii) For all $l \geq 2$, $|\lambda_l| \leq \lambda_1$.

An alternative extension of Theorem 7.7 can be formulated as follows.

Theorem 7.9. Assume that $\mathbf{A} \geq \mathbf{O}$, and $\mathbf{A}^m > \mathbf{O}$ for some positive integer m. Then all conditions (i)-(iv) of Theorem 7.7 apply to $\mathbf{A}$.

Proof. Let λ_1 be the dominant eigenvalue of $\mathbf{A}$ with associated eigenvector $\mathbf{x}_1$. Then the assertion follows immidiately from the fact that λ_1^m is the dominant eigenvalue of $\mathbf{A}^m$ with the same associated eigenvector.

♣

In many applications it is very useful to compute lower and upper bounds for the dominant eigenvalues λ_1 of nonnegative matrices. Such a result is presented in the following theorem.

Theorem 7.10. Let $\mathbf{A} = (a_{ij})$ be an $n \times n$ nonnegative matrix with dominant eigenvalue λ_1. Then

$$\min_j c_j \leq \lambda_1 \leq \max_j c_j, \qquad (7.13)$$

where

$$c_j = \sum_{i=1}^{n} a_{ij}, \, j = 1, 2, ..., n,$$

which is the sum of elements of column j of matrix $\mathbf{A}$.

Proof. Assume that the nonnegative eigenvector $\mathbf{x}_1 = (x_{1i})$ associated to λ_1 is normalized so that $\mathbf{1}^T \mathbf{x}_1 = 1$, where $\mathbf{1}^T = (1, 1, ..., 1)$. Multiply the eigenvector equation $\mathbf{A}\mathbf{x}_1 = \lambda_1 \mathbf{x}_1$ by $\mathbf{1}^T$ to get

$$\mathbf{1}^T \mathbf{A} \mathbf{x}_1 = \lambda_1 \mathbf{1}^T \mathbf{x}_1 = \lambda_1.$$

Observe that

$$\mathbf{1}^T \mathbf{A} = \left(c_1, c_2, ..., c_n \right),$$

and therefore

$$\lambda_1 = c_1 x_{11} + c_2 x_{12} + ... + c_n x_{1n}.$$

Since all components x_{1i} of $\mathbf{x}_1$ are nonnegative,

$$\lambda_1 \geq \left(\min_j c_j \right) \cdot \left(x_{11} + x_{12} + ... + x_{1n} \right) = \min_j c_j$$

and

$$\lambda_1 \leq \left(\max_j c_j \right) \cdot \left(x_{11} + x_{12} + ... + x_{1n} \right) = \max_j c_j.$$

♣

Corollary. Let $\mathbf{A} = (a_{ij})$ be an $n \times n$ nonnegative matrix with dominant eigenvalue λ_1. Then

$$\min_i r_i \leq \lambda_1 \leq \max_i r_i, \tag{7.14}$$

where

$$r_i = \sum_{j=1}^{n} a_{ij}, \quad i = 1, 2, ..., n,$$

which is the sum of elements of row i of matrix $\mathbf{A}$.

Proof. The eigenvalues of $\mathbf{A}$ and $\mathbf{A}^T$ are the same. Apply relation (7.13) to the transpose $\mathbf{A}^T$ of $\mathbf{A}$ to get the desired bounds.

♣

Relations (7.13) and (7.14) are illustrated in the following examples.

Example 7.5. In the case of matrix

$$\mathbf{A} = \begin{pmatrix} 1 & 2 \\ 4 & 3 \end{pmatrix},$$

$c_1 = c_2 = 5$, and therefore

$$\min_{j} c_j = \max_{j} c_j = 5.$$

Hence relation (7.13) implies that $\lambda_1 = 5$. This result can be also verified by solving the characteristic polynomial

$$\varphi(\lambda) = \det \begin{pmatrix} 1-\lambda & 2 \\ 4 & 3-\lambda \end{pmatrix} = (1-\lambda)(3-\lambda) - 8 = \lambda^2 - 4\lambda - 5.$$

The roots are $\lambda_1 = 5$ and $\lambda_2 = -1$.

♦

Example 7.6. Consider next matrix

$$\mathbf{A} = \begin{pmatrix} 1 & 0 & 0 \\ 1 & 1 & 4 \\ 0 & 1 & 3 \end{pmatrix},$$

where

$$c_1 = 2, c_2 = 2, c_3 = 7 \text{ and } r_1 = 1, r_2 = 6, r_3 = 4.$$

Therefore relations (7.13) and (7.14) imply that

$$2 \le \lambda_1 \le 7 \quad \text{and} \quad 1 \le \lambda_1 \le 6.$$

Note that from these inequalities we conclude that

$$2 \le \lambda_1 \le 6.$$

As a comparison we mention that the true value is

$$\lambda_1 = 2 + \sqrt{5} \approx 4.236.$$

♦

In many applications the nonnegativity of inverse matrices is assumed. In the next part of this section conditions will be introduced which guarantee that $\mathbf{A}^{-1} \ge \mathbf{0}$ for a given square matrix $\mathbf{A}$. We begin our analysis with a simple result.

Theorem 7.11. Let $\mathbf{B}$ be an $n \times n$ real matrix. The inverse $\mathbf{B}^{-1}$ exists and is nonnegative if and only if there exists an $n \times n$ real matrix $\mathbf{D}$ such that

(i) $\mathbf{D} \ge \mathbf{O}$;

(ii) $\mathbf{I} - \mathbf{DB} \ge \mathbf{O}$;

(iii) All eigenvalues of $\mathbf{I} - \mathbf{DB}$ are inside the unit circle.

Proof. (a) If $\mathbf{B}^{-1}$ exists and is nonnegative, then select $\mathbf{D} = \mathbf{B}^{-1}$. Then $\mathbf{I} - \mathbf{DB} = \mathbf{O}$, and therefore conditions (i), (ii) and (iii) are obviously satisfied.
(b) Assume now the existence of matrix $\mathbf{D}$ satisfying conditions (i), (ii) and (iii). Note first that condition (iii) and the results of the third application of Section 6.8 imply that $\mathbf{I} - (\mathbf{I} - \mathbf{DB}) = \mathbf{DB}$ is invertible. Therefore $\mathbf{D}$ and $\mathbf{B}$ are both nonsingular. From the same application we also know that

$$(\mathbf{DB})^{-1} = \mathbf{I} + (\mathbf{I} - \mathbf{DB}) + (\mathbf{I} - \mathbf{DB})^2 + \ldots \ge \mathbf{O},$$

since each term of the right hand side is nonnegative as the consequence of condition (ii). Hence, from (i) we conclude that

$$\mathbf{B}^{-1} = \left(\mathbf{D}^{-1}\mathbf{DB}\right)^{-1} = \left(\mathbf{DB}\right)^{-1}\mathbf{D} \geq \mathbf{O},$$

which completes the proof.

♣

Definition 7.8. An $n \times n$ real matrix $\mathbf{B} = (b_{ij})$ is called an *M-matrix*, if $\mathbf{D} = \mathrm{diag}\left(b_{11}, b_{22}, ..., b_{nn}\right)^{-1}$ satisfies conditions (i), (ii) and (iii) of Theorem 7.11.

Note that in the case of an *M*-matrix,

$$\mathbf{I} - \mathbf{DB} = \begin{pmatrix} 0 & -\dfrac{b_{12}}{b_{11}} & \cdots & -\dfrac{b_{1n}}{b_{11}} \\[2mm] -\dfrac{b_{21}}{b_{22}} & 0 & \cdots & -\dfrac{b_{2n}}{b_{22}} \\[2mm] \cdots & \cdots & \cdots & \cdots \\[2mm] -\dfrac{b_{n1}}{b_{nn}} & -\dfrac{b_{n2}}{b_{nn}} & \cdots & 0 \end{pmatrix}, \tag{7.15}$$

therefore $b_{ii} > 0$ for all i, and $b_{ij} \leq 0$ for all $j \neq i$. These conditions however do not imply that $\mathbf{B}$ is an *M*-matrix, since there is no guarantee in general that the eigenvalues of $\mathbf{I}$–$\mathbf{DB}$ are inside the unit circle.

Example 7.7. We will show that matrix

$$\mathbf{B} = \begin{pmatrix} 5 & -1 & -1 \\ -1 & 5 & -1 \\ -1 & -1 & 5 \end{pmatrix}$$

is an *M*-matrix. In this case,

$$\mathbf{D} = \text{diag}(5,5,5)^{-1} = \text{diag}\left(\frac{1}{5},\frac{1}{5},\frac{1}{5}\right),$$

and

$$\mathbf{I} - \mathbf{DB} = \begin{pmatrix} 0 & \dfrac{1}{5} & \dfrac{1}{5} \\[2mm] \dfrac{1}{5} & 0 & \dfrac{1}{5} \\[2mm] \dfrac{1}{5} & \dfrac{1}{5} & 0 \end{pmatrix}.$$

Conditions (i) and (ii) of Theorem 7.11. are obviously satisfied. By using relation (7.13) we see that $c_1 = c_2 = c_3 = \dfrac{2}{5}$, therefore for all eigenvalues λ_i of $\mathbf{I} - \mathbf{DB}$, $|\lambda_i| \leq \dfrac{2}{5}$. (The true eigenvalues are $\lambda_1 = \lambda_2 = -\dfrac{1}{5}$ and $\lambda_3 = \dfrac{2}{5}$.)

♦

In the conclusion of this section a new class of special matrices will be introduced, which has a very close relation to nonnegative matrices.

Definition 7.9. An $n \times n$ real matrix $\mathbf{A}$ is called a *Metzler-matrix*, if $a_{ij} \geq 0$ for all $i \neq j$.

Notice that Metzler-matrices have nonnegative elements outside the diagonal, and there is no constraint on the diagonal elements. A square matrix $\mathbf{A}$ is a Metzler matrix if and only if there is a real number α such that $\mathbf{A} + \alpha\mathbf{I} \geq 0$. This strong relation between nonnegative and Metzler matrices makes us able to extend the Perron-Frobenius theorem to Metzler matrices.

Theorem 7.12. Assume that $\mathbf{A}$ is an $n \times n$ real Metzler matrix. Then there exists a real eigenvalue λ_1 of $\mathbf{A}$ such that
(i) λ_1 has a nonnegative associated eigenvector;
(ii) For any other eigenvalue λ_l of $\mathbf{A}$, Re $\lambda_l < \lambda_1$.

Proof. (i) Select an α such that $\mathbf{A} + \alpha\mathbf{I} \geq 0$, and let μ_1 denote the Perron eigenvalue of $\mathbf{A} + \alpha\mathbf{I}$ with associated eigenvector $\mathbf{x}_1 \geq 0$. Then $\lambda_1 = \mu_1 - \alpha$ is an eigenvalue of $\mathbf{A}$ with associated eigenvector $\mathbf{x}_1$.
(ii) Let λ_l denote an eigenvalue of $\mathbf{A}$ such that $\lambda_l \neq \lambda_1$. Note that $\lambda_1 + \alpha$ is nonnegative and is the Perron eigenvalue of $\mathbf{A} + \alpha\mathbf{I}$, and for any other eigenvalue λ_l of $\mathbf{A}$, $\lambda_l + \alpha$ is also an eigenvalue of $\mathbf{A} + \alpha\mathbf{I}$ such that $\lambda_1 + \alpha \neq \lambda_l + \alpha$.
If $\lambda_l + \alpha$ is real, then from Theorem 7.8 we know that $|\lambda_l + \alpha| \leq \lambda_1 + \alpha$. If $\lambda_l + \alpha$ is nonnegative, then $\lambda_l + \alpha < \lambda_1 + \alpha$, which implies that $\lambda_l < \lambda_1$. If $\lambda_l + \alpha$ is negative, then obviously $\lambda_l + \alpha < \lambda_1 + \alpha$, that is $\lambda_l < \lambda_1$ again. Hence, Re $\lambda_l = \lambda_l < \lambda_1$.
If $\lambda_l = a + ib$ with $b \neq 0$, then from Theorem 7.8 we conclude that

$$|\lambda_l + \alpha| \leq \lambda_1 + \alpha,$$

that is,

$$|(a+\alpha) + ib| \leq \lambda_1 + \alpha.$$

This relation is equivalent to inequality

$$(a+\alpha)^2 + b^2 \leq (\lambda_1 + \alpha)^2,$$

which implies that

$$(a+\alpha)^2 < (\lambda_1 + \alpha)^2. \tag{7.16}$$

This inequality can be rewritten as

$$(a - \lambda_1)(a + 2\alpha + \lambda_1) < 0. \tag{7.17}$$

Note first that (7.16) implies that

$$|a + \alpha| < \lambda_1 + \alpha,$$

and therefore

$$a + 2\alpha + \lambda_1 > 0.$$

And hence, (7.17) holds only if $a - \lambda_1 < 0$, that is, when $a = \operatorname{Re} \lambda_l < \lambda_1$. Thus the proof is completed.

♣

Assume again that $\mathbf{A}$, is a Metzler matrix. We know that $\mathbf{A} + \alpha\mathbf{I}$ is nonnegative with some nonnegative α. It is also known, that the eigenvalues of $\mathbf{A} + \alpha\mathbf{I}$ are $\lambda_l + \alpha$ and the Perron eigenvalue of $\mathbf{A} + \alpha\mathbf{I}$ is $\lambda_1 + \alpha$, where λ_1 is the eigenvalue of $\mathbf{A}$, which satisfies properties (i) and (ii) of the above Theorem. If $\mathbf{A} = (a_{ij})$ and $\mathbf{A} + \alpha\mathbf{I} = (a'_{ij})$

$$a'_{ij} = \begin{cases} a_{ij} + \alpha & \text{if } i = j, \\ a_{ij} & \text{if } i \neq j, \end{cases}$$

Use relations (7.13) and (7.14) to this matrix to get inequalities

$$\min_j \left(c_j + \alpha \right) \leq \lambda_1 + \alpha \leq \max_j \left(c_j + \alpha \right),$$

and

$$\min_j \left(r_j + \alpha \right) \leq \lambda_1 + \alpha \leq \max_j \left(r_j + \alpha \right)$$

which imply that Theorem 7.10 and its corollary remain true for Metzler matrices.

Example 7.8. Consider matrix

$$\mathbf{A} = \begin{pmatrix} -2 & 1 \\ 2 & -1 \end{pmatrix},$$

where $c_1 = c_2 = 0$, and therefore $\lambda_1 = 0$. Note that the characteristic polynomial of $\mathbf{A}$ is

$$\varphi(\lambda) = \det \begin{pmatrix} -2-\lambda & 1 \\ 2 & -1-\lambda \end{pmatrix} = (-2-\lambda)(-1-\lambda) - 2 = \lambda^2 + 3\lambda.$$

The eigenvalues are $\lambda_1 - 0$ and $\lambda_2 - 3$.

♦

Example 7.9. Consider next matrix

$$A = \begin{pmatrix} -1 & 0 & 0 \\ 1 & 1 & 4 \\ 0 & 1 & 3 \end{pmatrix},$$

where

$$c_1 = 0,\ c_2 = 2,\ c_3 = 7 \quad \text{and} \quad r_1 = -1, r_2 = 6, r_3 = 4.$$

Therefore relations (7.13) and (7.14) imply that

$$0 \le \lambda_1 \le 7 \quad \text{and} \quad -1 \le \lambda_1 \le 6.$$

Hence, $0 \le \lambda_1 \le 6$. Note that the true value is

$$\lambda_1 = 2 + \sqrt{5} \approx 4.236.$$

♦

7.7 Applications

In this section several applications of special matrices are discussed.

1. In Section 3.8. we introduced cubic interpolating splines and derived a system of linear equations for the unknown coefficients. The system (3.31) is tridiagonal, so the special elimination method presented in Section 7.2 can be used to find the solution. The last application of Section 3.8 introduced the difference method for solving the boundary value problem (3.35). The application of the method requires the solution of linear equations (3.38), which can be rewritten as

$$y\left(x_{k-1}\right)\left(\frac{1}{h^2}-\frac{p_k}{2h}\right)+y\left(x_k\right)\left(q_k-\frac{2}{h^2}\right)+y\left(x_{k+1}\right)\left(\frac{1}{h^2}+\frac{p_k}{2h}\right)=r_k$$

for $k = 1, 2, \ldots, n-1$ with $y(x_0) = A$ and $y(x_n) = B$. This system is also tridiagional, and the special elimination method can be used in this case as well.

2. Consider a Markov chain with n states and the $n \times n$ transition matrix **P**. In Section 3.8 we have seen that **P** is a nonnegative matrix, and the sum of the elements in each column is equal to one. Therefore we can use Theorems 7.8 and 7.10 to see that the dominant eigenvalue of **P** is unity, and this eigenvalue has a nonnegative associated eigenvector. The eigenvalue equation with $\lambda_1 = 1$ and eigenvector **x** has the special form

$$\mathbf{Px} = 1 \cdot \mathbf{x}.$$

Since $\mathbf{x} \geq \mathbf{0}$ and $\mathbf{x} \neq \mathbf{0}$, $\sum_{k=1}^{n} x_k > 0$ and vector

$$\mathbf{x}^* = \frac{1}{\displaystyle\sum_{k=1}^{n} x_k} \cdot \mathbf{x}$$

satisfies relations

$$\mathbf{Px}^* = \mathbf{x}^*$$
$$x_1^* + x_2^* + \ldots + x_n^* = 1,$$

(7.18)

which is equivalent to relation (3.34). Hence **x*** is the steady state probability vector.

3. Consider now an economy with n products and a continuum of consumers of the same type with utility function

$$U(\mathbf{s}) = \mathbf{s}^T \mathbf{Qs} + \mathbf{b}^T \mathbf{s} + \mathbf{c},$$

(7.19)

where **Q** is a negative definite matrix. The representative consumer maximizes $U(\mathbf{s}) - \mathbf{p}^T\mathbf{s}$, where **p** is the price vector. This objective function is strictly concave in **s**, and it converges to $-\infty$ if any one of the components of **s** tends to ∞, since (7.19) is quadratic in all components s_i of **s** with the only quadratic term $q_{ii}s_i^2$ when all other terms are constants or linear in s_i. Therefore there is a unique vector $\mathbf{s} = \mathbf{d}(\mathbf{p})$ maximizing the objective function. This vector is called the demand function and can be obtained by simple differentiation. The gradient of $U(\mathbf{s}) - \mathbf{p}^T\mathbf{s}$ is given as $2\mathbf{Q}\mathbf{s} + \mathbf{b} - \mathbf{p}$, so the optimal **s** can be obtained by solving equation

$$2\mathbf{Q}\mathbf{s} + \mathbf{b} - \mathbf{p} = 0$$

which implies that

$$\mathbf{d}(\mathbf{p}) = \mathbf{s} = \frac{1}{2}\mathbf{Q}^{-1}\mathbf{p} - \frac{1}{2}\mathbf{Q}^{-1}\mathbf{b} \qquad (7.20)$$

resulting in a linear demand function.

4. Definite matrices have an important role in determining if a function is convex, or concave. Let $f : R^n \mapsto R$ be a twice continuously differentiable real valued function in an open domain $D \subseteq R^n$. Construct the Hessian matrix **H** of f in the following way. The (i, j) element of **H** is given by the second-order partial derivative $\partial^2 f / \partial x_i \partial x_j$. Hence

$$\mathbf{H}(\mathbf{x}) = \begin{pmatrix} \dfrac{\partial^2 f(\mathbf{x})}{\partial x_1^2} & \dfrac{\partial^2 f(\mathbf{x})}{\partial x_1 \partial x_2} & \cdots & \dfrac{\partial^2 f(\mathbf{x})}{\partial x_1 \partial x_n} \\[2ex] \dfrac{\partial^2 f(\mathbf{x})}{\partial x_2 \partial x_1} & \dfrac{\partial^2 f(\mathbf{x})}{\partial x_2^2} & \cdots & \dfrac{\partial^2 f(\mathbf{x})}{\partial x_2 \partial x_n} \\[2ex] \cdots & \cdots & \cdots & \cdots \\[2ex] \dfrac{\partial^2 f(\mathbf{x})}{\partial x_n \partial x_1} & \dfrac{\partial^2 f(\mathbf{x})}{\partial x_n \partial x_2} & \cdots & \dfrac{\partial^2 f(\mathbf{x})}{\partial x_n^2} \end{pmatrix}. \qquad (7.21)$$

Let now **x*** be an arbitraty point from D. It is well known from convex analysis, that if $\mathbf{H}(\mathbf{x}^*)$ is positive definite, then f is stictly convex in a neighborhood of **x***, and if $\mathbf{H}(\mathbf{x}^*)$ is negative definite, then f is stictly concave in a neighborhood of $\mathbf{x}^*$. In Section 7.5 we have discussed the conditions which have to be checked in order to verify the definiteness of the Hessian.

5. Let $\mathbf{f} : R^n \mapsto R^n$ be a continuously differentiable function on a convex, open domain $D \subseteq R^n$. We say that $\mathbf{f}$ is monotone, if for all $\mathbf{x}, \mathbf{y} \in D$,

$$(\mathbf{x} - \mathbf{y})^T \left(\mathbf{f}(\mathbf{x}) - \mathbf{f}(\mathbf{y}) \right) \geq 0. \tag{7.22}$$

In the one-dimensional case this kind of monotonicity coincides with the property that $\mathbf{f}$ is nondecreasing. Similarly we say that $\mathbf{f}$ is strictly monotone, if for all $\mathbf{x}, \mathbf{y} \in D$ and $\mathbf{x} \neq \mathbf{y}$,

$$(\mathbf{x} - \mathbf{y})^T \left(\mathbf{f}(\mathbf{x}) - \mathbf{f}(\mathbf{y}) \right) > 0. \tag{7.23}$$

In the one-dimensional case this kind of strict monotonicity coincides with the assumption that $\mathbf{f}$ is strictly increasing.

The Jacobian of $\mathbf{f}$ is the $n \times n$ matrix with (i, j) element $\partial f_i / \partial x_j$, that is,

$$\mathbf{J}(\mathbf{x}) = \begin{pmatrix} \dfrac{\partial f_1(\mathbf{x})}{\partial x_1} & \dfrac{\partial f_1(\mathbf{x})}{\partial x_2} & \cdots & \dfrac{\partial f_1(\mathbf{x})}{\partial x_n} \\ \dfrac{\partial f_2(\mathbf{x})}{\partial x_1} & \dfrac{\partial f_2(\mathbf{x})}{\partial x_2} & \cdots & \dfrac{\partial f_2(\mathbf{x})}{\partial x_n} \\ \cdots & \cdots & \cdots & \cdots \\ \dfrac{\partial f_n(\mathbf{x})}{\partial x_1} & \dfrac{\partial f_n(\mathbf{x})}{\partial x_2} & \cdots & \dfrac{\partial f_n(\mathbf{x})}{\partial x_n} \end{pmatrix}$$

where x_i and f_i denote the ith entries of $\mathbf{x}$ and $\mathbf{f}$, respectively. The monotonicity and strict monotonicity of $\mathbf{f}$ can be checked in the following way. If $\mathbf{J}(\mathbf{x}) + \mathbf{J}(\mathbf{x})^T$ is nonnegative definite for all $\mathbf{x} \in D$, then $\mathbf{f}$ is monotone, and if in addition, $\mathbf{J}(\mathbf{x}) + \mathbf{J}(\mathbf{x})^T$ is positive definite for all $\mathbf{x} \in D$, then $\mathbf{f}$ is strictly monotone.

Monotone and strictly monotone functions have an important role in may fields of optimization. As the conclusion of this application we show a classical result.
Consider a nonlinear complementarity problem

$$x^T \cdot f(x) = 0$$
$$x \geq 0, \ f(x) \geq 0$$

(7.24)

and assume that f is strictly monotone. Then the nonlinear complementarity problem must not have multiple solutions. Contrary to this assertion assume that x and z are both solutions, then

$$0 < (x-z)^T \left(f(x) - f(z) \right) = x^T f(x) - z^T f(x) - x^T f(z) + z^T f(z)$$
$$= -z^T f(x) - x^T f(z) \leq 0,$$

which is an obvious contradiction.

6. In this application *singular value decomposition* (SVD) of real matrices will be introduced and examined. Let A be a real $m \times n$ matrix. The existence of the SVD is guaranteed by the following theorem.

Theorem 7.13. There exist orthogonal matrices U and V such that

$$U^T AV = \begin{pmatrix} D & O \\ O & O \end{pmatrix},$$

where D is a nonsingular diagonal matrix with positive diagonal elements, which are arranged in nonincreasing order. The size of D is the rank of A.

Proof. From Section 7.5 we know that $A^T A$ is a positive semidefinite matrix with nonnegative eigenvalues.
Let $\lambda_1 \geq \lambda_2 \geq ... \geq \lambda_r > \lambda_{r+1} = ... = \lambda_n = 0$ denote the eigenvalues of $A^T A$. Denote the corresponding set of orthonormal eigenvectors by $\{v_1, v_2, ..., v_n\}$, which exists, since $A^T A$ is symmetric.

For $i = 1, 2, \ldots, n$,

$$\mathbf{A}^T \mathbf{A} \mathbf{v}_i = \sigma_i^2 \mathbf{v}_i,$$

where $\lambda_i = \sigma_i^2$ with a nonnegative σ_i. This equation implies that for all i and j,

$$\mathbf{v}_j^T \mathbf{A}^T \mathbf{A} \mathbf{v}_i = \sigma_i^2 \mathbf{v}_j^T \mathbf{v}_i = \begin{cases} 0 & \text{if } i \neq j \\ \sigma_i^2 & \text{if } i = j \end{cases} . \tag{7.25}$$

Notice also that from (5.25) with $i = j = r+1, \ldots, n$,

$$\mathbf{A} \mathbf{v}_i = 0 \quad \mathbf{v}_i = 0. \tag{7.26}$$

Introduce matrices

$$\mathbf{V}_1 = (\mathbf{v}_1, \ldots, \mathbf{v}_r), \quad \mathbf{V}_2 = (\mathbf{v}_{r+1}, \ldots, \mathbf{v}_n),$$

and the m-element vectors

$$\mathbf{u}_i = \frac{1}{\sigma_i} \mathbf{A} \mathbf{v}_i \quad (i = 1, 2, \ldots, r).$$

First we show that these vectors form an orthonormal system. From (7.25) we see that

$$\mathbf{u}_i^T \mathbf{u}_j = \frac{1}{\sigma_i} (\mathbf{A} \mathbf{v}_i)^T \frac{1}{\sigma_j} (\mathbf{A} \mathbf{v}_j) = \frac{1}{\sigma_i \sigma_j} (\mathbf{v}_i^T \mathbf{A}^T \mathbf{A} \mathbf{v}_j) = \begin{cases} 1 & \text{if } i = j \\ 0 & \text{if } i \neq j. \end{cases}$$

Define $\mathbf{U}_1 = (\mathbf{u}_1, \ldots, \mathbf{u}_r)$ and select $\mathbf{U}_2 = (\mathbf{u}_{r+1}, \ldots, \mathbf{u}_m)$ such that system $\{\mathbf{u}_1, \ldots, \mathbf{u}_r, \mathbf{u}_{r+1}, \ldots, \mathbf{u}_m\}$ is orthonormal. Vectors $\mathbf{u}_{r+1}, \ldots, \mathbf{u}_m$ can be selected as an orthonormal basis of the orthogonal complementary subspace of the subspace generated by the columns of $\mathbf{U}_1$. Notice that for any $j \geq r+1$ and $i \leq r$,

$$\mathbf{u}_j^T \mathbf{A} \mathbf{v}_i = \mathbf{u}_j^T \sigma_i \mathbf{u}_i = 0,$$

and for any $j \geq r + 1$ and $i \geq r + 1$,

$$\mathbf{u}_j^t \mathbf{A} \mathbf{v}_i = 0$$

as a consequence of relation (7.26).

Define matrices

$$\mathbf{U} = \left(\mathbf{U}_1, \mathbf{U}_2 \right) \quad \text{and} \quad \mathbf{V} = \left(\mathbf{V}_1, \mathbf{V}_2 \right).$$

Then

$$\mathbf{U}^T \mathbf{A} \mathbf{V} = \begin{pmatrix} \mathbf{u}_1^T \\ \mathbf{u}_2^T \\ \cdots \\ \mathbf{u}_m^T \end{pmatrix} \mathbf{A} \left(\mathbf{v}_1, \mathbf{v}_2, \ldots, \mathbf{v}_n \right) = \begin{pmatrix} \dfrac{1}{\sigma_1} \mathbf{v}_1^T \mathbf{A}^T \\ \cdots \\ \dfrac{1}{\sigma_r} \mathbf{v}_r^T \mathbf{A}^T \\ \mathbf{u}_{r+1}^T \\ \cdots \\ \mathbf{u}_m^T \end{pmatrix} \left(\mathbf{A}\mathbf{v}_1, \mathbf{A}\mathbf{v}_2, \ldots, \mathbf{A}\mathbf{v}_n \right)$$

$$= \begin{pmatrix} \dfrac{1}{\sigma_1}\sigma_1^2 & & & & \\ & \ddots & & & \\ & & \dfrac{1}{\sigma_r}\sigma_r^2 & & \\ & & & 0 & \\ & & & & \ddots \\ & & & & & 0 \end{pmatrix} = \begin{pmatrix} \mathbf{D} & \mathbf{O} \\ \mathbf{O} & \mathbf{O} \end{pmatrix}.$$

Notice that the both matrices $\mathbf{U}^T$ and $\mathbf{V}$ are invertible. Therefore

$$\text{rank}(\mathbf{A}) = \text{rank}(\mathbf{U}^T \mathbf{A}) = \text{rank}(\mathbf{U}^T \mathbf{A} \mathbf{V}) = \text{rank}(\mathbf{D}),$$

which completes the proof.

♣

Corollary. Matrix **A** can be factored as

$$\mathbf{A} = \mathbf{U} \begin{pmatrix} \sigma_1 & & & & \mathbf{O} \\ & \ddots & & & \\ & & \sigma_r & & \\ & & & \ddots & \\ \mathbf{O} & & & & \mathbf{O} \end{pmatrix} \mathbf{V}^T, \tag{7.27}$$

which is called the *singular value decomposition* of **A**, and the positive scalars $\sigma_1,\ldots,\sigma_r$ are called the *singular values* of **A**. Notice that **U** is $m \times m$, **V** is $n \times n$ and the second factor is $m \times n$.

Example 7.10. Consider the 3×2 real matrix

$$\mathbf{A} = \begin{pmatrix} 0 & 0 \\ 0 & 0 \\ 3 & 4 \end{pmatrix}.$$

Since

$$\mathbf{A}^T \mathbf{A} = \begin{pmatrix} 0 & 0 & 3 \\ 0 & 0 & 4 \end{pmatrix} \begin{pmatrix} 0 & 0 \\ 0 & 0 \\ 3 & 4 \end{pmatrix} = \begin{pmatrix} 9 & 12 \\ 12 & 16 \end{pmatrix},$$

the characteristic equation of $\mathbf{A}^T \mathbf{A}$ has the form

$$\varphi(\lambda) = \det \begin{pmatrix} 9 - \lambda & 12 \\ 12 & 16 - \lambda \end{pmatrix} = \lambda^2 - 25\lambda,$$

so the eigenvalues of $\mathbf{A}^T\mathbf{A}$ are $\lambda_1 = 25$ and $\lambda_2 = 0$. Therefore $\sigma_1 = 5$ and $\sigma_2 = 0$. Next, the associated eigenvectors will be determined. For eigenvalue λ_1, the associated eigenvector $\mathbf{v}_1 = (v_{1i})$ solves equation

$$\begin{pmatrix} 9 & 12 \\ 12 & 16 \end{pmatrix}\begin{pmatrix} v_{11} \\ v_{12} \end{pmatrix} = 25\begin{pmatrix} v_{11} \\ v_{12} \end{pmatrix},$$

that is,

$$-16v_{11} + 12v_{12} = 0$$
$$12v_{11} - 9v_{12} = 0.$$

A solution is vector

$$\begin{pmatrix} v_{11} \\ v_{12} \end{pmatrix} = \begin{pmatrix} 3 \\ 4 \end{pmatrix},$$

which can be normalized as

$$\mathbf{v}_1 = \begin{pmatrix} \dfrac{3}{5} \\ \dfrac{4}{5} \end{pmatrix}.$$

For eigenvalue λ_2, the associated eigenvector satisfies equation

$$\begin{pmatrix} 9 & 12 \\ 12 & 16 \end{pmatrix}\begin{pmatrix} v_{21} \\ v_{22} \end{pmatrix} = 0\begin{pmatrix} v_{21} \\ v_{22} \end{pmatrix},$$

that is,

$$9v_{21} + 12v_{22} = 0$$
$$12v_{21} + 16v_{22} = 0.$$

A solution is vector

$$\begin{pmatrix} v_{21} \\ v_{22} \end{pmatrix} = \begin{pmatrix} -4 \\ 3 \end{pmatrix},$$

which can be normalized as

$$\mathbf{v}_2 = \begin{pmatrix} -\dfrac{4}{5} \\ \dfrac{3}{5} \end{pmatrix}.$$

In our case $r = 1$ and

$$\mathbf{u}_1 = \frac{1}{5}\begin{pmatrix} 0 & 0 \\ 0 & 0 \\ 3 & 4 \end{pmatrix}\begin{pmatrix} \dfrac{3}{5} \\ \dfrac{4}{5} \end{pmatrix} = \frac{1}{5}\begin{pmatrix} 0 \\ 0 \\ 5 \end{pmatrix} = \begin{pmatrix} 0 \\ 0 \\ 1 \end{pmatrix}.$$

Since $\mathbf{u}_1$ is the third natural basis vector, we may select

$$\mathbf{u}_2 = \begin{pmatrix} 1 \\ 0 \\ 0 \end{pmatrix} \quad \text{and} \quad \mathbf{u}_3 = \begin{pmatrix} 0 \\ 1 \\ 0 \end{pmatrix}.$$

Thus,

$$\mathbf{U} = \begin{pmatrix} 0 & 1 & 0 \\ 0 & 0 & 1 \\ 1 & 0 & 0 \end{pmatrix} \quad \text{and} \quad \mathbf{V} = \begin{pmatrix} \dfrac{3}{5} & -\dfrac{4}{5} \\ \dfrac{4}{5} & \dfrac{3}{5} \end{pmatrix},$$

consequently

$$\mathbf{U}^T\mathbf{AV} = \begin{pmatrix} 0 & 0 & 1 \\ 1 & 0 & 0 \\ 0 & 1 & 0 \end{pmatrix}\begin{pmatrix} 0 & 0 \\ 0 & 0 \\ 3 & 4 \end{pmatrix}\begin{pmatrix} \dfrac{3}{5} & -\dfrac{4}{5} \\ \dfrac{4}{5} & \dfrac{3}{5} \end{pmatrix}$$

$$= \begin{pmatrix} 3 & 4 \\ 0 & 0 \\ 0 & 0 \end{pmatrix}\begin{pmatrix} \dfrac{3}{5} & -\dfrac{4}{5} \\ \dfrac{4}{5} & \dfrac{3}{5} \end{pmatrix} = \begin{pmatrix} 5 & 0 \\ 0 & 0 \\ 0 & 0 \end{pmatrix},$$

which is 3×2, and $\mathbf{D} = (5)$ is a 1×1 matrix.

♦

7. The next application will introduce the *pseudoinverses* of (not necessarily square) real matrices. Let $\mathbf{A}$ be a real $m \times n$ matrix.

Definition 7.10. The pseudoinverse of a real matrix $\mathbf{A}$ is an $n \times m$ matrix $\mathbf{X}$ satisfying the following conditions:
(i) $\mathbf{AXA} = \mathbf{A}$;
(ii) $\mathbf{XAX} = \mathbf{X}$;
(iii) $\mathbf{AX}$ is symmetric;
(iv) $\mathbf{XA}$ is symmetric.

In the special case when $\mathbf{A}$ is a nonsingular square matrix, $\mathbf{X} = \mathbf{A}^{-1}$ satisfies these relations showing that this concept of pseudoinverses of general matrices is a generalization of that of inverses of nonsingular square matrices.
The following theorem guarantees the existence and the uniqueness of the pseudoinverse of any real matrix $\mathbf{A}$.

Theorem 7.14. The pseudoinverse of any $m \times n$ real matrix $\mathbf{A}$ exists and is unique.

Proof. Consider the singular value decomposition (7.27) of $\mathbf{A}$, and define matrix

$$X = V \begin{pmatrix} \dfrac{1}{\sigma_1} & & & O \\ & \ddots & & \\ & & \dfrac{1}{\sigma_r} & \\ O & & & O \end{pmatrix} U^T, \qquad (7.28)$$

where the second factor is an $n \times m$ matrix. We can easily show that this matrix satisfies the conditions of Definition 7.10. By using the notation of Theorem 7.13 we have

(i)

$$\mathbf{AXA} = \mathbf{U} \begin{pmatrix} \mathbf{D} & \mathbf{O} \\ \mathbf{O} & \mathbf{O} \end{pmatrix} \mathbf{V}^T \mathbf{V} \begin{pmatrix} \mathbf{D}^{-1} & \mathbf{O} \\ \mathbf{O} & \mathbf{O} \end{pmatrix} \mathbf{U}^T \mathbf{U} \begin{pmatrix} \mathbf{D} & \mathbf{O} \\ \mathbf{O} & \mathbf{O} \end{pmatrix} \mathbf{V}^T = \mathbf{U} \begin{pmatrix} \mathbf{D} & \mathbf{O} \\ \mathbf{O} & \mathbf{O} \end{pmatrix} \mathbf{V}^T = \mathbf{A},$$

since $\mathbf{V}^T\mathbf{V}$ and $\mathbf{U}^T\mathbf{U}$ are equal to identity matrices of sizes m and n, respectively.

(ii)

$$\mathbf{XAX} = \mathbf{V} \begin{pmatrix} \mathbf{D}^{-1} & \mathbf{O} \\ \mathbf{O} & \mathbf{O} \end{pmatrix} \mathbf{U}^T \mathbf{U} \begin{pmatrix} \mathbf{D} & \mathbf{O} \\ \mathbf{O} & \mathbf{O} \end{pmatrix} \mathbf{V}^T \mathbf{V} \begin{pmatrix} \mathbf{D}^{-1} & \mathbf{O} \\ \mathbf{O} & \mathbf{O} \end{pmatrix} \mathbf{U}^T = \mathbf{V} \begin{pmatrix} \mathbf{D}^{-1} & \mathbf{O} \\ \mathbf{O} & \mathbf{O} \end{pmatrix} \mathbf{U}^T = \mathbf{X}.$$

(iii) $\mathbf{AX} = \mathbf{U} \begin{pmatrix} \mathbf{D} & \mathbf{O} \\ \mathbf{O} & \mathbf{O} \end{pmatrix} \mathbf{V}^T \mathbf{V} \begin{pmatrix} \mathbf{D}^{-1} & \mathbf{O} \\ \mathbf{O} & \mathbf{O} \end{pmatrix} \mathbf{U}^T = \mathbf{U} \begin{pmatrix} \mathbf{I} & \mathbf{O} \\ \mathbf{O} & \mathbf{O} \end{pmatrix} \mathbf{U}^T,$

where $\mathbf{I}$ is the $r \times r$ identity matrix, and the second factor is $m \times m$. This matrix is obviously symmetric.

(iv) $\mathbf{XA} = \mathbf{V} \begin{pmatrix} \mathbf{D}^{-1} & \mathbf{O} \\ \mathbf{O} & \mathbf{O} \end{pmatrix} \mathbf{U}^T \mathbf{U} \begin{pmatrix} \mathbf{D} & \mathbf{O} \\ \mathbf{O} & \mathbf{O} \end{pmatrix} \mathbf{V}^T = \mathbf{V} \begin{pmatrix} \mathbf{I} & \mathbf{O} \\ \mathbf{O} & \mathbf{O} \end{pmatrix} \mathbf{V}^T,$

where the second factor is now $n \times n$. This matrix is obviously symmetric.

We will next show that matrix (7.28) is the only pseudoinverse of **A**. Assume in contrary that matrix **Y** also satisfies the conditions of Definition 7.10. From conditions (i) and (iv) we see that

$$\mathbf{A}^T = (\mathbf{AYA})^T = \mathbf{A}^T\mathbf{Y}^T\mathbf{A}^T = (\mathbf{YA})^T \mathbf{A}^T = \mathbf{YAA}^T. \qquad (7.29)$$

Conditions (ii) and (iii) imply that

$$\mathbf{Y}^T = (\mathbf{YAY})^T = \mathbf{Y}^T\mathbf{A}^T\mathbf{Y}^T = (\mathbf{AY})^T \mathbf{Y}^T = \mathbf{AYY}^T. \qquad (7.30)$$

Since **X** is also a pseudoinverse, it also satisfies the above relation:

$$\mathbf{A}^T = \mathbf{XAA}^T \quad \text{and} \quad \mathbf{X}^T = \mathbf{AXX}^T, \qquad (7.31)$$

from which and equation (7.29) we have

$$\mathbf{O} = \mathbf{A}^T - \mathbf{A}^T = \mathbf{YAA}^T - \mathbf{XAA}^T = (\mathbf{Y} - \mathbf{X})\mathbf{AA}^T,$$

that is,

$$(\mathbf{Y} - \mathbf{X})\mathbf{AA}^T (\mathbf{Y} - \mathbf{X})^T = \mathbf{O}.$$

This equation implies (see Exercise 1.25) that

$$(\mathbf{Y} - \mathbf{X})\mathbf{A} = \mathbf{O}. \qquad (7.32)$$

Notice in addition that from equations (7.30) and (7.31) we see that

$$\mathbf{Y} - \mathbf{X} = (\mathbf{YY}^T - \mathbf{XX}^T)\mathbf{A}^T,$$

and therefore

$$(\mathbf{Y} - \mathbf{X})\mathbf{C} = \mathbf{O} \qquad (7.33)$$

for any matrix $\mathbf{C}$ the columns of which are orthogonal to the columns of $\mathbf{A}$. Finally, observe that equations (7.32) and (7.33) imply that $\mathbf{Y} - \mathbf{X} = \mathbf{O}$, that is $\mathbf{Y} = \mathbf{X}$. Hence, the pseudoinverse is unique.

♣

Example 7.11. Consider again the matrix

$$\mathbf{A} = \begin{pmatrix} 0 & 0 \\ 0 & 0 \\ 3 & 4 \end{pmatrix}.$$

From the previous example we know that

$$\mathbf{U} = \begin{pmatrix} 0 & 1 & 0 \\ 0 & 0 & 1 \\ 1 & 0 & 0 \end{pmatrix} \text{ and } \mathbf{V} = \begin{pmatrix} \dfrac{3}{5} & \dfrac{4}{5} \\ -\dfrac{4}{5} & \dfrac{3}{5} \end{pmatrix},$$

therefore

$$\mathbf{X} = \begin{pmatrix} \dfrac{3}{5} & -\dfrac{4}{5} \\ \dfrac{4}{5} & \dfrac{3}{5} \end{pmatrix} \begin{pmatrix} \dfrac{1}{5} & 0 & 0 \\ 0 & 0 & 0 \end{pmatrix} \begin{pmatrix} 0 & 0 & 1 \\ 1 & 0 & 0 \\ 0 & 1 & 0 \end{pmatrix}$$

$$= \begin{pmatrix} \dfrac{3}{25} & 0 & 0 \\ \dfrac{4}{25} & 0 & 0 \end{pmatrix} \begin{pmatrix} 0 & 0 & 1 \\ 1 & 0 & 0 \\ 0 & 1 & 0 \end{pmatrix} = \begin{pmatrix} 0 & 0 & \dfrac{3}{25} \\ 0 & 0 & \dfrac{4}{25} \end{pmatrix}.$$

As an exercise, we verify the conditions of Definition 7.10:

(i)

$$\mathbf{AXA} = \begin{pmatrix} 0 & 0 \\ 0 & 0 \\ 3 & 4 \end{pmatrix} \begin{pmatrix} 0 & 0 & \dfrac{3}{25} \\ 0 & 0 & \dfrac{4}{25} \end{pmatrix} \begin{pmatrix} 0 & 0 \\ 0 & 0 \\ 3 & 4 \end{pmatrix} = \begin{pmatrix} 0 & 0 \\ 0 & 0 \\ 3 & 4 \end{pmatrix} \begin{pmatrix} \dfrac{9}{25} & \dfrac{12}{25} \\ \dfrac{12}{25} & \dfrac{16}{25} \end{pmatrix} = \begin{pmatrix} 0 & 0 \\ 0 & 0 \\ 3 & 4 \end{pmatrix} = \mathbf{A};$$

(ii)

$$\mathbf{XAX} = \begin{pmatrix} 0 & 0 & \dfrac{3}{25} \\ 0 & 0 & \dfrac{4}{25} \end{pmatrix} \begin{pmatrix} 0 & 0 \\ 0 & 0 \\ 3 & 4 \end{pmatrix} \begin{pmatrix} 0 & 0 & \dfrac{3}{25} \\ 0 & 0 & \dfrac{4}{25} \end{pmatrix} = \begin{pmatrix} \dfrac{9}{25} & \dfrac{12}{25} \\ \dfrac{12}{25} & \dfrac{16}{25} \end{pmatrix} \begin{pmatrix} 0 & 0 & \dfrac{3}{25} \\ 0 & 0 & \dfrac{4}{25} \end{pmatrix}$$

$$= \begin{pmatrix} 0 & 0 & \dfrac{3}{25} \\ 0 & 0 & \dfrac{4}{25} \end{pmatrix} = \mathbf{X};$$

(iii) $\mathbf{AX} = \begin{pmatrix} 0 & 0 \\ 0 & 0 \\ 3 & 4 \end{pmatrix} \begin{pmatrix} 0 & 0 & \dfrac{3}{25} \\ 0 & 0 & \dfrac{4}{25} \end{pmatrix} = \begin{pmatrix} 0 & 0 & 0 \\ 0 & 0 & 0 \\ 0 & 0 & 1 \end{pmatrix};$

(iv) $\mathbf{XA} = \begin{pmatrix} 0 & 0 & \dfrac{3}{25} \\ 0 & 0 & \dfrac{4}{25} \end{pmatrix} \begin{pmatrix} 0 & 0 \\ 0 & 0 \\ 3 & 4 \end{pmatrix} = \begin{pmatrix} \dfrac{9}{25} & \dfrac{12}{25} \\ \dfrac{12}{25} & \dfrac{16}{25} \end{pmatrix}.$

Notice that both **AX** and **XA** are symmetric.

◆

7.8 Exercises

1. Which special matrix classes the following matrices belong to?

$$\begin{pmatrix} 0 & 0 & 0 \\ 0 & 0 & 0 \end{pmatrix}, \begin{pmatrix} 1 & 0 \\ 0 & 2 \end{pmatrix}, \begin{pmatrix} 1 & 1 \\ 0 & 1 \end{pmatrix}, \begin{pmatrix} 1 & 1 \\ 1 & 0 \end{pmatrix}, \begin{pmatrix} 1 & 0 \\ 0 & 1 \end{pmatrix}, \begin{pmatrix} 0 & 2 \\ -2 & 0 \end{pmatrix}, \begin{pmatrix} 1 & -1 \\ 1 & 1 \end{pmatrix}.$$

2. Characterize all real 2×2 symmetric matrices such that $\mathbf{A}^2 = \mathbf{A}$.

3. Assume that $\mathbf{A}^3 = \mathbf{O}$. Show that

$$(I - A)^{-1} = I + A + A^2.$$

4. If $A + B$ is symmetric, simplify the following expression:

$$\left(3(A + B)^T + A \right)^T - 3B^T.$$

5. Find all values of a and b such that matrix

$$\begin{pmatrix} a - b & a + b \\ a + 2b & b \end{pmatrix}$$

is symmetric.

6. Repeat Example 7.1 for system

$$\begin{aligned}
x_1 + 2x_2 & = 3 \\
x_1 + x_2 - x_3 & = 2 \\
x_2 + 2x_3 & = 1 \\
x_3 - x_4 & = 0
\end{aligned}$$

7. Find the characteristic polynomial of the coefficient matrix of the previous Exercise. (Use recursion (7.4)).

8. Prove that if A is symmetric, then aA is also symmetric.

9. Prove that if A all B are $n \times n$, symmetric, and $AB = BA$, then AB is symmetric.

10. Prove that if $A + B$ and $A - B$ are symmetric, then A and B are symmetric.

11. Prove that if A is skew symmetric, that A^2 is symmetric.

12. Is the sum of two triangular matrices also triangular?

13. Is matrix $\begin{pmatrix} i & 0 \\ 0 & 1 \end{pmatrix}$ unitary?

14. Prove that the inverse of a symmetric real matrix is symmetric.

15. Repeat Example 7.2 for matrix

$$A = \begin{pmatrix} 1 & 1 \\ 1 & 2 \end{pmatrix}.$$

16. Repeat Example 7.3 for matrix

$$A = \begin{pmatrix} \dfrac{1}{\sqrt{5}} & \dfrac{2}{\sqrt{5}} \\ \dfrac{2}{\sqrt{5}} & -\dfrac{1}{\sqrt{5}} \end{pmatrix}.$$

17. Check which of the following matrices are positive definite, negative definite, positive semidefinite, or negative semidefine:

a) $\begin{pmatrix} 1 & 1 \\ 1 & 1 \end{pmatrix}$;

b) $\begin{pmatrix} 1 & 2 \\ 2 & 4 \end{pmatrix}$;

c) $\begin{pmatrix} 8 & 1 \\ 1 & 8 \end{pmatrix}$;

d) $\begin{pmatrix} -10 & 0 & 1 \\ 0 & -12 & 0 \\ 1 & 0 & -14 \end{pmatrix}.$

18. Illustrate the Perron-Frobenius Theorem for matrices a), b), and c) of the previous Exercise.

19. Use Theorem 7.10 and its corollary to bound the Perron-eigenvalues of matrices a), b), and c) of the previous Exercise.

20. Repeat Example 7.7 for matrix

$$A = \begin{pmatrix} 10 & -1 & -1 \\ 0 & 8 & -1 \\ -1 & 0 & 6 \end{pmatrix}.$$

21. Let

$$A = U \begin{pmatrix} D & O \\ O & O \end{pmatrix} V^T$$

be the SVD of on $m \times n$ real matrix A. Prove that

$$V^T (A^T A) V = \mathrm{diag}(\sigma_1^2, ..., \sigma_r^2, 0, ..., 0),$$

which is $n \times n$, and

$$U^T (A A^T) U = \mathrm{diag}(\sigma_1^2, ..., \sigma_r^2, 0, ..., 0),$$

which is $m \times m$.

22. Assume that A is an $n \times n$ real symmetric matrix with eigenvalues $\lambda_i, i = 1, 2, ..., n$. Show that the singular values of A are $|\lambda_i|$ for all nonzero eigenvalues.

23. (Continuation of the previous Exercise) Prove that an $n \times n$ real matrix is nonsingular if and only if all its singular values are nonzero.

24. Find the SVD for matrix

$$A = \begin{pmatrix} 1 & 0 \\ 0 & 1 \\ 0 & 0 \end{pmatrix}.$$

25. Find the pseudoinverse of the matrix given in the previous Exercise.

26. Assume that matrix **A** has the following special structure:

$$A = \begin{pmatrix} A_1 & O \\ O & O \end{pmatrix}.$$

Find the singular value decomposition of **A**.

27. Find the pseudoinverse of the matrix of the previous Exercise.

28. What is the pseudoinverse of an $n \times n$ diagonal matrix where $a_{ii} \neq 0$ $(i = 1, 2, \ldots, k)$ and $a_{ii} = 0$ $(i = k+1, \ldots, n)$.

Chapter 8

Elements of Matrix Analysis

8.1 Introduction

In Chapter 2 we have already introduced some elements of matrix analysis such as lengths of vectors, which give the possibility to introduce, for example, convergence and to estimate the speed of convergence. In this chapter we will consider these issues is more detail. We will first generalize vector lengths by introducing vector and matrix norms, and their applications in convergence analysis and perturbation analysis will be then outlined.

8.2 Vector Norms

Let V denote the set of all n-element real (or complex) vectors.

Definition 8.1. A real valued function $\mathbf{x} \mapsto \|\mathbf{x}\|$ defined on V is called a vector norm, if it satisfies the following conditions:

(i) $\|\mathbf{x}\| \geq 0$ for all $\mathbf{x} \in V$, and $\|\mathbf{x}\| = 0$ if and only if $\mathbf{x} = \mathbf{0}$;

(ii) $\|a\mathbf{x}\| = |a| \cdot \|\mathbf{x}\|$ for all $\mathbf{x} \in V$ and scalar a;

(iii) $\|\mathbf{x} + \mathbf{y}\| \leq \|\mathbf{x}\| + \|\mathbf{y}\|$ for all $\mathbf{x}, \mathbf{y} \in V$.

Notice that these properties imply that

$$\|-\mathbf{x}\| = \|\mathbf{x}\|$$

and

$$\left| \|\mathbf{x}\| - \|\mathbf{y}\| \right| < \|\mathbf{x} - \mathbf{y}\|. \qquad (8.1)$$

The first relation is a consequence of property (ii) with the selection of $a = -1$. The second inequality can be proved as follows. From property (iii) we have

$$\|\mathbf{x}\| = \|\mathbf{y} + (\mathbf{x} - \mathbf{y})\| \le \|\mathbf{y}\| + \|\mathbf{x} - \mathbf{y}\|,$$

which implies that

$$\|\mathbf{x}\| - \|\mathbf{y}\| \le \|\mathbf{x} - \mathbf{y}\|. \qquad (8.2)$$

By interchanging $\mathbf{x}$ and $\mathbf{y}$, we see that

$$\|\mathbf{y}\| - \|\mathbf{x}\| \le \|\mathbf{y} - \mathbf{x}\| = \|-(\mathbf{x} - \mathbf{y})\| = \|\mathbf{x} - \mathbf{y}\|. \qquad (8.3)$$

If $\|\mathbf{x}\| \ge \|\mathbf{y}\|$, then from (8.2) we have

$$\left| \|\mathbf{x}\| - \|\mathbf{y}\| \right| = \|\mathbf{x}\| - \|\mathbf{y}\| \le \|\mathbf{x} - \mathbf{y}\|,$$

and if $\|\mathbf{x}\| < \|\mathbf{y}\|$, then from (8.3) we obtain relation

$$\left| \|\mathbf{x}\| - \|\mathbf{y}\| \right| = \|\mathbf{y}\| - \|\mathbf{x}\| \le \|\mathbf{x} - \mathbf{y}\|.$$

Hence inequality (8.1) is verified.

Three particular vector norms will be introduced in the next examples.

Example 8.1. Let $x_1, \ldots, x_n$ denote the elements of vector $\mathbf{x}$. Consider the vector norm

$$\|\mathbf{x}\|_\infty = \max_i \{|x_i|\},$$

which is called the "infinite-norm" of **x**. We can easily verify that conditions (i), (ii), and (iii) are satisfied.

(i) $\|\mathbf{x}\|_\infty \geq 0$ obviously, and $\|\mathbf{x}\|_\infty = 0$ if and only if all $x_i = 0$ which means that $\mathbf{x} = \mathbf{0}$;

(ii) For all i, $|ax_i| = |a| \cdot |x_i|$, which implies the condition;

(iii) For all i, $|x_i + y_i| \leq |x_i| + |y_i|$.

Select i_0 such that

$$|x_{i_0} + y_{i_0}| = \max_i \{|x_i + y_i|\}.$$

Then

$$\|\mathbf{x} + \mathbf{y}\|_\infty = |x_{i_0} + y_{i_0}| \leq |x_{i_0}| + |y_{i_0}| \leq \max_i \{|x_i|\} + \max_i \{|y_i|\}$$
$$= \|\mathbf{x}\|_\infty + \|\mathbf{y}\|_\infty.$$

For example, if $n = 4$ and $\mathbf{x} = (-1, 2, -4, 3)^T$, then $\|\mathbf{x}\|_\infty = 4$, since

$$4 = \max\{|-1|; |2|; |-4|; |3|\}.$$

♦

Example 8.2. Consider next the "one-norm" of vectors defined as

$$\|\mathbf{x}\|_1 = \sum_{i=1}^n |x_i|.$$

This quantity also satisfies the conditions of Definition 8.1, so it can also be considered as a vector norm. Conditions (i), (ii) are trivially satisfied, and (iii) can be proven as follows. For all i,

$$|x_i + y_i| \leq |x_i| + |y_i|.$$

The sum of this inequality for $i = 1, 2, \dots, n$ yields

$$\sum_i \left| x_i + y_i \right| \le \sum_i \left| x_i \right| + \sum_i \left| y_i \right|$$

which is equivalent to condition (iii) applied to this norm. For example, if $n = 4$ and $\mathbf{x} = \left(-1, 2, -4, 3\right)^T$, then $\left\| \mathbf{x} \right\|_1 = 1 + 2 + 4 + 3 = 10$.

$\blacklozenge$

Example 8.3. In Section 2.4 we have introduced the length of n-element vectors, which can also be considered as a vector norm

$$\left\| \mathbf{x} \right\|_2 = \sqrt{\left| x_1 \right|^2 + \left| x_2 \right|^2 + \dots + \left| x_n \right|^2}.$$

This norm is known as the "two-norm" or "Euclidean-norm". Properties (i) and (ii) are obviously satisfied, and condition (iii) has been verified as the Corollary of Theorem 2.10. For example, if $n = 4$ and $\mathbf{x} = \left(-1, 2, -4, 3\right)^T$, then $\left\| \mathbf{x} \right\|_2 = \sqrt{1^2 + 2^2 + 4^2 + 3^2} = \sqrt{30}$.

$\blacklozenge$

Notice that all these norms are special cases of the Hölder-norm:

$$\left\| \mathbf{x} \right\|_p = \left\{ \sum_{i=1}^n \left| x_i \right|^p \right\}^{1/p} \tag{8.4}$$

with some $p \ge 1$. A relation between the 1,2, and ∞-norms can be given as follows.

Theorem 8.1. For all $\mathbf{x} \in V$,

$$\left\| \mathbf{x} \right\|_2 \le \left\| \mathbf{x} \right\|_1 \le \sqrt{n} \left\| \mathbf{x} \right\|_2$$
$$\left\| \mathbf{x} \right\|_\infty \le \left\| \mathbf{x} \right\|_2 \le \sqrt{n} \left\| \mathbf{x} \right\|_\infty$$
$$\left\| \mathbf{x} \right\|_\infty \le \left\| \mathbf{x} \right\|_1 \le n \left\| \mathbf{x} \right\|_\infty.$$

Proof. The first parts of these relations are obvious, since

$$\|\mathbf{x}\|_2^2 = \sum_{i=1}^n |x_i|^2 \leq \left(\sum_{i=1}^n |x_i|\right)^2 = \|\mathbf{x}\|_1^2,$$

$$\|\mathbf{x}\|_\infty^2 = \max_i\{|x_i|^2\} \leq \sum_{i=1}^n |x_i|^2 = \|\mathbf{x}\|_2^2,$$

and

$$\|\mathbf{x}\|_\infty = \max_i\{|x_i|\} \leq \sum_{i=1}^n |x_i| = \|\mathbf{x}\|_1.$$

The second parts can be shown as follows:

$$\|\mathbf{x}\|_1^2 = \left(\sum_{i=1}^n |x_i|\right)^2 = \left(\sum_{i=1}^n 1 \cdot |x_i|\right)^2 \leq \left(\sum_{i=1}^n 1^2\right)\left(\sum_{i=1}^n |x_i|^2\right) = n \cdot \|\mathbf{x}\|_2^2,$$

where we used the Cauchy-Schwarz inequality.
Similarly,

$$\|\mathbf{x}\|_2^2 = \sum_{i=1}^n |x_i|^2 \leq n \cdot \max_i\{|x_i|^2\} = n \cdot \|\mathbf{x}\|_\infty^2,$$

and

$$\|\mathbf{x}\|_1 = \left(\sum_{i=1}^n |x_i|\right) \leq n \cdot \max_i\{|x_i|\} = n \cdot \|\mathbf{x}\|_\infty.$$

♣

Remark. We can easily prove that all vector norms are equivalent to each other, that is, if $\|...\|$ and $\|...\|^*$ are two vector norms in V, then there exist constants K_1, K_2 such that for all $\mathbf{x} \in V$,

$$K_1\|\mathbf{x}\| \leq \|\mathbf{x}\|^* \leq K_2\|\mathbf{x}\|.$$

Proof. Notice first that it is sufficient to prove that any vector norm is equivalent to $\|...\|_\infty$, since if

$$K_1 \parallel \mathbf{x} \parallel_\infty \le \parallel \mathbf{x} \parallel \le K_2 \parallel \mathbf{x} \parallel_\infty$$

and

$$\overline{K}_1 \parallel \mathbf{x} \parallel_\infty \le \parallel \mathbf{x} \parallel^* \le \overline{K}_2 \parallel \mathbf{x} \parallel_\infty$$

with positive constants $K_1, K_2, \overline{K}_1, \overline{K}_2$ then

$$\parallel \mathbf{x} \parallel \le K_2 \parallel \mathbf{x} \parallel_\infty \le K_2 \frac{1}{\overline{K}_1} \parallel \mathbf{x} \parallel^*$$

and

$$\parallel \mathbf{x} \parallel \ge K_1 \parallel \mathbf{x} \parallel_\infty \ge K_1 \frac{1}{\overline{K}_2} \parallel \mathbf{x} \parallel^*.$$

Relation (8.1) implies that any norm is a continous function of $\mathbf{x}$. Clearly, with any vector norm

$$\parallel \mathbf{x} \parallel = \left\| \sum_{i=1}^{n} x_i \mathbf{e}_i \right\| \le \sum_{i=1}^{n} |x_i| \cdot \parallel e_i \parallel \le \max_i |x_i| \cdot \sum_{i=1}^{n} \parallel \mathbf{e}_i \parallel = K_2 \parallel \mathbf{x} \parallel_\infty$$

with

$$K_2 = \sum_{i=1}^{n} \parallel \mathbf{e}_i \parallel$$

where $\mathbf{e}_i$ ($i = 1, 2, ..., n$) are the natural basis vectors introduced in Example 2.10. Define set $S = \left\{ \mathbf{x} \mid \max_i |x_i| = 1 \right\}$ which does not contain the origin and is compact. If $\mathbf{x}$ is any vector, then $\mathbf{z} = \mathbf{x} \cdot \dfrac{1}{\parallel x \parallel_\infty} \in S$, and since S is compact and the norm is continuous, there is $K_1 > 0$ such that for all $\mathbf{z} \in S$, $\parallel \mathbf{z} \parallel \ge K_1$. That is

$$\left\| \mathbf{x} \cdot \frac{1}{\|\mathbf{x}\|_{\infty}} \right\| \geq K_1$$

implying that

$$\|\mathbf{x}\| \geq K_1 \|\mathbf{x}\|_{\infty}$$

Thus

$$K_1 \|\mathbf{x}\|_{\infty} \leq \|\mathbf{x}\| \leq K_2 \|\mathbf{x}\|_{\infty}$$

for all $\mathbf{x} \in V$.

Let $\mathbf{A}$ be a nonsingular $n \times n$ real or complex matrix, and assume that $\|...\|$ is a given vector norm. Consider the mapping: $\mathbf{x} \mapsto \|\mathbf{A}\mathbf{x}\|$ for all n-element vectors $\mathbf{x}$. We can easily verify that this mapping also satisfies the conditions of Definition 8.1, that is, it can also be considered as a vector norm:

(i) $\|\mathbf{A}\mathbf{x}\| \geq 0$, since $\mathbf{A}\mathbf{x}$ is an n-element vector, and $\|\mathbf{A}\mathbf{x}\| = 0$ if and only if $\mathbf{A}\mathbf{x} = \mathbf{0}$, which holds if and only if $\mathbf{x} = \mathbf{0}$, since $\mathbf{A}$ is nonsingular.

(ii) $\|a\mathbf{A}\mathbf{x}\| = |a| \cdot \|\mathbf{A}\mathbf{x}\|$, since $\|...\|$ is a vector norm.

(iii) $\|\mathbf{A}(\mathbf{x} + \mathbf{y})\| = \|\mathbf{A}\mathbf{x} + \mathbf{A}\mathbf{y}\| \leq \|\mathbf{A}\mathbf{x}\| + \|\mathbf{A}\mathbf{y}\|$,

where we used the triangle inequality for norm $\|...\|$.

Notice, that in the case of a singular matrix $\mathbf{A}$ condition (i) might fail, since $\mathbf{A}\mathbf{x} = \mathbf{0}$ does not imply that $\mathbf{x} = \mathbf{0}$, therefore $\|\mathbf{A}\mathbf{x}\|$ has zero value for all nonzero vectors $\mathbf{x} \in N(\mathbf{A})$.

8.3 Matrix Norms

Consider now the set M of all $m \times n$ real (or complex) matrices.

Definition 8.2. A real-valued function $\mathbf{A} \mapsto \|\mathbf{A}\|$ defined on M is called a matrix norm, if it satisfies the following conditions:

(i) $\|\mathbf{A}\| \geq 0$ for all $\mathbf{A} \in M$, and $\|\mathbf{A}\| = 0$ if and only if $\mathbf{A} = \mathbf{O}$;

(ii) $\|a\mathbf{A}\| = |a| \cdot \|\mathbf{A}\|$ for all $\mathbf{A} \in M$ and scalar a;

(iii) For all $\mathbf{A}, \mathbf{B} \in M$,

$$\|\mathbf{A} + \mathbf{B}\| \le \|\mathbf{A}\| + \|\mathbf{B}\|.$$

A special class of matrix norms can be defined in the following way.

Definition 8.3. Let $\|\cdots\|$ be a given vector norm. For all $\mathbf{A} \in M$, define

$$\|\mathbf{A}\| = \max_{\mathbf{x} \neq 0} \frac{\|\mathbf{A}\mathbf{x}\|}{\|\mathbf{x}\|}, \tag{8.5}$$

which is called the *matrix norm subordinate* to the given vector norm.

In order to show that (8.5) defines a matrix norm, we have to verify that it satisfies the conditions of Definition 8.2:

(i) $\|\mathbf{A}\| \ge 0$, since $\|\mathbf{A}\mathbf{x}\| \ge 0$ and for $\mathbf{x} \neq 0$, $\|\mathbf{x}\| > 0$, furthermore $\|\mathbf{A}\| = 0$ if and only if $\|\mathbf{A}\mathbf{x}\| = 0$ for all $\mathbf{x} \neq \mathbf{0}$, which holds if and only if $\mathbf{A}$ is the zero matrix.

(ii) $\|a\mathbf{A}\| = \max\limits_{\mathbf{x} \neq 0} \dfrac{\|a\mathbf{A}\mathbf{x}\|}{\|\mathbf{x}\|} = \max\limits_{\mathbf{x} \neq 0} \dfrac{|a| \cdot \|\mathbf{A}\mathbf{x}\|}{\|\mathbf{x}\|} = |a| \cdot \max\limits_{\mathbf{x} \neq 0} \dfrac{\|\mathbf{A}\mathbf{x}\|}{\|\mathbf{x}\|} = |a| \cdot \|\mathbf{A}\|$;

(iii) For all $\mathbf{A}$ and $\mathbf{B} \in M$,

$$\|\mathbf{A} + \mathbf{B}\| = \max_{\mathbf{x} \neq 0} \frac{\|(\mathbf{A} + \mathbf{B})\mathbf{x}\|}{\|\mathbf{x}\|} \le \max_{\mathbf{x} \neq 0} \frac{\|\mathbf{A}\mathbf{x}\| + \|\mathbf{B}\mathbf{x}\|}{\|\mathbf{x}\|}$$

$$\le \max_{\mathbf{x} \neq 0} \frac{\|\mathbf{A}\mathbf{x}\|}{\|\mathbf{x}\|} + \max_{\mathbf{x} \neq 0} \frac{\|\mathbf{B}\mathbf{x}\|}{\|\mathbf{x}\|} = \|\mathbf{A}\| + \|\mathbf{B}\|.$$

Property (ii) implies that the matrix norm subordinate to a given vector norm can also be defined as

$$\|\mathbf{A}\| = \max\left\{\|\mathbf{A}\mathbf{x}\| \mid \mathbf{x} \in V, \|\mathbf{x}\| = 1\right\}$$

In the next three examples the subordinate matrix norms to the 1, 2, and ∞ vector norms will be determined.

Example 8.4. Consider first the ∞-vector norm. Assume that

$$\|\mathbf{x}\|_\infty = \max_i\left\{|x_i|\right\} = 1.$$

Then

$$\|\mathbf{A}\mathbf{x}\|_\infty = \max_i\left|\sum_{j=1}^n a_{ij} x_j\right| \le \max_i \sum_{j=1}^n |a_{ij}| \cdot |x_j| \le \max_i \sum_{j=1}^n |a_{ij}|. \qquad (8.6)$$

Therefore

$$\|\mathbf{A}\|_\infty \le \max_i \sum_{j=1}^n |a_{ij}|.$$

Assume that maximum is obtained for $i = i_0$ on the right hand side. Select vector **x** such that

$$x_j = \begin{cases} 0 & \text{if } a_{i_0 j} = 0 \\[2mm] \dfrac{\overline{a_{i_0 j}}}{|a_{i_0 j}|} & \text{if } a_{i_0 j} \ne 0, \end{cases}$$

where overbar denotes complex conjugate. In the real case we select

$$x_j = \begin{cases} 0 & \text{if } a_{i_0 j} = 0 \\ 1 & \text{if } a_{i_0 j} > 0 \\ -1 & \text{if } a_{i_0 j} < 0. \end{cases}$$

Then

$$a_{i_0 j} x_j = \left| a_{i_0 j} \right|,$$

therefore in inequality (8.6) we have equality. Hence, the subordinate matrix norm to the ∞ vector norm is the following:

$$\left\| \mathbf{A} \right\|_\infty = \max_i \sum_{j=1}^n \left| a_{ij} \right|, \tag{8.7}$$

which is sometimes called the *row-norm* of matrix $\mathbf{A}$.

$\blacklozenge$

Example 8.5. Consider next the 1-vector norm. Assume that

$$\left\| \mathbf{x} \right\|_1 = \sum_{j=1}^n \left| x_j \right| = 1.$$

Then

$$\left\| \mathbf{A}\mathbf{x} \right\|_1 = \sum_{i=1}^m \left| \sum_{j=1}^n a_{ij} x_j \right| \le \sum_{i=1}^m \sum_{j=1}^n \left| a_{ij} x_j \right|$$

$$= \sum_{j=1}^n \sum_{i=1}^m \left| a_{ij} \right| \cdot \left| x_j \right| = \sum_{j=1}^n \left(\left| x_j \right| \sum_{i=1}^m \left| a_{ij} \right| \right) \tag{8.8}$$

$$\le \sum_{j=1}^n \left| x_j \right| \cdot \left(\max_j \sum_{i=1}^m \left| a_{ij} \right| \right) = \max_j \sum_{i=1}^m \left| a_{ij} \right|.$$

Therefore,

$$\left\| \mathbf{A} \right\|_1 \le \max_j \sum_{i=1}^m \left| a_{ij} \right|.$$

Assume next that maximum occurs for $j = j_0$. Select vector $\mathbf{x} = \mathbf{e}_{j_0}$, that is,

$$x_j = \begin{cases} 1 & \text{if} \quad j = j_0 \\ 0 & \text{otherwise.} \end{cases}$$

Then in inequality (8.8) we have equality, therefore

$$\|\mathbf{A}\|_1 = \max_j \sum_{i=1}^m |a_{ij}|, \tag{8.9}$$

which is sometimes called the *column-norm* of matrix $\mathbf{A}$.

◆

Example 8.6. Consider next the 2-vector norm

$$\|\mathbf{x}\|_2 = \sqrt{|x_1|^2 + |x_2|^2 + \ldots + |x_n|^2} = \sqrt{\overline{\mathbf{x}}^T \mathbf{x}},$$

where overbar denotes complex conjugate. Assume that $\|\mathbf{x}\|_2 = 1$. Then

$$\|\mathbf{Ax}\|_2^2 = \overline{\mathbf{x}}^T \overline{\mathbf{A}}^T \mathbf{Ax} = \overline{\mathbf{x}}^T \cdot \mathbf{Hx},$$

where matrix $\mathbf{H}$ is an $n \times n$ self-adjoint matrix. Theorem 7.2 implies that all eigenvalues of $\mathbf{H}$ are real. Let $\lambda_1 \geq \lambda_2 \geq \ldots \geq \lambda_n$ denote the eigenvalues of $\mathbf{H}$. Notice that $\mathbf{H}$ is positive semidefinite, since with any vector $\mathbf{x}$

$$\overline{\mathbf{x}}^T \mathbf{Hx} = \overline{\mathbf{x}}^T \overline{\mathbf{A}}^T \mathbf{Ax} = \left(\overline{\mathbf{Ax}} \right)^T (\mathbf{Ax}) = \|\mathbf{Ax}\|_2^2 \geq 0,$$

therefore $\lambda_1 \geq 0$. From Section 7.4 we know that there is a unitary matrix $\mathbf{U}$ such that

$$\mathbf{H} = \overline{\mathbf{U}}^T \mathbf{DU},$$

where $\mathbf{D}$ is a diagonal matrix with diagonal elements $\lambda_1, \ldots, \lambda_n$. Then

$$\|\mathbf{Ax}\|_2^2 = \overline{\mathbf{x}}^T \overline{\mathbf{U}}^T \mathbf{DUx} = \overline{\mathbf{z}}^T \mathbf{Dz} = \sum_{i=1}^n \lambda_i |z_i|^2 \leq \lambda_1 \sum_{i=1}^n |z_i|^2,$$

where $\mathbf{z} = \mathbf{Ux}$. Since $\mathbf{U}$ is unitary, $\|\mathbf{z}\|_2 = \|\mathbf{Ux}\|_2 = \|\mathbf{x}\|_2 = 1$. Therefore

$$\|\mathbf{Ax}\|_2^2 \leq \lambda_1,$$

that is,

$$\|\mathbf{Ax}\|_2 \leq \sqrt{\lambda_1}.$$

Select $\mathbf{x}$ as an eigenvector of $\mathbf{H}$ associated to λ_1 such that $\|\mathbf{x}\|_2 = 1$. Then

$$\|\mathbf{A}\mathbf{x}\|_2^2 = \overline{\mathbf{x}}^T \mathbf{H}\mathbf{x} = \overline{\mathbf{x}}^T \lambda_1 \mathbf{x} = \lambda_1 \|\mathbf{x}\|_2^2 = \lambda_1,$$

which shows that

$$\|\mathbf{A}\|_2 = \sqrt{\lambda_1}. \tag{8.10}$$

which is called the *Euclidean-norm* of matrix $\mathbf{A}$.

♦

In the following example we introduce a matrix norm, which differs from the ones shown in the previous examples. In many practical applications it is used instead of the Euclidean matrix norm, since it is easy to obtain.

Example 8.7. For an $m \times n$ real (or complex) matrix introduce the norm

$$\|\mathbf{A}\|_F = \left\{ \sum_{i=1}^{m} \sum_{j=1}^{n} |a_{ij}|^2 \right\}^{1/2},$$

which is called the *Frobenius-norm* of $\mathbf{A}$. Notice that it is the 2-vector norm of the $m \cdot n$ element vector

$$\mathbf{a} = \left(a_{11}, \dots, a_{1n}, \; a_{21}, \dots, a_{2n}, \dots, a_{m1}, \dots, a_{mn} \right)^T,$$

therefore it satisfies the properties of Definition 8.2. We will next show that for all $\mathbf{A}$,

$$\|\mathbf{A}\|_F \geq \|\mathbf{A}\|_2.$$

Assume that $\|\mathbf{x}\|_2 = 1$, then using the Cauchy-Schwarz inequality (Theorem 2.10) we have

$$\|\mathbf{A}\mathbf{x}\|_2^2 = \sum_{i=1}^{m} \left| \sum_{j=1}^{n} a_{ij} x_j \right|^2 \le \sum_{i=1}^{m} \left\{ \sum_{j=1}^{n} |a_{ij}|^2 \cdot \sum_{j=1}^{n} |x_j|^2 \right\}$$

$$= \left(\sum_{i=1}^{m} \sum_{j=1}^{n} |a_{ij}|^2 \right) \left(\sum_{j=1}^{n} |x_j|^2 \right) = \|\mathbf{A}\|_F^2 \,,$$

since the second factor equals one. Therefore

$$\|\mathbf{A}\|_2 = \max \left\{ \|\mathbf{A}\mathbf{x}\|_2 \mid \|\mathbf{x}\|_2 = 1 \right\} \le \|\mathbf{A}\|_F \,.$$

◆

A very useful property of the above introduced matrix norms is given next.

Theorem 8.2. For any real or complex matrices $\mathbf{A}$, $\mathbf{B}$ such that $\mathbf{A} \cdot \mathbf{B}$ is defined,

$$\|\mathbf{A} \cdot \mathbf{B}\|_p \le \|\mathbf{A}\|_p \cdot \|\mathbf{B}\|_p \quad (p = 1,\, 2,\, \infty,\, F)$$

Proof. Assume first that $p = 1$, 2, or ∞. Then for any vector $\mathbf{x}$ such that $\mathbf{B} \cdot \mathbf{x}$ exists and $\|\mathbf{x}\|_p = 1$,

$$\|(\mathbf{A} \cdot \mathbf{B})\mathbf{x}\|_p = \|\mathbf{A} \cdot (\mathbf{B} \cdot \mathbf{x})\|_p \le \|\mathbf{A}\|_p \cdot \|\mathbf{B} \cdot \mathbf{x}\|_p$$

$$\le \|\mathbf{A}\|_p \cdot \|\mathbf{B}\|_p \cdot \|\mathbf{x}\|_p = \|\mathbf{A}\|_p \cdot \|\mathbf{B}\|_p \,,$$

which implies the assertion. If $p = F$, then assuming that $\mathbf{A}$ is $m \times n$ and $\mathbf{B}$ is $n \times r$,

$$\|\mathbf{A} \cdot \mathbf{B}\|_F^2 = \sum_{i=1}^{m} \sum_{j=1}^{r} \left| \sum_{l=1}^{n} a_{il} b_{lj} \right|^2 \le \sum_{i=1}^{m} \sum_{j=1}^{r} \left(\sum_{l=1}^{n} |a_{il}|^2 \cdot \sum_{l=1}^{n} |b_{lj}|^2 \right)$$

$$= \left(\sum_{i=1}^{m} \sum_{l=1}^{n} |a_{il}|^2 \right) \left(\sum_{l=1}^{n} \sum_{j=1}^{r} |b_{lj}|^2 \right) = \|\mathbf{A}\|_F^2 \cdot \|\mathbf{B}\|_F^2 \,.$$

♣

Remark. The assertion of the theorem holds for any subordinate matrix norm.

Corollary. Let $\mathbf{A}$ be an $n \times n$ real or complex matrix. Then for $p = 1, 2, \infty, F$ and $k = 1, 2, \ldots,$

$$\left\| \mathbf{A}^k \right\|_p < \left(\left\| \mathbf{A} \right\|_p \right)^k. \tag{8.11}$$

Theorem 8.3. Let $\mathbf{A}$ be any real or complex $n \times n$ matrix, and assume that λ is an eigenvalue of $\mathbf{A}$. Then

$$|\lambda| \leq \|\mathbf{A}\|_p \quad (p = 1, 2, \infty, F).$$

Proof. Let $p = 1, 2$ or ∞, and assume that $\mathbf{x} \neq \mathbf{0}$ is an eigenvector of $\mathbf{A}$ associated to λ. Then the eigenvector equation implies that

$$|\lambda| \cdot \|\mathbf{x}\|_p = \|\lambda \mathbf{x}\|_p = \|\mathbf{A}\mathbf{x}\|_p \leq \|\mathbf{A}\|_p \cdot \|\mathbf{x}\|_p.$$

Since $\mathbf{x} \neq \mathbf{0}$, $\|\mathbf{x}\|_p \neq 0$. Divide this inequality by $\|\mathbf{x}\|_p$ to obtain the assertion. For $p = F$, the assertion follows from the fact that $|\lambda| \leq \|\mathbf{A}\|_2 \leq \|\mathbf{A}\|_F$.

♣

Remark 1. This theorem provides simple bounds for the eigenvalues of real or complex matrices:

$$|\lambda| \leq \max_i \sum_{j=1}^n |a_{ij}|;$$

$$|\lambda| \leq \max_j \sum_{i=1}^n |a_{ij}|;$$

$$|\lambda| \leq \left\{ \sum_{i=1}^n \sum_{j=1}^n |a_{ij}|^2 \right\}^{1/2}.$$

Example 8.8. Consider the 3×3 real matrix

$$A = \begin{pmatrix} 1 & 2 & 1 \\ 1 & 1 & 1 \\ 1 & 1 & 1 \end{pmatrix}.$$

Notice that

$$\|A\|_1 = \max\{1+1+1;\ 2+1+1;\ 1+1+1\} = 4;$$
$$\|A\|_\infty = \max\{1+2+1;\ 1+1+1;\ 1+1+1\} = 4;$$

and

$$\|A\|_F = \{1+4+1+1+1+1+1+1+1\}^{1/2} = \sqrt{12}.$$

Therefore for all eigenvalues λ of A, $|\lambda| \le \sqrt{12} \approx 3.464$. The true eigenvalues are:

$$\lambda_1 = \frac{3-\sqrt{13}}{2} = -0.303, \quad \lambda_2 = \frac{3+\sqrt{13}}{2} = 3.303, \quad \lambda_3 = 0.$$

$\blacklozenge$

Remark 2. For nonnegative matrices, in Theorem 7.10 we presented lower and upper bounds for the largest eigenvalue. The upper bounds are the ∞-norms of the matrices. Therefore the above theorem generalizes this upper bound for any $n \times n$ real or complex matrices.

A nice and practically important refinement of the previous theorem for the 1-norm and ∞-norm is presented in the following result, which is known as the *Gerschgorin-disk theorem*.

Theorem 8.4. Let A be a real or complex $n \times n$ matrix, and

$$r_i = \sum_{\substack{j=1 \\ j \ne i}}^{n} |a_{ij}| \qquad (i = 1, 2, \dots, n).$$

For all i, define the disk

$$B_i = \left\{ z \, \big| \, |z - a_{ii}| \leq r_i \right\}$$

on the complex plane. Then all eigenvalues of $\mathbf{A}$ can be found in the union $B_1 \cup B_2 \cup ... \cup B_n$ of those disks.

Proof. The eigenvector equation can be written as

$$\lambda x_i = \sum_{j=1}^{n} a_{ij} x_j \quad (i = 1, 2, ..., n), \tag{8.12}$$

where λ is an eigenvalue of $\mathbf{A}$ and $\mathbf{x} = (x_i)$ is an associated eigenvector to λ. Let i_0 be selected so that

$$\left| x_{i_0} \right| = \max_i |x_i|,$$

then $x_{i_0} \neq 0,$ and from equation (8.12) we have

$$\lambda = a_{i_0 i_0} + \sum_{\substack{j=1 \\ j \neq i_0}}^{n} a_{i_0 j} \frac{x_j}{x_{i_0}},$$

which implies that

$$\left| \lambda - a_{i_0 i_0} \right| \leq \sum_{\substack{j=1 \\ j \neq i_0}}^{n} \left| a_{i_0 j} \right|,$$

that is, $\lambda \in B_{i_0}$. Hence λ is in the union of all disks B_i.

♣

Corollary. Since the eigenvalues of $\mathbf{A}$ and $\mathbf{A}^T$ are the same, we can apply the theorem to $\mathbf{A}^T$ to obtain new Gerschgorin disks. For $j = 1, 2, ..., n$ let

$$s_j = \sum_{\substack{i=1 \\ i \neq j}}^{n} |a_{ij}|,$$

and

$$D_j = \left\{ z \,\middle|\, \left| z - a_{jj} \right| \le s_j \right\}.$$

Then all eigenvalues of $\mathbf{A}$ can be found in the union $D_1 \cup D_2 \cup ... \cup D_n$ of these disks. The application of the theorem and its corollary is illustrated in the next example.

Example 8.9. Consider again the 3×3 matrix

$$\mathbf{A} = \begin{pmatrix} 1 & 2 & 1 \\ 1 & 1 & 1 \\ 1 & 1 & 1 \end{pmatrix},$$

which was the subject of the previous example. In this case, $a_{11} = a_{22} = a_{33} = 1$, $r_1 = 3$, $r_2 = r_3 = 2$, $s_1 = s_3 = 2$, $s_2 = 3$. Therefore

$$B_1 = D_2 = \left\{ z \,\middle|\, \left| z - 1 \right| < 3 \right\},$$
$$B_2 = B_3 = D_1 = D_3 = \left\{ z \,\middle|\, \left| z - 1 \right| < 2 \right\}.$$

The union of these disk is B_1, which therefore contains all eigenvalues of $\mathbf{A}$. This domain and the true eigenvalues are shown in Figure 8.1.

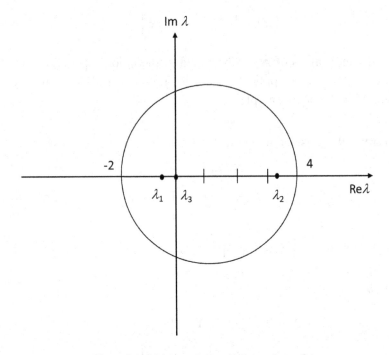

Figure 8.1. Disk containing all eigenvalues of **A**

◆

8.4 Applications

In this section some useful applications of the properties of vector and matrix norms will be presented.

1. In our first application we derive a *bound for the inverses* of matrices of a special class. Assume that $\|...\|$ is a subordinate matrix norm such that $\|\mathbf{I}\|=1$, where **I** is the identity matrix. Let **A** be a real or complex $n \times n$ matrix, and assume that $\|\mathbf{A}\|<1$. We will prove that $\mathbf{I} - \mathbf{A}$ is nonsingular, and

$$\left\|(\mathbf{I}-\mathbf{A})^{-1}\right\| \le \frac{1}{1-\|\mathbf{A}\|}. \tag{8.13}$$

Notice first that the assumption implies that for all eigenvalues λ_i of **A**,

$|\lambda_i| < 1$ therefore the eigenvalues $1 - \lambda_i$ of $\mathbf{I} - \mathbf{A}$ are all nonzero. Thus, $\mathbf{I} - \mathbf{A}$ is nonsingular. From Application 3 of Section 6.8 we know that

$$\left(\mathbf{I} - \mathbf{A}\right)^{-1} = \mathbf{I} + \mathbf{A} + \mathbf{A}^2 + \mathbf{A}^3 + \dots.$$

Property (iii) of Definition 8.2 and Theorem 8.2 imply that

$$\left\|\left(\mathbf{I} - \mathbf{A}\right)^{-1}\right\| \le \|\mathbf{I}\| + \sum_{k=1}^{\infty} \|\mathbf{A}^k\| \le 1 + \sum_{k=1}^{\infty} \|\mathbf{A}\|^k = \frac{1}{1 - \|\mathbf{A}\|}.$$

2. In this application the *relative errors of inverses* of approximating matrices will be examined. Assume that $\mathbf{A}$ is nonsingular, and $\mathbf{A} - \mathbf{E}$ is a small perturbation of $\mathbf{A}$ such that $\|\mathbf{A}^{-1}\mathbf{E}\| < 1$. We will now prove that $\mathbf{A} - \mathbf{E}$ is also nonsingular and

$$\frac{\left\|\left(\mathbf{A} - \mathbf{E}\right)^{-1} - \mathbf{A}^{-1}\right\|}{\|\mathbf{A}^{-1}\|} \le \frac{\|\mathbf{A}^{-1}\mathbf{E}\|}{1 - \|\mathbf{A}^{-1}\mathbf{E}\|}. \tag{8.14}$$

Notice first that

$$\mathbf{A} - \mathbf{E} = \mathbf{A}\left(\mathbf{I} - \mathbf{A}^{-1}\mathbf{E}\right).$$

The first factor is nonsingular by our assumption, and the second factor is nonsingular as the consequence of the previous application. Thus, $\mathbf{A} - \mathbf{E}$ is nonsingular. From the above equation we have

$$\left(\mathbf{A} - \mathbf{E}\right)^{-1} = \left(\mathbf{I} - \mathbf{A}^{-1}\mathbf{E}\right)^{-1} \mathbf{A}^{-1} = \left(\mathbf{I} + \left(\mathbf{A}^{-1}\mathbf{E}\right) + \left(\mathbf{A}^{-1}\mathbf{E}\right)^2 + \dots\right)\mathbf{A}^{-1}.$$

That is,

$$\left(\mathbf{A} - \mathbf{E}\right)^{-1} - \mathbf{A}^{-1} = \sum_{k=1}^{\infty} \left(\mathbf{A}^{-1}\mathbf{E}\right)^k \mathbf{A}^{-1}.$$

Using property (iii) of Definition 8.2 and Theorem 8.2 we obtain the inequality

$$\left\|\left(\mathbf{A}-\mathbf{E}\right)^{-1}-\mathbf{A}^{-1}\right\| \leq \sum_{k=1}^{\infty}\left\|\left(\mathbf{A}^{-1}\mathbf{E}\right)^{k}\mathbf{A}^{-1}\right\| \leq \left(\sum_{k=1}^{\infty}\left\|\mathbf{A}^{-1}\mathbf{E}\right\|^{k}\right)\left\|\mathbf{A}^{-1}\right\|$$

$$=\frac{\left\|\mathbf{A}^{-1}\mathbf{E}\right\|}{1-\left\|\mathbf{A}^{-1}\mathbf{E}\right\|}\left\|\mathbf{A}^{-1}\right\|,$$

which implies the assertion. Notice that inequality (8.14) gives an upper bound for the relative error of inverse of an approximating matrix, if we consider $\mathbf{A}$ as the exact matrix and $\mathbf{A}-\mathbf{E}$ as its approximation.

3. Our next application is concerned with the solutions of systems of *linear equations with approximating data*. Assume that linear system $\mathbf{Ax} = \mathbf{b}$ is to be solved with an $n \times n$ nonsingular matrix $\mathbf{A}$. Let $\mathbf{x}$ denote the solution. We discuss first the effect of the perturbation in $\mathbf{b}$. Assume that $\mathbf{b} \neq \mathbf{0}$ and it is approximated by $\mathbf{b} + \mathbf{e}$, where $\mathbf{e}$ can be considered as the error term. Let $\mathbf{x}^*$ denote the solution of the approximating system, and let $\varepsilon = \mathbf{x}^* - \mathbf{x}$. Since $\mathbf{x}$ solves the exact system,

$$\mathbf{Ax} = \mathbf{b},$$

and since $\mathbf{x}^*$ is the solution of the approximating system,

$$\mathbf{A}\left(\mathbf{x}+\varepsilon\right) = \mathbf{b}+\mathbf{e}.$$

Subtract the two equations to see that

$$\mathbf{A}\varepsilon = \mathbf{e},$$

that is,

$$\varepsilon = \mathbf{A}^{-1}\mathbf{e},$$

which implies that

$$\left\|\varepsilon\right\| \leq \left\|\mathbf{A}^{-1}\right\| \cdot \left\|\mathbf{e}\right\|.$$

It is clear that

$$\left\|\mathbf{b}\right\| = \left\|\mathbf{Ax}\right\| \leq \left\|\mathbf{A}\right\| \cdot \left\|\mathbf{x}\right\|.$$

Combining the last two inequalities, we have

$$\frac{\|\varepsilon\|}{\|x\|} \le \frac{\|A^{-1}\|\|e\|}{\|b\|/\|A\|} = \|A\|\cdot\|A^{-1}\|\cdot\frac{\|e\|}{\|b\|}.$$

The quantity $\|A\|\cdot\|A^{-1}\|$ is called the *condition number* of matrix A. This inequality shows that the relative change in the solution is bounded by the condition number times the relative change in the right hand side vector. Therefore small condition number indicates that small relative change in the right hand side vector implies only a small relative change in the solution. On the other hand, if the condition number is large, then even a small relative change in the right hand side vector might imply a drastic relative change in the solution. The condition number is denoted by Cond(A), and the above inequality can be restated as

$$\frac{\|\varepsilon\|}{\|x\|} \le \text{Cond}(A)\cdot\frac{\|e\|}{\|b\|}. \tag{8.15}$$

Assume next that the coefficient matrix A is approximated by $A+E$, where E is a small matrix, but the right hand side remains the same. Let x^* denote again the solution of the approximating equation and let $\varepsilon = x^* - x$. Subtract equations

$$(A+E)(x+\varepsilon) = b$$

and

$$Ax = b$$

to get

$$(A+E)\varepsilon = -Ex,$$

which implies that

$$\varepsilon = -A^{-1}E(x+\varepsilon).$$

Therefore

$$\|\boldsymbol{\varepsilon}\| \le \|\mathbf{A}^{-1}\| \cdot \|\mathbf{E}\| (\|\mathbf{x}\| + \|\boldsymbol{\varepsilon}\|) = \frac{\mathrm{Cond}(\mathbf{A})}{\|\mathbf{A}\|} \|\mathbf{E}\| (\|\mathbf{x}\| + \|\boldsymbol{\varepsilon}\|).$$

Solving this inequality for $\|\boldsymbol{\varepsilon}\|$ we have

$$\|\boldsymbol{\varepsilon}\| \le \frac{\mathrm{Cond}(\mathbf{A}) \cdot \dfrac{\|\mathbf{E}\|}{\|\mathbf{A}\|} \|\mathbf{x}\|}{1 - \mathrm{Cond}(\mathbf{A}) \cdot \dfrac{\|\mathbf{E}\|}{\|\mathbf{A}\|}},$$

where we have to assume that

$$\mathrm{Cond}(\mathbf{A}) \frac{\|\mathbf{E}\|}{\|\mathbf{A}\|} = \|\mathbf{A}^{-1}\| \cdot \|\mathbf{E}\| < 1. \tag{8.16}$$

From the above inequality we conclude that

$$\frac{\|\boldsymbol{\varepsilon}\|}{\|\mathbf{x}\|} \le \frac{\mathrm{Cond}(\mathbf{A}) \dfrac{\|\mathbf{E}\|}{\|\mathbf{A}\|}}{1 - \mathrm{Cond}(\mathbf{A}) \dfrac{\|\mathbf{E}\|}{\|\mathbf{A}\|}}. \tag{8.17}$$

This inequality shows that if Cond(**A**) is small, then small relative change in **A** indicates only small relative change in the solution. If Cond(**A**) is large, then **E** has to be very small in order to guarantee that the validity of condition (8.16) holds. Even in this case, the relative change in **x** might be very large.

Consider next the case, when both **A** and **b** are approximated. Similarly to the previous cases one can easily prove that

$$\frac{\|\boldsymbol{\varepsilon}\|}{\|\mathbf{x}\|} \le \frac{\mathrm{Cond}(\mathbf{A})\dfrac{\|\mathbf{E}\|}{\|\mathbf{A}\|}}{1 - \mathrm{Cond}(\mathbf{A})\dfrac{\|\mathbf{E}\|}{\|\mathbf{A}\|}} \left(\frac{\|\mathbf{E}\|}{\|\mathbf{A}\|} + \frac{\|\mathbf{e}\|}{\|\mathbf{b}\|}\right). \tag{8.18}$$

This inequality can be interpreted in the same way as it was done in the previous cases.

Example 8.10. Consider equations

$$2x_1 + 6x_2 = 8$$
$$2x_1 + 6.00001x_2 = 8.00001$$

which have a unique solution $x_1 = x_2 = 1$. The coefficient matrix is

$$\mathbf{A} = \begin{pmatrix} 2 & 6 \\ 2 & 6.00001 \end{pmatrix},$$

with inverse

$$\mathbf{A}^{-1} = 50,000 \begin{pmatrix} 6.00001 & -6 \\ -2 & 2 \end{pmatrix}.$$

Using the row-norm of matrices we can see that

$$\|\mathbf{A}\| = 8.00001 \text{ and } \|\mathbf{A}^{-1}\| = 50,000 \cdot 12.00001 = 600,000.5$$

implying that the condition number of $\mathbf{A}$ equals

$$\|\mathbf{A}\| \cdot \|\mathbf{A}^{-1}\| = 4,800,010,$$

which is a very large number in comparison to the matrix elements.

Consider next a small perturbation of the system

$$2x_1 + 6x_2 = 8$$
$$2x_1 + 5.99999x_2 = 8.00002$$

where the unique solution is $x_1 = 10$, $x_2 = -2$. The large discrepancy between the solutions of the two systems is due to the large value of the condition number of $\mathbf{A}$.

4. In Section 3.8 we have discussed the *least squares method* to find the best polynomial fit for a given data set. In this application we will consider this problem under more general conditions. We will offer a solution, and two alternative formulations will be presented based on other than the 2-vector norm.

Assume that the value of a quantity y depends on variables $x_1, x_2, \ldots, x_n$. For example, $x_1, x_2, \ldots, x_n$ may represent the energy usage, manpower, technology level, etc. of a firm, with y being the output value of the firm in a certain time period. It is assumed that simultaneous values of these variables are measured, for example, at different time periods. Let the values of the kth ($k = 1, 2, \ldots, N$) measurement be denoted as

$$x_1^{(k)}, x_2^{(k)}, \ldots, x_n^{(k)}, y^{(k)}.$$

The most simple functional relation is linear, so for the sake of simplicity assume that we are looking for a function of the form

$$y = c_0 + c_1 x_1 + c_2 x_2 + \ldots + c_n x_n,$$

where the coefficients $c_0, c_1, \ldots, c_n$ are unknown. We will determine these values based on the condition that they should provide best overall fit with respect to the measurement data. The function values provide the vector

$$\mathbf{y} = \begin{pmatrix} y^{(1)} \\ y^{(2)} \\ \vdots \\ y^{(N)} \end{pmatrix},$$

where the kth element is the kth measurement for y.

With fixed values of $c_0, c_1, \ldots, c_n$, the linear function gives the vector

$$
\begin{pmatrix}
c_0 + c_1 x_1^{(1)} + \ldots + c_n x_n^{(1)} \\
c_0 + c_1 x_1^{(2)} + \ldots + c_n x_n^{(2)} \\
\cdots\cdots\cdots\cdots \\
c_0 + c_1 x_1^{(N)} + \ldots + c_n x_n^{(N)}
\end{pmatrix},
$$

where the kth element shows what would be the function value at the kth measurement if the relation were linear and $c_0, c_1, \ldots, c_n$ were the correct coefficient values. This vector can be rewritten as $\mathbf{Xc}$ where

$$
\mathbf{X} =
\begin{pmatrix}
1 & x_1^{(1)} & \ldots & x_n^{(1)} \\
1 & x_1^{(2)} & \ldots & x_n^{(2)} \\
\cdots & \cdots & \cdots & \cdots \\
1 & x_1^{(N)} & \ldots & x_n^{(N)}
\end{pmatrix}
\quad \text{and} \quad
\mathbf{c} =
\begin{pmatrix}
c_0 \\
c_1 \\
c_2 \\
\vdots \\
c_n
\end{pmatrix}.
$$

Hence, a logical way to select the values of $c_0, c_1, \ldots, c_n$ that minimize the overall discrepancy between vectors $\mathbf{y}$ and $\mathbf{Xc}$. This concept can be mathematically formulated as minimizing $\|\mathbf{Xc} - \mathbf{y}\|$, where $\|...\|$ is a given vector norm. Depending on the norm selection different approximation methods are obtained.

The selection of the 2-vector norm is known as the *least squares method*. The objective function can be rewritten as

$$
\|\mathbf{Xc} - \mathbf{y}\|_2 = \sqrt{(\mathbf{Xc} - \mathbf{y})^T (\mathbf{Xc} - \mathbf{y})}
$$
$$
= \sqrt{\left(\mathbf{c}^T \mathbf{X}^T - \mathbf{y}^T \right)(\mathbf{Xc} - \mathbf{y})} = \sqrt{\mathbf{c}^T \mathbf{X}^T \mathbf{Xc} - 2\mathbf{c}^T \mathbf{X}^T \mathbf{y} + \mathbf{y}^T \mathbf{y}}.
$$

This norm is minimal if and only if the expression under the square root is minimal. Consider therefore the $(n + 1)$-variable function

$$
g(\mathbf{c}) = \mathbf{c}^T \mathbf{X}^T \mathbf{Xc} - 2\mathbf{c}^T \mathbf{X}^T \mathbf{y} + \mathbf{y}^T \mathbf{y}.
$$

The gradient vector of g is the following:

$$\nabla g(\mathbf{c}) = 2\mathbf{X}^T \mathbf{X}\mathbf{c} - 2\mathbf{X}^T \mathbf{y},$$

and the Hessian matrix is

$$\mathbf{H}(\mathbf{c}) = 2\mathbf{X}^T \mathbf{X}.$$

Since $\mathbf{X}^T\mathbf{X}$ is positive semidefinite, function g is convex. Therefore the stationary points provide global optimum. Equating the gradient to zero leads to the so called *normal equations*.

$$\mathbf{X}^T \mathbf{X}\mathbf{c} = \mathbf{X}^T \mathbf{y}. \qquad (8.19)$$

Notice that this is a system of linear equations with coefficient matrix $\mathbf{X}^T\mathbf{X}$ and right hand side vector $\mathbf{X}^T\mathbf{y}$.

In the special case of $n = 1$, the above problem reduces to *linear regression*. In this case,

$$\mathbf{X} = \begin{pmatrix} 1 & x^{(1)} \\ 1 & x^{(2)} \\ \cdots & \cdots \\ 1 & x^{(N)} \end{pmatrix} \quad \text{and} \quad \mathbf{y} = \begin{pmatrix} y^{(1)} \\ y^{(2)} \\ \cdots \\ y^{(N)} \end{pmatrix},$$

therefore the normal equations can be written as

$$\begin{pmatrix} 1 & 1 & \cdots & 1 \\ x^{(1)} & x^{(2)} & \cdots & x^{(N)} \end{pmatrix} \begin{pmatrix} 1 & x^{(1)} \\ 1 & x^{(2)} \\ \cdots & \cdots \\ 1 & x^{(N)} \end{pmatrix} \begin{pmatrix} c_0 \\ c_1 \end{pmatrix} = \begin{pmatrix} 1 & 1 & \cdots & 1 \\ x^{(1)} & x^{(2)} & \cdots & x^{(N)} \end{pmatrix} \begin{pmatrix} y^{(1)} \\ y^{(2)} \\ \cdots \\ y^{(N)} \end{pmatrix}.$$

Simple calculation shows that this equation is equivalent to the system

$$N c_0 + \left(\sum_{k=1}^{N} x^{(k)} \right) c_1 = \sum_{k=1}^{N} y^{(k)},$$

$$\left(\sum_{k=1}^{N} x^{(k)} \right) c_0 + \left(\sum_{k=1}^{N} x^{(k)2} \right) c_1 = \sum_{k=1}^{N} x^{(k)} y^{(k)}.$$

In order to simplify these equations introduce the following notation:

$$\overline{x} = \frac{1}{N}\sum_{k=1}^{N} x^{(k)}, \overline{y} = \frac{1}{N}\sum_{k=1}^{N} y^{(k)}, \overline{x^2} = \frac{1}{N}\sum_{k=1}^{N} x^{(k)2},$$

and

$$\overline{xy} = \frac{1}{N}\sum_{k=1}^{N} x^{(k)}y^{(k)}.$$

Notice that these quantities are the averages of the values of $x^{(k)}, y^{(k)}, x^{(k)2}$ and the products $x^{(k)}y^{(k)}$, respectively. Dividing both equations by N we have

$$c_0 + \overline{x}\,c_1 = \overline{y},$$
$$\overline{x}\,c_0 + \overline{x^2}\,c_1 = \overline{xy}.$$

Subtracting the $\overline{x}$-multiple of the first equation from the second equation eliminates c_0:

$$c_1\left(\overline{x^2} - \overline{x}^2\right) = \overline{xy} - \overline{x}\cdot\overline{y},$$

which implies that

$$c_1 = \frac{\overline{xy} - \overline{x}\cdot\overline{y}}{\overline{x^2} - \overline{x}^2}. \tag{8.20}$$

From the first equation we conclude that

$$c_0 = \overline{y} - \overline{x}\cdot c_1. \tag{8.21}$$

The least squares method is illustrated in the next examples.

Example 8.11. Consider the following data values: $N = 5$,

$$X = \begin{pmatrix} 1 & -2 & 4 \\ 1 & -1 & 1 \\ 1 & 0 & 0 \\ 1 & 1 & 1 \\ 1 & 2 & 4 \end{pmatrix} \quad \text{and} \quad y = \begin{pmatrix} 0 \\ 1 \\ 2 \\ 1 \\ 0 \end{pmatrix}.$$

That is, the measurements are as follows:

$$x_1^{(1)} = -2, \quad x_2^{(1)} = 4, \quad y^{(1)} = 0,$$
$$x_1^{(2)} = -1, \quad x_2^{(2)} = 1, \quad y^{(2)} = 1,$$
$$x_1^{(3)} = 0, \quad x_2^{(3)} = 0, \quad y^{(3)} = 2,$$
$$x_1^{(4)} = 1, \quad x_2^{(4)} = 1, \quad y^{(4)} = 1,$$
$$x_1^{(5)} = 2, \quad x_2^{(5)} = 4, \quad y^{(5)} = 0.$$

The best fitting quadratic polynomial $c_0 + c_1 x + c_2 x^2$ will be determined. Simple calculation shows that

$$X^T X = \begin{pmatrix} 1 & 1 & 1 & 1 & 1 \\ -2 & -1 & 0 & 1 & 2 \\ 4 & 1 & 0 & 1 & 4 \end{pmatrix} \begin{pmatrix} 1 & -2 & 4 \\ 1 & -1 & 1 \\ 1 & 0 & 0 \\ 1 & 1 & 1 \\ 1 & 2 & 4 \end{pmatrix} = \begin{pmatrix} 5 & 0 & 10 \\ 0 & 10 & 0 \\ 10 & 0 & 34 \end{pmatrix},$$

$$X^T y = \begin{pmatrix} 1 & 1 & 1 & 1 & 1 \\ -2 & -1 & 0 & 1 & 2 \\ 4 & 1 & 0 & 1 & 4 \end{pmatrix} \begin{pmatrix} 0 \\ 1 \\ 2 \\ 1 \\ 0 \end{pmatrix} = \begin{pmatrix} 4 \\ 0 \\ 2 \end{pmatrix}.$$

Hence the normal equations can be written as

$$\begin{pmatrix} 5 & 0 & 10 \\ 0 & 10 & 0 \\ 10 & 0 & 34 \end{pmatrix} \begin{pmatrix} c_0 \\ c_1 \\ c_2 \end{pmatrix} = \begin{pmatrix} 4 \\ 0 \\ 2 \end{pmatrix}.$$

The application of the elimination method gives the solution

$$c_0 = \frac{58}{35}, \quad c_1 = 0, \quad \text{and} \quad c_2 = -\frac{3}{7}.$$

Hence, the least squares function is:

$$y = \frac{58}{35} - \frac{3}{7}x^2.$$

$\blacklozenge$

Example 8.12. Consider next a simple regression problem. Assume that a function $y = c_0 + c_1 x$ is to be determined with data: $N = 4$,

$$\begin{aligned} x^{(1)} &= 0 & y^{(1)} &= 0, \\ x^{(2)} &= 1 & y^{(2)} &= 0, \\ x^{(3)} &= 2 & y^{(3)} &= 1, \\ x^{(4)} &= 3 & y^{(4)} &= 1. \end{aligned}$$

In this case,

$$\overline{x} = \frac{3}{2}, \ \overline{x^2} = \frac{7}{2}, \ \overline{y} = \frac{1}{2}, \quad \text{and} \quad \overline{xy} = \frac{5}{4},$$

therefore equations (8.20) and (8.21) imply that

$$c_1 = \frac{\dfrac{5}{4} - \dfrac{3}{2} \cdot \dfrac{1}{2}}{\dfrac{7}{2} - \dfrac{9}{4}} = \frac{2}{5},$$

and

$$c_0 = \frac{1}{2} - \frac{2}{5} \cdot \frac{3}{2} = -\frac{1}{10}.$$

Hence, the regression line is the following:

$$y = -\frac{1}{10} + \frac{2}{5}x.$$

The original data points and the regression line are illustrated in Figure 8.2.

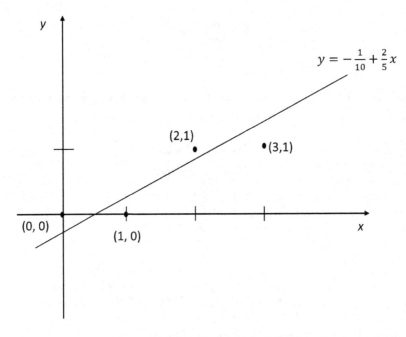

Figure 8.2. Illustration of a regression line

♦

An alternative approach to least squares is derived if we select the ∞-vector norm. In this case the maximal discrepancy between the components of vectors $\mathbf{Xc}$ and $\mathbf{y}$ is minimized. Mathematically this concept can be formulated as the unconstrained optimization problem:

$$\text{minimize} \quad \max_k \left\{ \left| c_0 + c_1 x_1^{(k)} + \ldots + c_n x_n^{(k)} - y^{(k)} \right| \right\}. \tag{8.22}$$

If E denotes the objective function, then this nonlinear optimization problem can be rewritten as a linear programming problem:

$$\text{minimize } E \tag{8.23}$$

$$\text{subject to} \quad \left. \begin{array}{l} c_0 + c_1 x_1^{(k)} + \ldots + c_n x_n^{(k)} + E \geq y^{(k)} \\ c_0 + c_1 x_1^{(k)} + \ldots + c_n x_n^{(k)} - E \leq y^{(k)} \end{array} \right\} \left(1 \leq k \leq N \right),$$

where the unknowns are $c_0, c_1, \ldots, c_n$, and E. The solution of this linear programming problem provides the best fitting linear function.

Example 8.13. Consider again the problem of Example 8.11. Based on the data values, the linear programming formulation has now the form:

$$\text{minimize } E$$

subject to

$$c_0 - 2c_1 + 4c_2 + E \geq 0,$$
$$c_0 - 2c_1 + 4c_2 - E \leq 0,$$
$$c_0 - c_1 + c_2 + E \geq 1,$$
$$c_0 - c_1 + c_2 \quad E \leq 1,$$
$$c_0 \qquad\qquad + E \geq 2,$$
$$c_0 \qquad\qquad - E \leq 2,$$
$$c_0 + c_1 + c_2 + E \geq 1,$$
$$c_0 + c_1 + c_2 - E \leq 1,$$
$$c_0 + 2c_1 + 4c_2 + E \geq 0,$$
$$c_0 + 2c_1 + 4c_2 - E \leq 0.$$

◆

Another alternative approach to least squares is derived if we select the 1-vector norm. In this case the sum of the absolute discrepancies between the components of vectors $\mathbf{Xc}$ and $\mathbf{y}$ is minimized. This concept can be mathematically formulated as the unconstrained optimization problem

$$\text{minimize} \sum_{k=1}^{N} \left| c_0 + c_1 x_1^{(k)} + \dots + c_n x_n^{(k)} - y^{(k)} \right|, \tag{8.24}$$

which can be rewritten as the following linear programming problem:

$$\text{minimize} \sum_{k=1}^{N} E_k \tag{8.25}$$

subject to $\left.\begin{array}{l} c_0 + c_1 x_1^{(k)} + \dots + c_n x_n^{(k)} + E_k \geq y^{(k)} \\ c_0 + c_1 x_1^{(k)} + \dots + c_n x_n^{(k)} - E_k \leq y^{(k)} \end{array}\right\} (1 \leq k \leq N),$

where E_k denotes the kth term of the objective function (8.24).

Example 8.14. In the case of the data values of the previous example, problem (8.25) has the form:

minimize

$$\sum_{k=1}^{5} E_k$$

subject to

$$c_0 - 2c_1 + 4c_2 + E_1 \geq 0$$
$$c_0 - 2c_1 + 4c_2 - E_1 \leq 0$$
$$c_0 - c_1 + c_2 + E_2 \geq 1$$
$$c_0 - c_1 + c_2 - E_2 \leq 1$$
$$c_0 \qquad\qquad + E_3 \geq 2$$
$$c_0 \qquad\qquad - E_3 \leq 2$$
$$c_0 + c_1 + c_2 + E_4 \geq 1$$
$$c_0 + c_1 + c_2 - E_4 \leq 1$$
$$c_0 + 2c_1 + 4c_2 + E_5 \geq 0$$
$$c_0 + 2c_1 + 4c_2 - E_5 \leq 0.$$

◆

In many applications in economics, for example, in computing production functions, linear relations are usually replaced by function forms as

$$y = a_0 x_1^{a_1} x_2^{a_2} ... x_n^{a_n},$$

where the constants $a_0, a_1, a_2, ..., a_n$ are the unknowns. The most commonly used technique to find the values of these unknowns is based on the following linearization idea. Take the logarithms of both sides to have

$$\log y = \log a_0 + a_1 \log x_1 + a_2 \log x_2 + ... + a_n \log x_n.$$

That is, $\log y$ is a linear function of $\log x_1, ..., \log x_n$. In the first step of the procedure, the logarithms of all data values are taken, and in the second step, the best linear fit is determined based on the new data set consisting of the logarithm values.

8.5 Exercises

1. Prove that if $\mathbf{x}, \mathbf{y} \in \mathbf{R}^n$ are orthogonal, then

$$\left\| \mathbf{x} + \mathbf{y} \right\|_2^2 = \left\| \mathbf{x} \right\|_2^2 + \left\| \mathbf{y} \right\|_2^2.$$

2. Let $\mathbf{A} = (a_{ij})$ be an $m \times n$ real matrix. Define

$$\left\| \mathbf{A} \right\| = \max \left\{ \left| a_{ij} \right| \mid 1 \le i \le m, 1 \le j \le n \right\}.$$

   Is this quantity a matrix norm?

3. Let

$$\mathbf{A} = \begin{pmatrix} 2 & 1 & 1 \\ 1 & 2 & 1 \\ 1 & 1 & 2 \end{pmatrix}.$$

   Give the particular forms of the induced vector norms $\mathbf{x} \mapsto \left\| \mathbf{Ax} \right\|_p$ for $p = 1, 2, \infty$.

4. Give an $n \times n$ real matrix $\mathbf{A}$ such that $\left\| \mathbf{A} \right\|_1 < 1$ but $\left\| \mathbf{A} \right\|_2 > 1$ and

$$\left\| \mathbf{A} \right\|_\infty > 1.$$

5. Give an $n \times n$ real matrix $\mathbf{A}$ such that $\left\| \mathbf{A} \right\|_2 < 1$ but $\left\| \mathbf{A} \right\|_1 > 1$ and

$$\left\| \mathbf{A} \right\|_\infty > 1.$$

6. Give an $n \times n$ real matrix $\mathbf{A}$ such that $\left\| \mathbf{A} \right\|_\infty < 1$ but $\left\| \mathbf{A} \right\|_1 > 1$ and

$$\left\| \mathbf{A} \right\|_2 > 1.$$

7. Repeat Example 8.8 for matrix

$$A = \begin{pmatrix} 3 & 1 & 1 \\ 1 & 3 & 1 \\ 1 & 1 & 3 \end{pmatrix}.$$

8. Repeat Example 8.9 for the matrix of the previous Exercise.

9. Select matrix

$$A = \begin{pmatrix} 0.01 & 0.01 & 0.01 \\ 0.01 & 0.01 & 0.01 \\ 0.01 & 0.01 & 0.01 \end{pmatrix},$$

and compare the right and left hand sides of inequality (8.13) with the ∞ matrix norm.

10. Repeat the previous exercise with the 1-matrix norm.

11. Select $A = \varepsilon \cdot I$, where $0 < \varepsilon < 1$ is a given constant. Show that inequality (8.13) becomes an equality for any subordinate matrix norm.

12. Select

$$A = \begin{pmatrix} 3 & 1 & 1 \\ 1 & 3 & 1 \\ 1 & 1 & 3 \end{pmatrix} \quad \text{and} \quad E = \begin{pmatrix} 0.01 & 0.01 & 0.01 \\ 0.01 & 0.01 & 0.01 \\ 0.01 & 0.01 & 0.01 \end{pmatrix},$$

and compare the right and left hand sides of inequality (8.14) with the ∞ matrix norm.

13. Repeat the previous Exercise with the 1-matrix norm.

14. Select $A = I$ and $E = \varepsilon \cdot I$ with $0 < \varepsilon < 1$. Show that in this special case inequality (8.14) becomes an equation for any subordinate norm.

15. Show that in the special case of $\mathbf{A} = \mathbf{I}$, inequality (8.15) becomes an equality for any subordinate matrix norm.

16. Illustrate inequality (8.17) in the special case of $\mathbf{A} = \mathbf{I}$ and $\mathbf{E} = \delta \cdot \mathbf{I}$ (with $0 < \delta < 1$) if the norm is subordinate.

17. Prove that for any $n \times n$ real matrix $\mathbf{A}$, $\left\| \mathbf{A}^T \mathbf{A} \right\|_2 = \left\| \mathbf{A} \right\|_2^2$.

18. Let $\mathbf{A}$ be an $n \times n$ real matrix with columns $\mathbf{a}_1, \mathbf{a}_2, ..., \mathbf{a}_n$. Prove that

$$\left\| \mathbf{A} \right\|_F^2 = \sum_{i=1}^{n} \left\| \mathbf{a}_i \right\|_2^2.$$

19. Let $\mathbf{A}$ be an $n \times n$ real symmetric matrix with eigenvalues $\lambda_1, ..., \lambda_n$. Assume that the eigenvalues are ordered so that $|\lambda_1| \geq |\lambda_2| \geq ... \geq |\lambda_n|$. Prove that

$$\left\| \mathbf{A} \right\|_2 = |\lambda_1|.$$

20. Repeat Example 8.11 with data values

$$\mathbf{X} = \begin{pmatrix} 1 & -1 & 1 & -1 \\ 1 & 0 & 0 & 0 \\ 1 & 1 & 1 & 1 \\ 1 & 2 & 4 & 8 \end{pmatrix}, \quad \text{and} \quad \mathbf{y} = \begin{pmatrix} 1 \\ 0 \\ 1 \\ 16 \end{pmatrix}.$$

21. Repeat Example 8.12 with data values

$$x^{(1)} = -2 \quad y^{(1)} = 1$$
$$x^{(2)} = 1 \quad y^{(2)} = 2$$
$$x^{(3)} = 0 \quad y^{(3)} = 4$$
$$x^{(4)} = 1 \quad y^{(4)} = 5$$
$$x^{(5)} = 2 \quad y^{(5)} = 7.$$

22. Repeat Example 8.13 with data values of Exercise 20.

23. Repeat Example 8.14 with data values of Exercise 20.

24. Let $\mathbf{B}$ be a given $n \times n$ real nonsingular matrix. Consider mapping $\mathbf{A} \mapsto \|\mathbf{BA}\|$ for $n \times n$ real matrices $\mathbf{A}$ with some matrix norm. Is this mapping a norm of $\mathbf{A}$?

25. Let $\mathbf{B}$ and $\mathbf{C}$ be given $n \times n$ real nonsingular matrices. Consider mapping $\mathbf{A} \mapsto \|\mathbf{BAC}\|$ for $n \times n$ real matrices $\mathbf{A}$ with some matrix norm. Is this mapping a norm of $\mathbf{A}$?

Bibliography

Argyros, I. K. and F. Szidarovszky (1993) The Theory and Applications of Iteration Methods. CRC Press, Boca Raton/London/Tokyo.

Baumol, W. J. (1970) Economic Dynamics. (3rd edition) Macmillan, New York.

Bellman, R. E. (1970) Introduction to Matrix Analysis. (3rd edition) McGraw-Hill Publ. Co., New York.

Friedman, J. W. (1977) The Theory of Games and Oligopoly. North Holland, Amsterdam.

Gandolfo, G. (1971) Mathematical Methods and Models in Economic Dynamics. North Holland, Amsterdam.

Gantmacher, F. R. (1959) The Theory of Matrices. (Vols. 1 and 2) Chelsea, New York.

Goldberg, S. (1958) Introduction to Difference Equations. John Wiley & Sons, New York.

Golub, G. and C. Van Loan (1983) Matrix Computations. The John Hopkins Univ. Press, Baltimore, Md.

Halmos, P. R. (1958) Finite Dimensional Vector Spaces. Van Nostrand, Princeton, N. J.

Henderson, J. M. and R. E. Quandt (1971) Microeconomic Theory: A Mathematical Approach. (2nd edition) McGraw-Hill, New York.

Herstein, I. (1964) Topics in Algebra. Blaisdell Publ. Co., New York.

Kaplan, W. (1952) Advanced Calculus. Addison-Wesley Publ. Co., Reading, Mass.

Lancester, P. (1969) Theory of Matrices. Academic Press, New York.

Leontief, W. (1973) Input-Output Economics. Oxford Univ. Press, New York.

Liu, D. and F. Szidarovszky (1990) Block-M-Matrices and Their Properties. Pure Math. and Appl., Ser. B, Vol. 1, No. 2-3, pp. 99-107.

Luenberger, D. G. (1979) Introduction to Dynamic Systems: Theory, Models, and Applications. John Wiley & Sons, New York.

Mirsky, L. (1990): An Introduction to Linear Algebra, Courier Dover Books on Mathematics, Mineola, New York.

Molnar, S. (1990): A Special Decomposition of Linear Systems, Belgian Journal of Operations Research, Statistics, and Computer Science, Vol 29., No 4, pp. 4-37

Molnar S., Szidarovszky F. (1992): Some notes on Cournot-oligopolies with sequential adjustments, Pure Math and Applications, Series B, Vol. 3., No-34, pp. 289-293.

Nikaido, H. (1968) Convex Structures and Economic Theory. Academic Press, New York/London.

Noble, B. (1968) Applied Linear Algebra. Prentice-Hall, Englewood Cliffs, N. J.

Okuguchi, K. (1976) Expectations and Stability in Oligopoly Models. Springer-Verlag, Berlin/Heidelberg/New York.

Okuguchi, K. and F. Szidarovszky (1990) The Theory of Oligopoly with Multi-Product Firms. Springer-Verlag, Berlin/Heidelberg/New York.

Ortega, J. M. and W. C. Rheinboldt (1970) Iterative Solutions of Nonlinear Equations in Several Variables. Academic Press, New York.

Ross, Sh. M. (1987) Introduction to Probability and Statistics for Engineers and Scientists. J. Wiley & Sons, New York/Toronto.

Rugh, W.J. (1996) Linear System Theory. (2nd edition) Prentice-Hall, Upper Saddle River, N.J.

Stewart, G. (1973) Introduction to Matrix Computations. Academic Press, New York.

Strang, G. (1976) Linear Algebra and Its Applications. Academic Press, New York.

Szidarovszky, F. (1989) On Non-Negative Solvability of Nonlinear Input-Output Systems. Econ. Letters, Vol. 30, pp. 319-321.

Szidarovszky, F. and A. T. Bahill (1992) Linear Systems Theory. CRC Press, Boca Raton/London/Tokyo.

Szidarovszky, F. and K. Okuguchi (1989) A Non-Differentiable Input-Output Model. Math. Social Sci., Vol. 18, pp. 187-190.

Szidarovszky, F. and S. Molnár (1986) Game Theory with Engineering Applications (in Hungarian) Műszaki Könyvkiadó, Budapest.

Szidarovszky, F. and S. Molnár (1994) Learning in a Dynamic Producer-Consumer Market. Appl. Math. and Comp., Vol. 62, pp. 223-233.

Szidarovszky, F. and S. Molnár (1994) On Discrete Dynamic Producer-Consumer Markets. Keio Econ. Studies, Vol. XXXI, No. 2, pp. 51-63.

Szidarovszky, F. and S. Molnár (1995) A Note on Extrapolative Expectations in a Dynamic Producer-Consumer Market. Keio Econ. Studies, Vol. XXXII, No. 2, pp. 71-73.

Szidarovszky, F. and S. Yakowitz (1978) Principles and Procedures of Numerical Analysis. Plenum Press, New York/London.

Yakowitz, S. and Szidarovszky F. (1989): An Introduction to Numerical Computations (2nd ed.), Macmillan Publishing Company, New York.

Index

Printed in the United States
by Baker & Taylor Publisher Services